Biomass and Green Chemistry

Sílvio Vaz Jr.
Editor

Biomass and Green Chemistry

Building a Renewable Pathway

Editor
Sílvio Vaz Jr.
Brazilian Agricultural Research
Corporation, National Research Center
for Agroenergy (Embrapa Agroenergy)
Embrapa Agroenergia, Parque Estação
Biológica
Brasília, DF, Brazil

ISBN 978-3-319-88308-3 ISBN 978-3-319-66736-2 (eBook)
https://doi.org/10.1007/978-3-319-66736-2

Softcover reprint of the hardcover 1st edition 2017

Printed on acid-free paper

This Springer imprint is published by Springer Nature
The registered company is Springer International Publishing AG
The registered company address is: Gewerbestrasse 11, 6330 Cham, Switzerland

Preface

The need to develop renewable raw materials as a substitute for oil in industrial chemistry has been shown to be a strategic challenge for the twenty-first century. In this context, the use of plant biomass can be consolidated both as an alternative of using cheaper and less polluting raw materials and as a model of aggregation of economic value to the agro-industrial chains. Green chemistry (GC), based on 12 principles, emerged in the 1990s as a new philosophy in academia and industry to break the old paradigms of chemistry, such as generation of large amounts of waste and intensive use of petrochemicals, through a holistic view of processes in laboratories and industries. In the case of plant biomass, the 7th GC principle – *use of renewable raw materials* – stands out as a great strategic opportunity for segments related to several areas of GC worldwide. Biomass is a renewable source of a large variety of bioproducts, and GC principles can be applied for its exploitation to promote sustainable processes and products. In this volume, the application of GC principles, especially in biomass conversion processes, is discussed with the aim to demonstrate their feasibility.

This book is composed of eight chapters dedicated to understanding the relationship between vegetable biomass and GC (Chap. 1): the bioproducts and their processes for saccharide (Chap. 2), oleaginous (Chap. 3), starch (Chap. 4), and lignocellulosic (Chap. 5) biomass. Furthermore, the potentiality of microalgae as feedstock (Chap. 6) and the potentiality of enzymes for conversion processes for sugars (Chap. 7) are discussed. In finalizing, Chap. 8 describes sustainability aspects of biomass according to the GC approach.

Brasília, Brazil

Sílvio Vaz Jr.

Contents

Contributors

Abla Alzagameem Faculty of Environment and Natural Sciences, Brandenburg University of Technology BTU Cottbus-Senftenberg, Cottbus, Germany

Department of Natural Sciences, Bonn-Rhein-Sieg University of Applied Sciences, Rheinbach, Germany

Mercedes Ballesteros Department of Energy, Biofuels Unit, CIEMAT, Madrid, Spain

Bruna de Barros Correia Faculdade de Engenharia Mecânica, Universidade Estadual de Campinas (Unicamp), Campinas, SP, Brazil

Simone P. Favaro Embrapa Agroenergia, Parque Estação Biológica, Brasília, DF, Brazil

Luz Marina Flórez Pardo Departamento de Energética y Mecánica, Universidad Autónoma de Occidente, Cali, Colombia

Mario Giordano Dipartimento di Scienze della Vita e dell'Ambiente, Università Politecnica delle Marche, Ancona, Italy

Università Politecnica delle Marche and Shantou University Joint Algal Research Center, Shantou, Guangdong, China

Francisco Gírio Bioenergy Unit, LNEG, Lisbon, Portugal

Alan T. Jensen Instituto de Química, Universidade de Brasília, Brasília, DF, Brazil

Birgit Kamm Faculty of Environment and Natural Sciences, Brandenburg University of Technology BTU Cottbus-Senftenberg, Cottbus, Germany

Kompetenzzentrum Holz GmbH, Linz, Austria

Basma El Khaldi-Hansen Department of Natural Sciences, Bonn-Rhein-Sieg University of Applied Sciences, Rheinbach, Germany

Jorge Enrique López Galán Escuela de Ingeniería Química, Universidad del Valle, Cali, Colombia

Tatiana Lozano Ramírez Departamento de Sistemas de Producción, Universidad Autónoma de Occidente, Cali, Colombia

Fabricio Machado Instituto de Química, Universidade de Brasília, Brasília, DF, Brazil

Pedro Gerber Machado Faculdade de Engenharia Mecânica, Universidade Estadual de Campinas (Unicamp), Campinas, SP, Brazil

Susana Marques Bioenergy Unit, LNEG, Lisbon, Portugal

Anderson M.M.S. Medeiros Instituto de Química, Universidade de Brasília, Brasília, DF, Brazil

Cesar H.B. Miranda Embrapa Agroenergia, Parque Estação Biológica, Brasília, DF, Brazil

Antonio D. Moreno Department of Energy, Biofuels Unit, CIEMAT, Madrid, Spain

Camila Ortolan Fernandes de Oliveira Faculdade de Engenharia Mecânica, Universidade Estadual de Campinas (Unicamp), Campinas, SP, Brazil

Margit Schulze Department of Natural Sciences, Bonn-Rhein-Sieg University of Applied Sciences, Rheinbach, Germany

Joaquim E.A. Seabra Faculdade de Engenharia Mecânica, Universidade Estadual de Campinas (Unicamp), Campinas, SP, Brazil

Roger A. Sheldon Molecular Science Institute, School of Chemistry, University of the Witwatersrand, Johannesburg, South Africa

Department of Biotechnology, Delft University of Technology, Delft, The Netherlands

Itânia P. Soares Embrapa Agroenergia, Parque Estação Biológica, Brasília, DF, Brazil

Sílvio Vaz Jr. Brazilian Agricultural Research Corporation, National Research Center for Agroenergy (Embrapa Agroenergy), Embrapa Agroenergia, Parque Estação Biológica, Brasília, DF, Brazil

Arnaldo Walter Faculdade de Engenharia Mecânica, Universidade Estadual de Campinas (Unicamp), Campinas, SP, Brazil

Qiang Wang Key Laboratory of Algal Biology, Institute of Hydrobiology, the Chinese Academy of Sciences, Wuhan, Hubei, China

---

The original version of the List of Contributors was revised.
An erratum to this chapter can be found at https://doi.org/10.1007/978-3-319-66736-2_9

Chapter 1
Biomass and the Green Chemistry Principles

Sílvio Vaz Jr.

Abstract The need to develop renewable raw materials for industrial chemistry as a substitute for oil has been shown to be a strategic challenge for the twenty-first century. In this context, the use of plant biomass can be construed as both the alternative of using cheaper and less polluting raw materials and as a model of aggregation of economic value to the agro-industrial chains. Green chemistry (GC), based on 12 principles, emerged in the 1990s as a new philosophy in both academia and industry to break old paradigms of chemistry such as the generation of large amounts of waste and the intensive use of petrochemicals through a holistic view of processes in laboratories and industries. In the case of plant biomass, the seventh principle—*use of renewable raw materials*—is notable as a great strategic opportunity for segments related to several areas of GC worldwide. Thereby, biomass is a renewable source of a large variety of bioproducts, and green chemistry principles can be applied for its exploitation to promote sustainable processes and products. In this chapter, the application of GC principles, especially in conversion processes for biomass, is discussed with the aim to demonstrate their feasibility.

Keywords Plant biomass • Green processes • Biomass conversion

1.1 Introduction

The need to develop renewable raw materials for industrial chemistry as a substitute for oil has been shown to be a strategic challenge for the twenty-first century. In this context, the use of different types of plant biomass—starch, lignocellulosic, oleaginous, and saccharide—can be considered as an alternative for using cheaper, less polluting raw materials and as a model of aggregation of economic value to the agro-industrial chains, such as soybeans, sugarcane, corn, and forests. These lines of action may, above all, contribute to the sustainability of a wide range of

S. Vaz Jr. (✉)
Brazilian Agricultural Research Corporation, National Research Center for Agroenergy (Embrapa Agroenergy), Embrapa Agroenergia, Parque Estação Biológica, Brasília, DF, Brazil
e-mail: silvio.vaz@embrapa.br

S. Vaz Jr. (ed.), *Biomass and Green Chemistry*,
https://doi.org/10.1007/978-3-319-66736-2_1

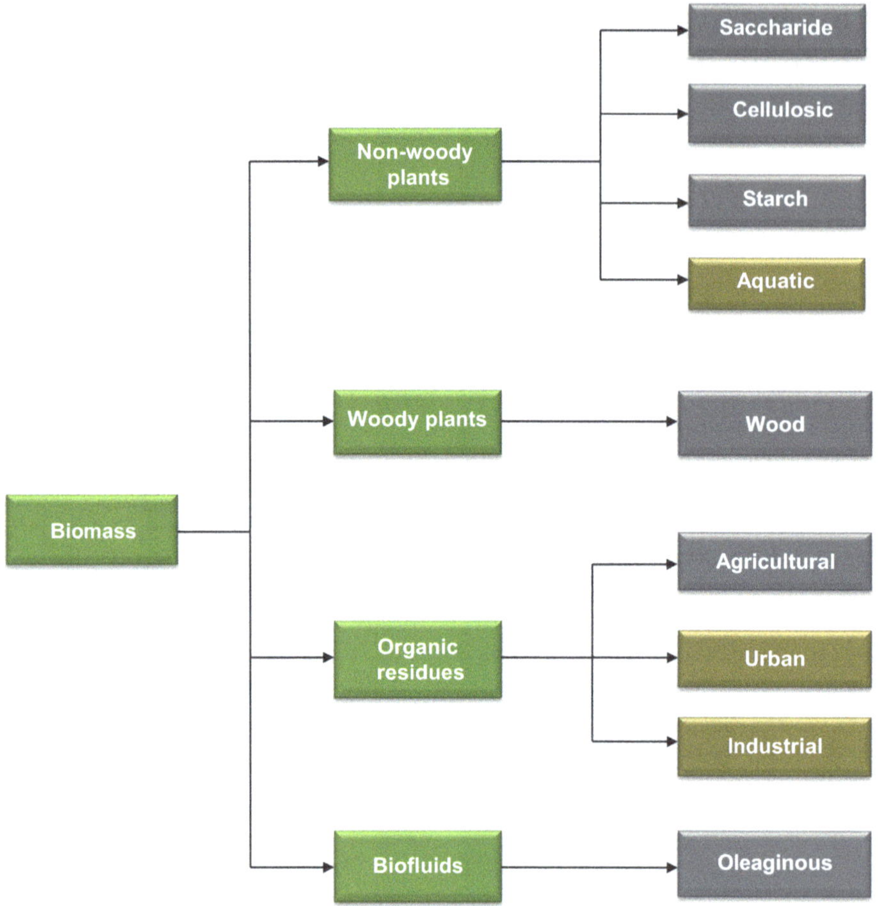

Fig. 1.1 Sources of biomass. *Gray boxes* represent the most used biomass types for industrial and research and development activities (Vaz Jr. 2014)

chemicals, especially organic chemicals (e.g., organic acids, esters, alcohols, sugars, phenolics), which are widely used in today's society.

The great heterogeneity and consequent chemical complexity of plant biomass provides the raw material for end products such as energy, food, chemicals, pharmaceuticals, and materials. As commented, we can highlight four types of plant biomass of great economic interest, to which we now turn our attention: oil crops or oleaginous, saccharides (or sugary), starchy, and lignocellulosic. Soybean (*Glycine max*) and palm oil (*Elaeis guineensis*) are examples of oil plant species; sugarcane (*Saccharum* spp.) and sorghum (*Sorghum bicolor* (L.) Moench) are biomass saccharides; maize (*Zea mays*) is a starchy biomass; and bagasse, straw, and wood biomass are lignocellulosic biomass (Vaz Jr. 2016). Figure 1.1 shows the classification of the sources of plant biomass.

Green chemistry (GC) emerged in the 1990s in countries such as the United States and England, spreading rapidly throughout the world as a new philosophy in both academia and industry and breaking old paradigms of chemistry, such as the generation of large amounts of waste and the intensive use of petrochemicals, through a holistic view of processes in laboratories and industries (Anastas and Kirchhoff 2002). This approach, described in the 12 principles, proposes to consider, among other aspects, the reduction of waste generation, atomic and energy economy, and the use of renewable raw materials (Anastas and Warner 1998).

The use of renewable raw materials is an extremely strategic issue for large biomass producer countries, such as Brazil, the United States, Germany, and France. These raw materials, the agro-industrial biomass, are an abundant and cheap feedstock for the transformation processes of chemistry or the conversion processes applied to biomass, which are biocatalytic, chemocatalytic, fermentative, and thermochemical.

Thus, the use of biomass through chemistry opens up new possibilities of business and wealth generation for a large number of countries, as well as promoting a less negative impact on the environment and the sustainability of biomass chains.

Chemical compounds are the products with the highest potential to add value into a generic biomass chain, given the importance of the conventional chemical industry and the fine chemical industry in different sectors of the economy. It is possible to highlight compounds that can be used as building blocks, synthetic intermediates, polymers, and specialties, among others; such ideas can be greatly explored by biorefineries (Kamm et al. 2006). On the other hand, the need to develop technologies to obtain these products presents considerable bottlenecks to be overcome related to technical, scientific, and market issues.

The 12 fundamental principles of GC are as follows (ACS Green Chemistry Institute 2017):

1. **Prevention**
   It is better to prevent waste than to treat or clean up waste after it has been created.
2. **Atom Economy**
   Synthetic methods should be designed to maximize the incorporation of all materials used in the process into the final product.
3. **Less Hazardous Chemical Syntheses**
   Wherever practicable, synthetic methods should be designed to use and generate substances that possess little or no toxicity to human health and the environment.
4. **Designing Safer Chemicals**
   Chemical products should be designed to effect their desired function while minimizing their toxicity.
5. **Safer Solvents and Auxiliaries**
   The use of auxiliary substances (e.g., solvents, separation agents) should be made unnecessary wherever possible and innocuous when used.
6. **Design for Energy Efficiency**

Energy requirements of chemical processes should be recognized for their environmental and economic impacts and should be minimized. If possible, synthetic methods should be conducted at ambient temperature and pressure.

7. **Use of Renewable Feedstocks**
   A raw material or feedstock should be renewable rather than depleting whenever technically and economically practicable.
8. **Reduce Derivatives**
   Unnecessary derivatization (use of blocking groups, protection/deprotection, temporary modification of physical/chemical processes) should be minimized or avoided if possible, because such steps require additional reagents and can generate waste.
9. **Catalysis**
   Catalytic reagents (as selective as possible) are superior to stoichiometric reagents.
10. **Design for Degradation**
    Chemical products should be designed so that at the end of their function they break down into innocuous degradation products and do not persist in the environment.
11. **Real-Time Analysis for Pollution Prevention**
    Analytical methodologies need to be further developed to allow for real-time, in-process monitoring and control before the formation of hazardous substances.
12. **Inherently Safer Chemistry for Accident Prevention**
    Substances and the form of a substance used in a chemical process should be chosen to minimize the potential for chemical accidents, including releases, explosions, and fires.

These concepts, which also refer to clean production and green innovations, are already relatively widespread for industrial applications, particularly in countries with a well-developed chemical industry and with strict controls over the emission of pollutants. The concepts are based on the assumption that chemical processes with potential for negative impact on the environment will be replaced by less polluting or nonpolluting processes. Clean technology, reduction of pollutants at source, environmental chemistry, and GC are denominations that have emerged and were minted during the past two decades to translate concerns for chemical sustainability (Sheldon 2014).

In the case of plant biomass, the seventh principle—the use of renewable raw materials—stands out as a great strategic opportunity for several areas of GC worldwide. Examples of market segments that may be positively impacted by GC and the use of biomass are these:

- Polymers and materials for various applications
- Chemical commodities such as monomers
- Pharmaceuticals, cosmetics, and hygiene products
- Fine chemicals (agrochemicals, catalysts, etc.) and specialties
- Fuels and energy

In this way, it is possible to observe the great range of opportunities associated with the use of plant biomass for chemistry, which is better observed in Fig. 1.2.

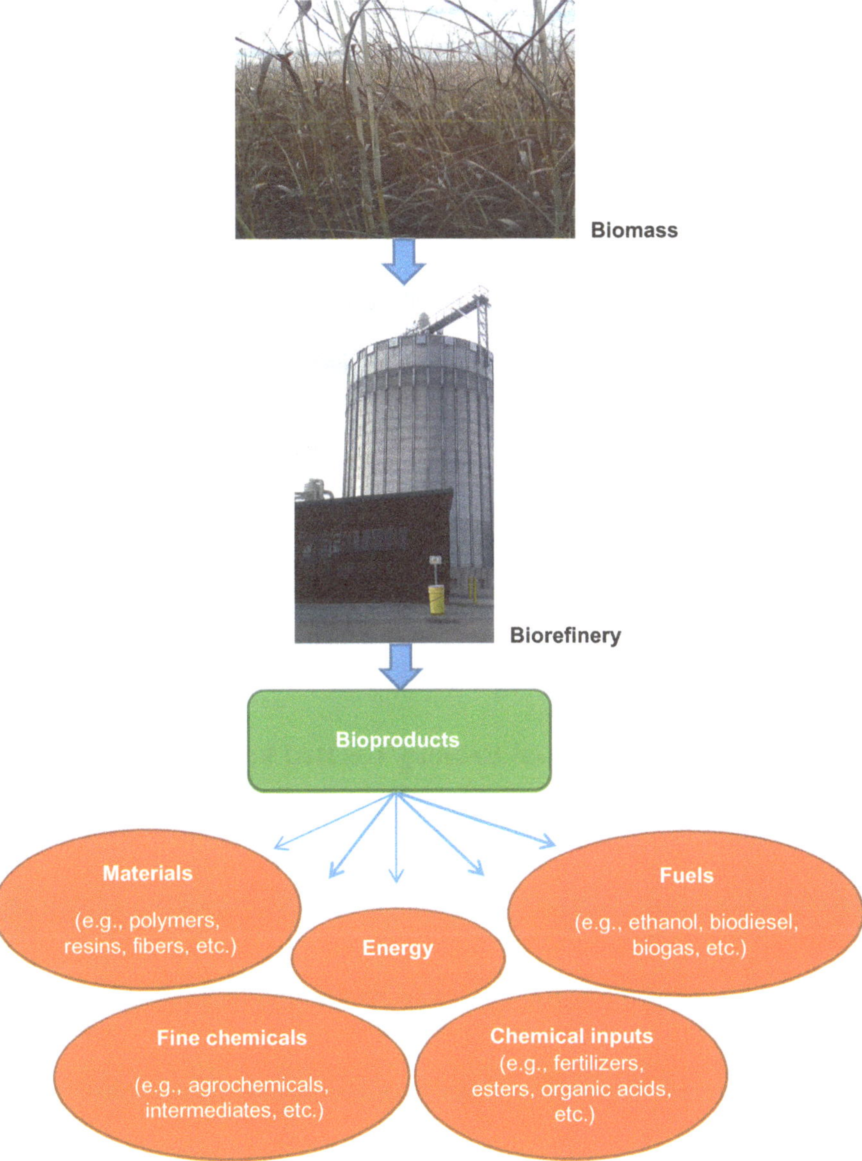

Fig. 1.2 Products that can be obtained from plant biomass by means of a biorefinery and their conversion processes

1.2 Exploring the Green Chemistry Principles in Biomass Conversion

The application of the 12 principles at the same time in a process is the ideal situation but it is not always possible. Raw materials, working conditions, budget restriction, and other factors could be limiting. On the other hand, the application of all principles could cause a process to become not feasible.

We can see that those principles were formulated and more easily applied for two main areas: analytical chemistry and industrial chemistry. Both areas include processes in which principles such as prevention (principle 1), safer solvents and auxiliaries (principle 5), and inherently safer chemistry (principle 12) are feasible for direct application, differing from R&D activities. Of course, R&D is not excluded from GC because green processes and products can be developed based mainly on the use of renewable feedstock (principle 7).

Table 1.1 shows the application of the 12 principles in biomass exploitation. We can see how to apply each principle and what is the expected result from this application to promote green processes for biomass conversion.

Figure 1.3 shows a process that uses lignocellulosic biomass as feedstock (e.g., wood residues). Some principles are suggested to make the process more "green."

We can state these highlights from Fig. 1.3:

- Fractionation step: better for principles 1, 5, 6, 7, 11, and 12.
- Depolymerization step: better for principles 1, 5, 6, 7, 11, and 12.
- Synthesis step: better for principles 1, 2, 3, 4, 5, 6, 7, 8, 9, 10, 11, and 12.

1.3 Aspects of Industrial Ecology Related to Biomass Processing

Industrial ecology explores, by means of a holistic vision, a certain industrial process according to its relationship with the environment (extraction, processing, fabrication, use, disposal, etc.). It is very useful to understand the biomass transformation and have a direct synergy with GC. Figure 1.4 shows a good example of a generic conversion process from the aspects of industrial ecology. Certainly, the most important point to keep in mind is the residue generation and how to recycle it; energy consumption is also an important aspect as it can improve the sustainability of the process, mainly because of the environmental component.

Table 1.1 Use of green chemistry (GC) principles in biomass exploitation to promote green processes

GC principle	How to apply to biomass and their processes and products	Expected results
Prevention	Agro-industry can generate much waste, mainly lignocellulosic residues (e.g., sugarcane bagasse); waste reuse to reclaim chemicals, materials, and energy should be considered to explore biomass chains	Reduction of environmental issues and increase of profits
Atom economy	Optimize the use of reagents during synthesis, for example, to produce biodiesel, furfural, etc.	Waste and cost reduction
Less hazardous chemical syntheses	Change toxic solvents (e.g., hexane, benzene) to oxygenated solvents	Decrease in negative impacts on health and environment
Designing safer chemicals	Develop molecules that are environmentally and health friendly as petrochemical substitutes, which is one of the advantages of biomass as the renewable raw material	
Safer solvents and auxiliaries	The use of water as solvent should be prioritized. However, as biomass has a high chemical heterogeneity, separation is needed; membranes are good alternatives	
Design for energy efficiency	Conversion processes based on enzymes can work in mild conditions	Cost and environmental impact reduction
Use of renewable feedstocks	Intrinsic	Intrinsic
Reduce derivatives	Change a certain analytical technique (e.g., gas chromatography) based on derivative formation to another without this need (e.g., liquid chromatography) or decrease the number of steps in a synthetic route	Cost and environmental impact reduction
Catalysis	Use enzymes and heterogeneous catalysts in synthetic routes	
Design for degradation	Increase the "degradability" of products; e.g., change petrochemical plastics to biobased plastics such as starch-derived or polylactic acid	Decrease in negative impacts on health and environment
Real-time analysis for pollution prevention	Use spectroscopic probes (e.g., Raman and Fourier-transform infrared spectroscopy (FTIR)) in field and industry instead of laboratory techniques	Cost and environmental impact reduction
Inherently safer chemistry for accident prevention	Pay attention to the physicochemical properties of substances to be handled; these can be accessed in the safety data sheet (SDS). Good laboratory practices (GLP) and good production practices (GPP) should be implemented in laboratory and industry	Decrease in negative impacts on health and environment Cost reduction

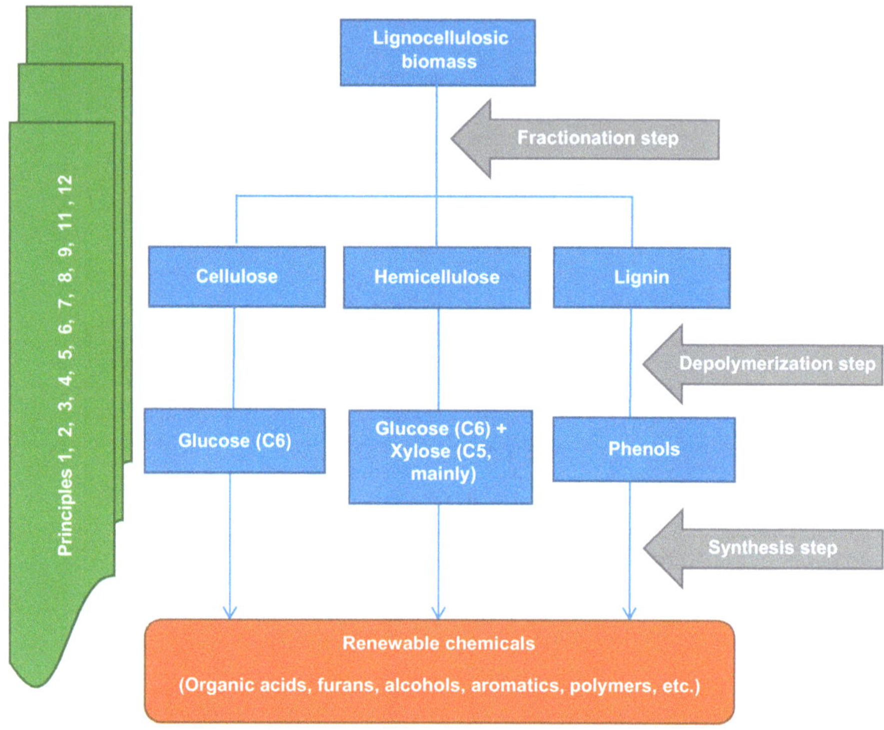

Fig. 1.3 Application of 11 GC principles (*left*) in a process of three steps (*gray arrows*) to obtain renewable chemicals

1.4 Conclusions

Biomass is a renewable source of a large variety of bioproducts, and green chemistry principles can be applied for its exploitation to promote sustainable processes and products. It is fostered by the fact that the use of biomass as feedstock for the chemical industry is one of the most representative principles.

Unfortunately, it is not easy to apply all 12 principles, and we need to choose those appropriate on the basis of biomass type, process characteristics, and products. Principles 1 (prevention), 5 (safer solvents and auxiliaries), and 12 (inherently safer chemistry) can be considered the most feasible for direct application in processes for biomass conversion.

Furthermore, industrial ecology and GC for biomass are closely related to improving environmental aspects of the conversion processes and promoting their sustainability.

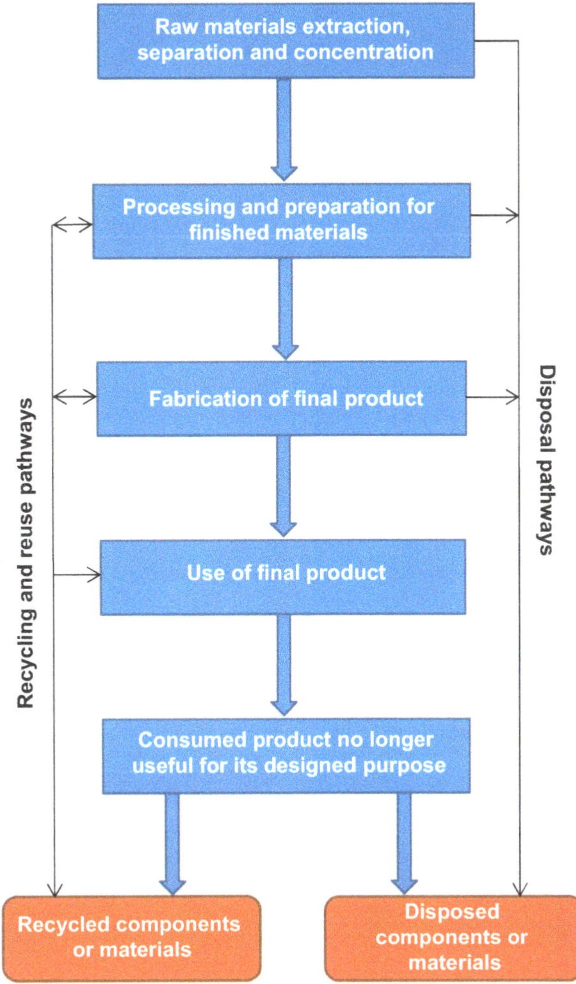

Fig. 1.4 Flowchart of a complete industrial ecosystem (*Source:* Adapted from Manahan (2000))

References

ACS Green Chemistry Institute (2017) 12 principles of green chemistry. Homepage: https://www.acs.org/content/acs/en/greenchemistry/what-is-green-chemistry/principles/12-principles-of-green-chemistry.html. Accessed Jan 2017

Anastas PT, Kirchhoff MM (2002) Origins, current status and future challenges of green chemistry. ACC Chem Res 35:686–694

Anastas PT, Warner JC (1998) Green chemistry: theory and practice. Oxford University Press, New York

Kamm B, Gruber PR, Kamm M (2006) Biorefineries – industrial processes and products: status quo and future directions, 1st edn. Wiley-CH, Weinheim

Manahan S (2000) Environmental chemistry. CRC Press, Boca Raton

Sheldon RA (2014) Green and analytical manufacture of chemicals from biomass: state of the art. Green Chem 16:950–963

Vaz S Jr (2014) Analytical techniques for the chemical analysis of plant biomass and biomass products. Anal Methods 6:8094–8105

Vaz S Jr (ed) (2016) Analytical techniques and methods for biomass. Springer Nature, Cham

Chapter 2
Saccharide Biomass for Biofuels, Biomaterials, and Chemicals

Luz Marina Flórez Pardo, Jorge Enrique López Galán, and Tatiana Lozano Ramírez

Abstract This chapter is a description of the main applications of saccharides in industry for obtaining energetic and nonenergetic products by means of the biorefinery concept allied to green chemistry principles. A biorefinery seeks to use the entire biomass completely, exploiting polysaccharides, proteins, and lignin in various manufacturing processes, to obtain food, pharmaceutical products, biomaterials, bioproducts, and biopolymers, in addition to energetic products, in a sustainable manner. After analyzing demand aspects, costs, transformation technology to be used, and possibilities for the molecule to be a source for many technological applications, the most promising saccharide applications are succinic acid, bioethanol, and 3-HP (3-hydroxypropionic acid).

Keywords Bioplastics • Biochemical • Biorefinery

2.1 Introduction

Biomass is a renewable resource, as it is part of the natural and repetitive cycles of nature's processes (on a human scale) and is derived from plants starting with the process of photosynthesis, transforming solar light into energy that accumulates in the cells in the form of organic matter. This action is essential to maintain ecological equilibrium and allows biological and soil diversity to be preserved and enriched.

L.M. Flórez Pardo (✉)
Departamento de Energética y Mecánica, Universidad Autónoma de Occidente, Cali, Colombia
e-mail: lmflorez@uao.edu.co

J.E. López Galán
Escuela de Ingeniería Química, Universidad del Valle, Cali, Colombia

T. Lozano Ramírez
Departamento de Sistemas de Producción, Universidad Autónoma de Occidente, Cali, Colombia

S. Vaz Jr. (ed.), *Biomass and Green Chemistry*,
https://doi.org/10.1007/978-3-319-66736-2_2

Since the beginning of the twenty-first century, in Europe, the United States (US), some Latin-American countries, and most recently some Asian countries, projects are being driven by the use of sugars and starch products for carburant bioethanol, and the use of oils and fats from different origins for biodiesel production. These were first-generation projects that were developed from good agricultural management, a known logistic for raw material transportation, and the implementation of technologies already developed for alimentary applications that had excellent performance. But their execution required support from states or governments for exemption of biofuels from taxes if they were to be competitive to gasoil and diesel from fossil origins. Then, environmentalist movements against the first-generation biofuels appeared because of the competition between the use of raw materials to feed the population and their direct bioenergetic use. This conflict drove strong research efforts, from the beginning of this century, to use lignocellulosic materials to produce ethanol, and microalgae and used kitchen or residual oils for biodiesel production, which originated the second, third, and fourth generations of biofuels.

When biomass is seen as a complete system in which every advantage is taken, the concept of biorefinery appears, a term that has been used since the end of the 2000–2010 decade. This concept makes the use of biomass for production of second-, third-, and fourth-generation biofuels more economically sustainable and viable. This concept is in accord with the definition made recently by IEA Bioenergy (Task 42: Biorefineries: adding value to the sustainable utilization of biomass), referring to bioenergy as "the sustainable treatment of biomass in a spectrum of commercial and energy products" (International Energy Agency 2009).

Latin American countries in the north have abundant biomass as the result of their strategic position at the Earth's equator. This situation provides solar radiation 365 days per year, with 12-h daily cycles of light, abundant rain, and warm temperatures above 24 °C, giving a tropical setting with one of the highest rates of biodiversity. This characteristic influences crop production; for instance, sugarcane cultivated in Colombia has one of the highest performances worldwide, approximately 120–140 ton ha^{-1} per year, of which 40–60% corresponds to agricultural cane residues in a humid base (approximately 13 million tons), and in Brazil this residue can be nearly 300 million tons. These results are significantly close to those of other studies. For example, Central America (another tropical region) reports 60% of field residue, of which 20–40% (w/w) is process residue (BUN-CA 2002).

The biorefinery concept includes an extended range of technologies that are capable of separating biomass from its basic components (carbohydrates, proteins, triglycerides), which can be converted to products with added value (energetic or not energetic). Products can be intermediary or final products, in food, materials, chemicals, or energy (defined as fuel, electricity, or heat). A biorefinery can use every kind of biomass, including wood, agricultural crops, organic residues (vegetal, animal, industrial, or urban), and aquatic biomass (e.g., algae).

An important characteristic for biorefinery implementation is sustainability, which should be evaluated over the entire production chain. This evaluation has to consider possible externalities such as competition between food and biomass resources, impacts on water use and quality, soil changes (fertility) resulting from

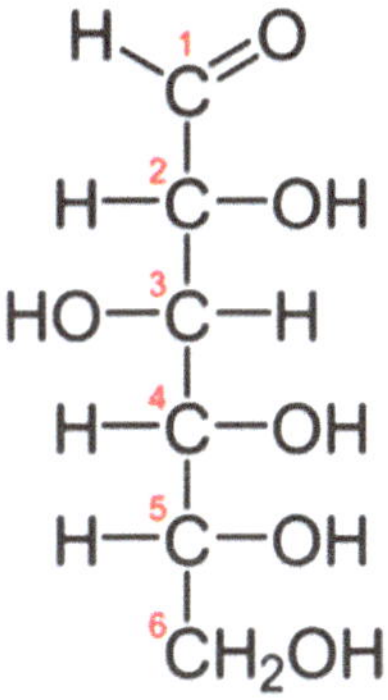

Fig. 2.1 Glucose structure

continuous use, the net balance of greenhouse gases, negative consequences on biodiversity, potential toxic risks, and energetic efficiency. A biorefinery is a concept that depends on continuous innovation, presenting opportunities for all sectors. The construction of a biologically based economy (bioeconomy) has the capacity to face actual difficulties and also provides an environmentally benign industry.

Thus, this chapter focuses on reviewing the application of sugars for bioenergy and non-bioenergy sectors in order to integrate them into a process by the biorefinery concept.

2.2 Chemical Constitution and Physicochemical Properties of Saccharides

Sugars are substances that have the particularity of giving a sweet sensation in the mouth. They are classified as carbohydrates because they contain carbon, hydrogen, and oxygen in their structure. Simple sugars or monosaccharides are classified depending on their number of carbons: as trioses if they have a three-carbon base, as tetroses if they have four, pentoses if they have five, and hexoses if they have six carbons. The most abundant monosaccharide is glucose. As shown in Fig. 2.1, this sugar, as any other, has a chiral structure, all carbons in positions two to five having different substitute radicals, which originate various stereoisomers that show mirror symmetry. Usually, these are called L- (levogyre, levorotatory) and D- (dextrogyre, dextrorotatory). Chemically they behave in a similar manner, even if they are not biologically similar (Pratt and Cornely 2012).

The more common C5–C6 sugars with these structures are D(+)-glucose, D(+)-xylose, and L(+)-arabinose, the physicochemical properties of which are shown in Table 2.1.

As noted in Table 2.1, there is little difference in the properties among these different sugars, except for molecular mass and water solubility.

Table 2.1 Physical and chemical properties of the principal monosaccharides

	Saccharide[a]	D(+)-Glucose[b]	D(+)-Xylose[c]	L(+)-Arabinose[d]
Appearance	White translucent crystals	Crystals or colorless to white crystalline powder	White powder	White powder
Water density at 20 °C (kg m^{-3})	1587	1540	1525	1585
Molar mass (g mol^{-1})	342.3	180.1	150.1	150.1
Fusion point (K)	459	419	419–425	431–436
Water solubility (g 100 ml^{-1} at 298 K)	203.9	91	55	83.4

[a]www.reactivosmeyer.com.mx/pdf/materias/hds_6710.pdf
[b]http://iio.ens.uabc.mx/hojas-seguridad/1593[1]glucosa%20anhidrida.pdf
[c]https://www.applichem.com/fileadmin/datenblaetter/A2241_es_ES.pdf
[d]https://www.carlroth.com/downloads/sdb/es/5/SDB_5118_ES_ES.pdf

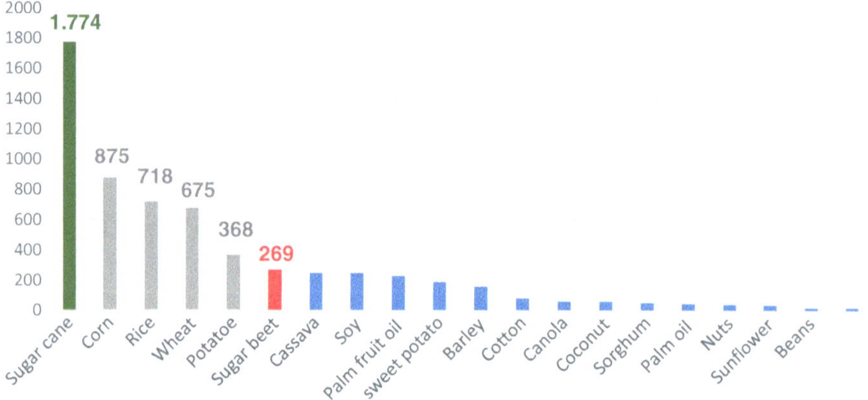

Fig. 2.2 World production of agricultural goods in 2013 (in million tons) (Adapted from Asocaña (2015))

2.3 Economic Aspects: Global Production, Availability, and Distribution of Sucrose

Sugarcane and sugar beet production are the dominant agricultural activity in tropical and subtropical regions around the world. More than 117 countries produce sugar, ethanol, honeys, bagasse, and derived products from these crops, resulting from State policies implemented in various countries that recognize the positive socioeconomic impact associated with the cultivation and processing of cane. Overall, world production of fresh cane in 2013 was 1774 million tons, from 26 million harvested hectares (Food and Agriculture Organization of the United Nations 2017), which bypasses the combined production of maize (875 million tons) and rice (718 million tons) (Fig. 2.2). During the 2014–2015 period,

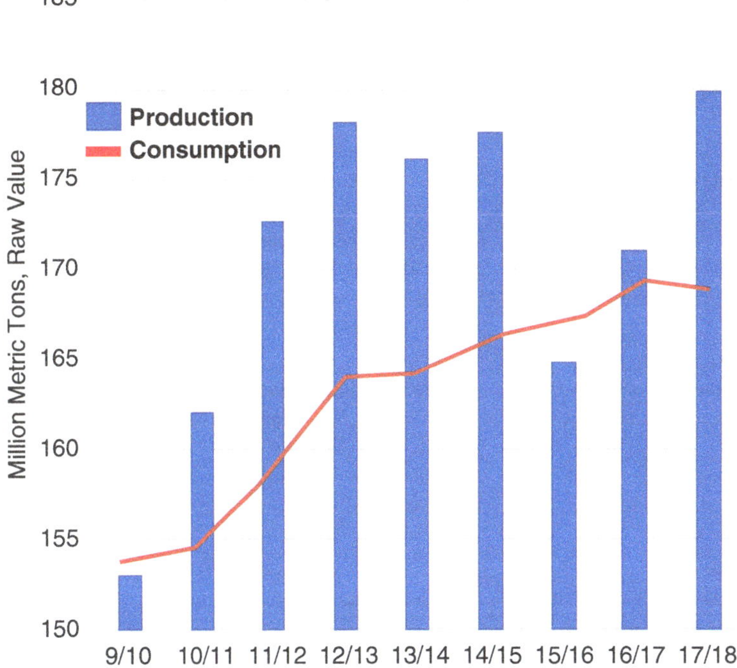

Fig. 2.3 Global consumption and production of sugar during 2009–2018 (*Source:* Adapted from U.S. Department of Agriculture (2017))

sugarcane production increased to 1900 million tons, quadrupling 1965 production (Statistica 2015). Brazil was the first producer (768 million tons), followed by India (341 million tons), China (128 million tons), Thailand (100 million tons), and Pakistan (63 million tons). Other producer countries are Mexico, Colombia, Indonesia, Philippines, and the US (Food and Agriculture Organization of the United Nations 2017). Sugar beet production is much lower, about 14% of cane production in 2014, of which France, US, Germany, and Russia are the main producers worldwide (Statistica 2016).

Worldwide average performance is about 69.8 ton ha^{-1} per year, but it varies from region to region, depending on such factors as weather potential, level of agricultural management, crop cycle, soil quality, and number of harvests (Food and Agriculture Organization of the United Nations 2012). There is a long-term tendency for average global performance to grow about 4 ton ha^{-1} per year. Colombia has the highest production of sugarcane in the world, at 120–140 ton ha^{-1} per year. An important factor on increasing production of sugarcane per hectare is the adoption of new varieties and increasing harvest age, which is maintained over 12.5 months.

Figure 2.3 shows that from 2010 to 2015 sugarcane production increased by 12%, whereas its consumption was 8.8%, creating an oversupply that lowered sugar

Table 2.2 Agricultural and economic value of sugarcane and its products and subproducts

Economic sector	Added-value products from sugarcane residues and the sugarcane industry
Food	Sweeteners, drinks, edible fats and oils, proteins
Health	Chemicals, antibiotics, anti-cholesterol
Fertilizers, compost, food, seeds, concentrates, grass, pesticides	Variety of foods, seeds, concentrates for animals, grass
Industry	Solvents, plastics, bioplastics, chemical products, alcohol-based compounds, anticorrosive compounds, surface-active compounds, biocides
Electric energy, biogas, bagasse fuel, alcohol	Bagasse as fuel, biogas, cogeneration of energy, ethanol from bagasse and sugars
Transport	Ethanol
Education and culture	Books, notebooks, broadsheets, paper
Construction/housing	Boards, alternating current/voltage conduits, decorative plating
Light industry	Textile, wax, bitumen, carbon paper, chemical products
Communication	Isolating materials
Heavy industry	Grout for casting mold
Human resources development	Job creation for rural areas

Source: Solomon (2011)

prices and discouraged its production between 2015 and 2016. It is also important to take into account that after 2010, diverse projects on fuel ethanol were created, particularly in South America, so that sugarcane production was diverted to fuel production. It is expected by projections that for the 2017-2018 that production will reach 180 million metric tons of saccharide, 4.7% more than expected consumption for this same period.

2.4 Main Products: Considering Energy, Biofuels, Biomaterials, and Chemicals

Only after the 1973 world energy crisis did scientists and technicians realize the value of sugarcane and its subproducts and co-products, as it is considered as one of the best plants for converting solar energy to biomass and sugar.

Sugarcane is a versatile crop, being a rich source of food (saccharides, syrup), fiber (cellulose), animal food (green leaves, buds), fuels and chemicals (bagasse, molasses, alcohol), and fertilizers and filter cake. Therefore, almost every country in the world that produces sugar has realized that although its production is profitable, it is better to develop added-value products as obtained by sugarcane industry diversification and the use of its by-products, instead of depending on only one product.

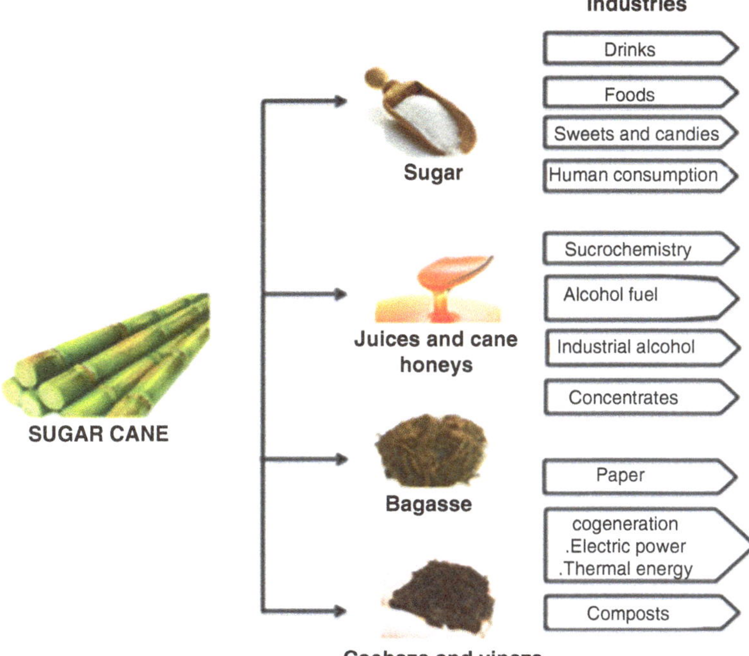

Fig. 2.4 Products and by-products of sugarcane

Sugarcane and its derivatives are raw materials for more than 25 industries: some of the most important are shown in Table 2.2.

Of all the sugarcane products and by-products shown in Table 2.2, those having the highest economic value are sugar (sucrose), juices/honeys (mostly molasses), bagasse, and filter cake (Fig. 2.4). Moreover, other residues produced during harvest and production that have less commercial value, such as green leaves, wax, and ashes, need to be valorized.

In addition to the applications already mentioned, saccharide is the base of the sucrochemical industry, from which bioproducts and other compounds of sugars are obtained and developed.

Bioproducts are made from biological compounds or renewable materials. There is no universally accepted definition for these products, but they can be described as the one that is fabricated in a sustainable manner, total or partially, from renewable resources. Thus, it includes all the processes, from raw material production to every step into the final product fabrication, in addition to processes of investigation, development, and commercialization.

Typically, bioproducts are divided in two categories: bioenergetic and non-bioenergetic. Those categories can be separated into seven segments. Table 2.3 shows the categorization and segmentation of bioproducts.

Table 2.3 Categorization and segmentation of bioproducts

Category	Segment	Products
Bioenergetic	Biofuels	Liquid fuels, such as ethanol, methanol, biobutanol, fuel oil, and biodiesel for means of transport
	Bioenergy	Solid/liquid biomass for combustion to generate heat and electricity
	Biogas	Gaseous fuel, as biogas, biomethane, and synthetic gas to generate heat and electricity
Non-bioenergetic	Bioplastics/biopolymers	Bioplastics from vegetable oils and C6 and C5 sugars
	Biocompounds	Fabricated from agricultural and forest fibers (hemp, linen), that can be used for panels and part production for vehicles, for example
	Biochemicals	Basic chemical products and industrial specialties, including grouts, paints, lubricants, solvents
	Biomedicine	*Pharmaceutical products*: antibodies and vaccines produced by genetically modified plants; medical compounds from a natural origin *Cosmetic products*: soaps, body creams, lotions

Source: Adapted from Gobina (2014)

Among compounds derived from saccharide, ethanol and butanol are important as bioenergetics. From the non-bioenergetics, the most significant are bioplastics/biopolymers and biochemicals. Other products are derived from lignocellulosics, which are discussed in Chap. 5. Hereafter, we examine more deeply the more relevant compounds for the sucrochemical industry.

2.4.1 Energetics Products

Biofuels are those compounds such as alcohols, ethers, and esters that come from sweetened juices or organic compounds of sugar-based cellulosic extracted from plants, cultures, or biomass residues. Since the 1970s, biofuels have been promoted to replace part of the consumption of traditional fossil fuels such as coal, oil, and natural gas. The best developed and most used biofuels are bioethanol and biodiesel; other alternatives are biopropanol and biobutanol, which have been until now less commercially valuable.

According to the study by Gobina (2014), during the period between 2010 and 2011, between 28.8 and 30.6 billion gallons (BG) of biofuels were sold worldwide. His projections show that global consumption of these biofuels could reach 50.9 BG

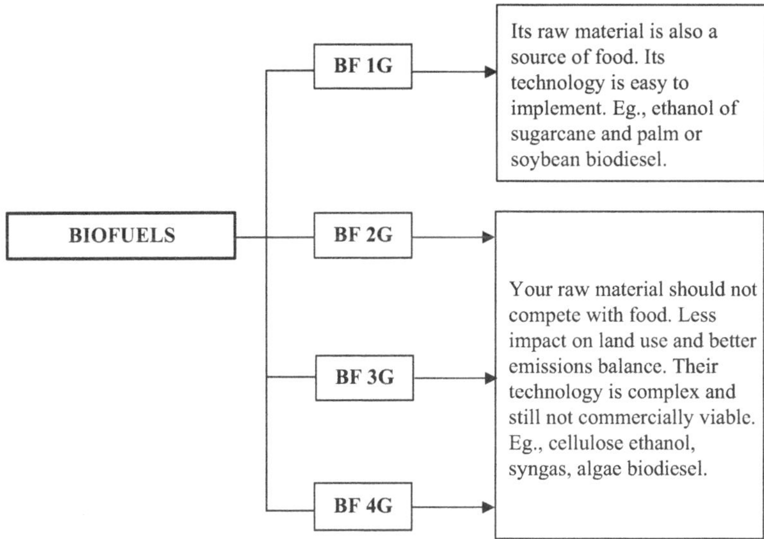

Fig. 2.5 Classification of biofuels (BF) by generations

in 2018, compared to 34.3 BG in 2013. This increase corresponds to an annual growth rate of 8.2% during the forecast period. Energy bioproducts, therefore, can help countries achieve their goals of a secure, reliable, and affordable energy policy, to expand, access, and promote development.

Although the increase is surprising, the production of neither ethanol nor biodiesel is expected to double for the period from 2013 to 2021 (Gobina 2014). Production volumes are expected to decline below the estimated demand of 71.8 billion gallons because of lack of access to low-cost raw materials and difficulties in obtaining financing (Mussattoa et al. 2010). However, this trend is not global, but is particular to each country and the policies implemented in each one. It also depends on the price of oil and the availability of biomass in each region. In Colombia during 2015, 468 million liters bioethanol were produced (Fedebiocombustibles 2017), far below major world powers producers of this biofuel, such as the USA (9000 million gallons) and Brazil (6472.2 million gallons), which covered 89% of the total market in the world in 2008 (Mussatto et al. 2010).

Biofuels can be made from different sources and have different levels of complexity in their processing, depending on the technology required. This variation allows classifying biofuels in four different generations (Fig. 2.5), of which the first two are explained next.

First-generation biofuels (BF 1G) are also called agro-fuels because they come from crops that are used for human consumption, directly or indirectly, and use conventional technologies, which are already produced and marketed in significant quantities by different countries. The US is currently the largest producer of bioethanol, 19.8 billion liters per year, with maize as the main raw material. Sugarcane is used as the main raw material in Brazil, currently the second largest producer in the world (17.8 billion liters per year). The European Union produces 3.44 trillion liters of bioethanol, mainly from sugar beet and starch crops (Food and Agriculture Organization of the United Nations 2007). This ethanol is obtained by processes of fermentation of sugar solutions in which pH value, temperature, type and concentration of inoculum, and quantity and type of micro- and macronutrients are controlled, to obtain musts with an alcohol concentration of 10% (v/v). After that, ethanol is purified by distillation and dehydrated by molecular sieves or pervaporation. Anhydrous bioethanol is thus obtained with a moisture content of only 0.4% (v/v), ready to be used as biofuel. The size of the global ethanol market can be around 86 million ton year^{-1}, of which only 18% is destined for applications other than for energy (Choi et al. 2015).

Second-generation and higher biofuels (BF 2G, 3G, 4G) emerged as the response to the largest critique of first-generation fuels: the dilemma between food and fuels. These biofuels are obtained from biomass that cannot be used as food and are produced with technological innovations that allow a more ecologically sound and more advanced production than that used nowadays. As they are obtained from nonalimentary raw materials, the sources for such biofuels can be cultivated in marginal lands that are not used for growing food, or can be obtained from residues from agricultural, forest, or agro-industrial activities. In that sense, they allow a higher diversification with new raw materials, new technologies, and new final products, promoting agricultural and agro-industrial development.

It has been found that biomass coming from cellulose can be a basic raw material for the production of second-generation biofuels. Biomass from cellulose generates cellulosic bioethanol, which allows the use of sawmill waste, reorienting and expanding the forestry industry to diversify the uses of wood and to protect forests from clearing for agricultural and farming uses.

Second-generation biofuels are in a preoperational stage, but they are not yet produced on a large scale. High manufacturing costs imply that they cannot be commercialized and that these biofuels need incentives and subventions from governments to be able to expand. Demonstration factories of cellulosic ethanol already exist in Sweden, Spain, Germany, US, and Canada. In Latin America, Brazil has a plant under construction in the São Paulo area (Sanford et al. 2016). However, the initial issue represents a potential development to help reduce CO_2 and diminish the consequences of greenhouse gases, braking global warming.

Some companies such as Braskem of Brazil, Dow, and Solvay have announced projects to produce ethylene from ethanol. In the Brazilian case, the plant of Braskem has a production capacity of polyethylene (PE) of 200 ton year^{-1} and a consumption of bioethanol of 462 million liters year^{-1} (Haro et al. 2013). Thereby, ethanol is projected as a commodity for the production of polyethylenes, styrene monomer, and ethylene oxide, from which biopolymers, rubbers, and polyesters are mostly obtained (Fig. 2.6). The world market for ethylene derived from petroleum is booming, with a growth rate around 4.5%. Production is expected to be close to 208.5 million tons by 2017 as the result of the new plants that will go into operation between 2016 and 2017 in the US and China, presently the world leaders for this product (Carvajal medios B2B 2017). This step can give a good margin of growth to

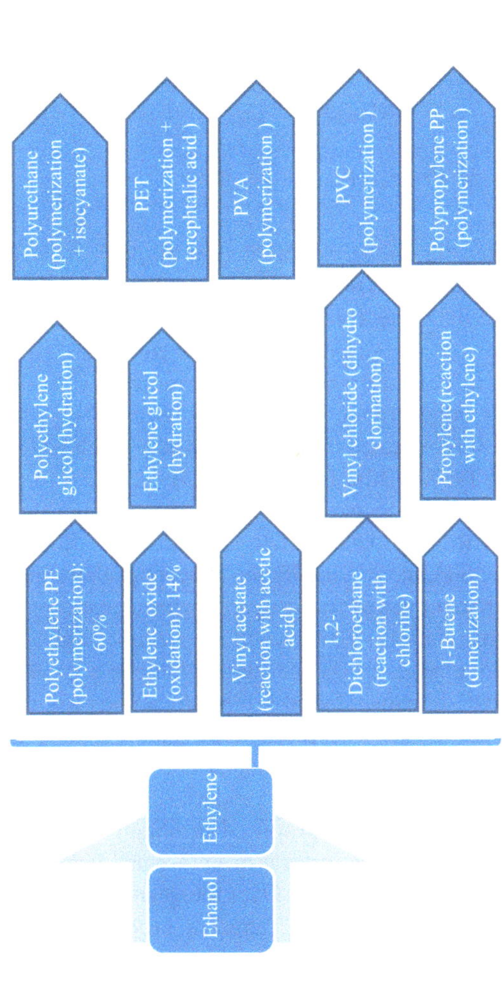

Fig. 2.6 Main products derived from ethylene. Terms in parentheses correspond to the transformation process. *PET* polyethylene terephthalate, *PVA* polyvinyl acetate, *PVC* polyvinylchloride (Adapted from Jong et al. (2007) and Choi et al. (2015))

the bioethylene that is produced from ethanol through a process of dehydration. This process is developed by introducing water and ethanol into a fluidized-bed reactor operating at a temperature of 613 K and a pressure of 0.48 MPa with activated alumina as catalyst. The effluent from this reactor is cooled and compressed to remove the water. The resulting gas then passes through two rectification columns to separate the C3–C4, ethane and propylene, to produce an ethylene with a purity of 99.99% (Haro et al. 2013; Bozell and Petersen 2010). It should be noted that the main product of ethylene is PE, with a participation of 60%, followed by polyethylene terephthalate (PET), with 14% (see Fig. 2.6).

2.4.2 Nonenergetic Products

BCC Research estimates that the world demand of bioproducts will increase at an annual growth rate of 12.6% in 2013 to attain $700.7 billon in 2018, when it will reach a penetration rate of 5.5%. This growth rate is higher than that expected for energetic products, which are expected to increase to an annual growth rate of 8.5% to attain $227.9 billon in 2018, as opposed to an estimated $151.3 billon in 2013. The global market for nonenergetic products is significantly higher; it attained $236.3 billion in 2013. It is expected that for 2018 it may reach $472.8 billon, growing at an annual rate of 14.4% (BCC Research 2014).

Regarding the principal products that can be obtained from the sucrochemical industry, it is important to note that the State Department of Energy of the United States made a study in 2004, identifying by the concept of biorefinery seven potential products that could be produced from sugars by fermentation processes (Werpy and Petersen 2004). This platform was formed by succinic acid, aspartic acid, and 3-HP (3-hydroxypropionic acid) by oxaloacetate, glutamic acid of α-ketoglutarate, itaconic acid from citrate, bioethanol, and bio-hydrocarbons from acetyl-CoA and lactic acid from pyruvate. By chemical means the evaluation of another seven products was also contemplated, such as the production of levulinic acid by acid hydrolysis of saccharides, glucaric acid by oxidation, and sorbitol by glucose hydrogenation. This last process also proposed to obtain arabitol from arabinose and xylitol from xylose. Finally, the furfural of xylose and furfural hydroxymethyl of fructose pass by a dehydration process. Over the years, and after making technical and economic feasibility studies, the range of products was reduced and L-aspartic acid, L-glutamic acid, itaconic acid, and glucaric acid were discarded from the list, because of the limited size of the market, low growth rates, and restrictions on the applications of each product. A large portion of those products comes from fermentation processes, and others from chemical processes (Werpy and Petersen 2004).

2.4.2.1 Bioplastics/Biopolymers

Plastics have become a fundamental part of modern development, to the point that most of the objects we buy are made or packaged with this material, or have at least one plastic part, or have used components of this type in their production or transport. With continued growth for more than 50 years, world production in 2013 reached 299 million tons, an increase of 3.9% as compared to 2012 (PlasticsEurope 2014).

These plastics are generally synthetic and manufactured by polymerization of petroleum derivatives. They are diverse and very versatile in their applications, which indicates that more than 700 types of plastics are produced currently, such as polystyrene, nylon, polyurethane, polyvinyl chloride (PVC), silicones, epoxy resins, and polyamides. Nevertheless, this versatility is accompanied by several environmental disadvantages: inability to biodegrade, high resistance to corrosion and water, and greenhouse gas emissions. Although recycling programs for plastic materials are increasingly efficient and economical, they are applied on residues already generated and, in addition, recycling is not an alternative for all plastics that come from the petrochemical industry. As a result, a significant percentage of residues are discarded as waste, generating serious environmental problems.

Despite this discouraging scenario, one of the alternatives that is gaining strength is the replacement of plastic by bioplastics because they do not constitute a single class of polymers but rather a family of materials with different properties and range of applications. In general terms, the European Bioplastics Association (European Bioplastics 2017) states a bioplastic has one or both of the following characteristics:

- Biological basis materials: these are called biopolymers or materials originated totally or partially from biomass (renewable resources). The biomass used for bioplastics is derived from corn, sugarcane, or cellulose.
- Biodegradable: biodegradation is the ability of a material to decompose into CO_2, methane, water, and organic components, or even biomass, in which the predominant mechanism is the enzymatic action of microorganisms. This property does not depend on the origin of the material, but on its chemical structure.

The classification of bioplastics according to their origin, as biopolymers or from petrochemical biodegradables, can be observed in Fig. 2.7.

An investigation by Jason Chen in June 2014 for BCC Research (Chen 2014) predicted an annual growth rate for bioplastics of 32.7% from 2014 to 2019. In this study, Asia and South America are expected to account for about 90% of the production capacity of bioplastics in 2019 compared to 65% in 2013.

The sugarcane fiber present in residues such as bagasse, leaves, and buds has great potential for development in the field of bioplastics. The degree of commercialization of bioplastics based on renewable resources such as fibers has reached different levels. Some of these have already achieved a certain level of maturity that allows them to be competitive in comparison to petroleum-based plastics. The differences in prices that insulated them from common resins have been significantly

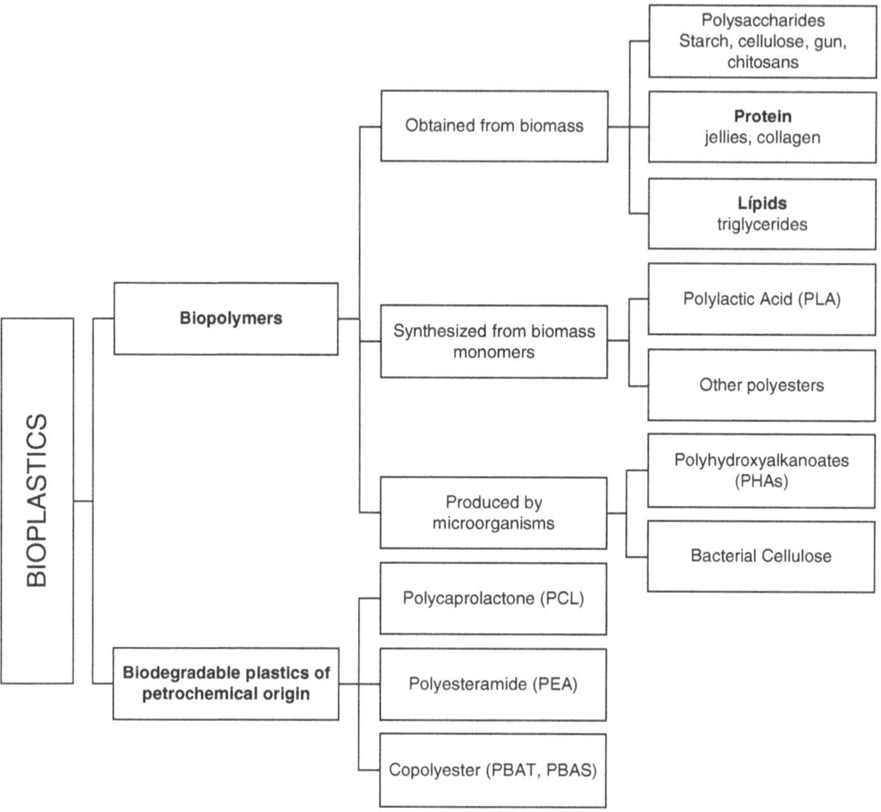

Fig. 2.7 Classification of bioplastics according to their origin

reduced. Furthermore, factors such as the high cost of oil, increasing advances in the processing of these types of materials, the growing number of consumers concerned about protection of the environment, and the creation of new government legislation have allowed consolidation of concrete applications of bioplastics at commercial levels.

Production capacity for biopolymers will triple from 5.7 million tons in 2014 to almost 17 million tons in 2020. Data show a 10% growth rate from 2012 to 2013 and as much as 11% from 2013 to 2014. Bioplastics consolidated in the market will grow at a slower pace because they have already been established for some time. In addition, competition during past years has increased, particularly, for example, for polymers derived from cellulose and starch besides polylactic acid (PLA), which will show a slower growth rate of 26.6% and 18.1% for the first two and 16.7% for the latter. In contrast, bioplastics such as polyethylene (Bio-PE) and polyethylene terephthalate (PET) have noted a significant increase during the past few years. It is expected that from 2014 to 2019, annual growth rates of 38% and 76.5%, respectively,

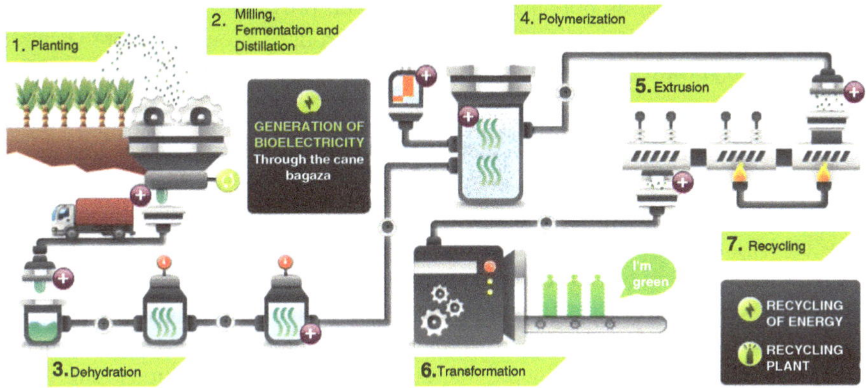

Fig. 2.8 Ecological production of green polyethylene (green PE) (*Source:* Adapted from Braskem (2017))

can be reached (Aeschelmann and Carus 2015). Some examples of bioplastics commercialized in the market are described next.

The renowned Coca-Cola Company launched a new plastic bottle made from materials derived from sugarcane. Since 2009, Coca-Cola has initiated the implementation of this type of new technique to elaborate its containers in the program *PlantBottle* (for PET), with an initial vegetable composition of 30%. Since its introduction, more than 35 million bottles have been distributed in almost 40 countries. The use of these containers has avoided annual emissions equivalent to more than 315,000 metric tons of carbon dioxide. In June 2015, the first 100% sugarcane bottle made was presented in Italy (Procaña 2015).

After years of research and development, Braskem, a Brazilian petrochemical company, launched the green ethylene plant in September 2010, as mentioned previously. This event marked the beginning of the production of *I'm green* polyethylene on a commercial scale, which ensured the company's global leadership position in the field of biopolymers (Fig. 2.8). This industrial plant was constructed by investments of $290 million and has a production capacity of 200,000 ton year^{-1} of green ethylene (Braskem 2017).

Green PE from Braskem is a plastic produced from a renewable raw material source, sugarcane ethanol. Braskem receives ethanol from its suppliers mainly through railways. When ethanol arrives at the ethylene plant, it goes through a dehydration process and is transformed into green ethylene. Then, green ethylene goes to the polymerization step for its transformation into green polyethylene by using the same equipment as used to produce conventional polyethylene (Fig. 2.8). The process and quality of the final product are maintained without the need to invest in new machinery and equipment. For the same reason, it can also be recycled within the company.

In addition to the ethanol mentioned in the previous section, there are other compounds derived from sugars from which plastics can be obtained through different chemical routes, such as polyurethane, polyesters, polyamides, polyacrylamides, and PLA. These compounds, which are used to produce monomers, are succinic acid, 3-hydroxypropionic acid, and lactic acid, explained in detail next.

Succinic Acid

Succinic acid is obtained by the fermentation of sugars (lactose, xylose, arabinose, glucose) by microorganisms such as *Escherichia coli*, *Actinobacillus succinogenes*, *Corynebacterium crenatum*, and *Basfia succiniciproducens* at 37 °C and a pH value between 2.7 and 6.5 (Moussa et al. 2016; Alexandri et al. 2017; Jiang et al. 2017). It is a dicarboxylic acid of four carbons that serves as the base to be transformed in various plastics by applying diverse processes of chemical transformation. In this way, 1,4-butanediol is obtained by hydrogenation. From this process, polyurethanes and polyesters are obtained using polymerization processes with various acids (terephthalic, dicarboxylic) or alkylene oxides. It is also possible to obtain polyamides after reacting succinic acid with ammonia to produce 1,4-butanediamine by hydrogenation, which after polymerization in acid medium produces the plastic (Choi et al. 2015). The size of the succinic acid market is between 30,000 and 38,000 ton year^{-1}. Several companies are currently manufacturing the acid, such as Reverdia, Myriam, Succinity, and Bioamber-Mitsubishi (Alexandri et al. 2017). The expectations for this commodity are to reach an annual growth of 18.7% (Moussa et al. 2016), although most comes from the petrochemical sector (95%) (Gallezot 2012).

3-Hydroxypropionic Acid (3-HP)

The biological production of 3-HP is recent. It is obtained from glucose via malonyl CoA (pathway II), which is expressed in microorganisms of the genus *Lactobacillus* and recombinant *E. coli*. Thus, the glucose is converted to lactic acid and then to 3-HP in a concentration approximately 49 g l^{-1} for 69 h with a 0.46 mol mol^{-1} yield relative to the sugar used (Kumar et al. 2013).

From 3-HP, 1,3-propanediol is obtained by hydrogenation to produce polyester by polymerization. If this polymerization takes place in the presence of terephthalic acid, PTT (polytrimethylene terephthalate) is obtained. Polyacrylamides, polyamides, and polyacrylonitriles are also obtained from the acrylic acid obtained by dehydration of 3-PH. Several companies, such as Cargil-BASF-Novozyme, Opxbio-Dow Chemical, and Perstorp, are already preparing the commercialization of several bio-based products, mainly acrylic acid derivatives with the largest market size, at about 5.5 million ton year^{-1} in 2013 (Choi et al. 2015).

Lactic Acid

Lactic acid is an organic acid with a very important market in the chemical, plastic, pharmaceutical, cosmetic, and food preservation industries (Murillo et al. 2016). PLA, a bioplastic that replaces polyethylene, polystyrene, or polyethylene terephthalate (Gerssen-Gondelach et al. 2014), is obtained from this acid through a polymerization process. About 90% of lactic acid is produced by fermentation from sugars of five or six carbons (Diaz et al. 2017; Murillo et al. 2016), but it has certain limitations in its application as the result of the long reaction times, technical difficulties in the stages of neutralization and purification, and the search of economic raw materials to make it profitable (Abdel-Rahman et al. 2011).

2.4.2.2 Biochemicals

We consider, under this heading, biochemicals as the set of chemicals produced from biomass. Several factors are driving the chemical industry to change its raw materials from petroleum to renewable counterparts for the production of organic chemicals: realization of finite oil resources, the harmful effects of oil derivatives on the environment, the availability of renewable resources, and the latest advances in processing technologies.

It is relevant that throughout the manufacture of chemicals, and in fact for the duration of the product's life cycle, the energy demand must be minimized, the processes applied must be safer, and the use of hazardous chemicals and their production avoided, as preconized by green chemistry (GC) principles. The final product must be nontoxic, and degradable into safe chemicals with minimal waste production; again, GC is closer to these criteria.

Expectations of the market for bio-based chemicals will grow as a result of business initiatives that favor the use of sustainable and renewable raw materials. The biochemical industry was valued at almost $92 billion in 2010. The market reached a value of $155.7 billion in 2013, equivalent to a compound annual rate increase of 19.2%. BCC Research forecasts that the market for biomass-derived chemicals will reach $331.2 billion in 2018, corresponding to an annual composite rate of 16.3% (Gobina 2014).

In 2004, the Pacific Northwest National Laboratory (PNNL) and the National Renewable Energy Laboratory (NREL) identified 12 classes of products that could be derived from sugars, as saccharose, to support the production of fuels and energy in an integrated biorefinery both economically and technically. Further, they could identify the challenges and barriers of associated (chemical or biological) conversion processes (Werpy and Petersen 2004). These products are considered "building blocks" or chemical platforms, that is, they have the potential to produce multiple derivatives. Table 2.4 shows the products, the different ways of conversion, and some potential applications.

The products listed in the table represent a small part of the biochemicals studied by different research groups around the world. Several aspects including high oil prices, consumer preference for products from renewable resources, business commitments, mandates, and government support are driving efforts to broaden the spectrum of biochemicals.

In the mid- to long term, biochemical products will share the market with petroleum-based chemicals and eventually replace them as competitively priced products after overcoming the technological challenges they face today, in some cases, from the lack of friendly technologies and green transformations.

According to research done by the U.S. Department of Agriculture (2017), the degree of research, development, and commercialization of chemicals based on

Table 2.4 Main applications of biochemical drivers

Chemical products	Conversion process	Potential applications
1,4-Diacids (succinic, fumaric, malic)	Biological	Green solvents, fibers (such as Lycra), water-soluble polymers, antifreeze, fuel additives, flavorings
2,5-Furandicarboxylic acid	Chemical	PET with new properties (bottles, films, containers), new polyesters, nylon
3-Hydroxypropionic acid	Biological	Contact lenses, diapers, absorbable sutures
Aspartic acid	Biological Chemical	Detergents, absorbent polymers, corrosion inhibitors, water treatment systems
Glucaric acid	Chemical	Solvents, nylon with different properties, detergents
Glutamic acid	Biological	Flavors, monomers for polyesters and polyamides
Itaconic acid	Biological	Paints, adhesives, lubricants, herbicides
Levulinic acid	Chemical	Oxygenated fuels, solvents, synthetic gums, cigarettes
3-Hydroxybutyrolactone	Chemical	Solvents
Glycerol	Biological chemical	Disinfectants, lubricants, cosmetic products, polyester fibers with new/better properties, antifreeze, polyurethane resins
Sorbitol	Chemical	Antifreezes, water-soluble polymers, cosmetic products, sweetener in dietetic foods, source of alcohol in manufacture of resins
Xylitol/arabinitol	Chemical	Flavors, antifreezes, chewing gums, cough syrups, toothpaste, mouthwashes, hard candies

Source: Xu et al. (2008)

renewable resources is at different levels. The most highly developed are derivatives of 1,4-diacids (succinic, fumaric, malic), as well as xylitol, sorbitol, and *y*-3-HP. The world demand for the latter acid was approximately $832 million in 2013, and it is expected that by the year 2018 this figure will reach $900 million (Evans 2011).

2.5 Conclusions

Some authors consider that the integration of bio-based products into the production of bioenergy from biomass is still in its first stages of growth. Nevertheless, companies such as Cargil-BASF-Novozyme, Reverdia, Myriam, and Dupont as well as Braskem are making great economic and technical efforts to bring this idea into reality.

There are huge potentials for bio-based products that can be obtained from sugars, but at the moment, those with a more developed platform from the commercial and technical aspects are succinic acid with a market of 30,000–38,000 tons $year^{-1}$

and acrylic acid, derived from 3-HP, with a market of 5.5 million tons year^{-1}, in addition to biopolymers that can be obtained from bioethanol. However, the expected horizon, in both the short term and medium term, is an increase in the market for these types of renewable chemicals.

References

Abdel-Rahman MA, Tashiro Y, Sonomoto K (2011) Lactic acid production from lignocellulose-derived sugars using lactic acid bacteria: overview and limits. J Biotechnol 156:286–301

Aeschelmann F, Carus M (2015) Markets. Available at: http://www.bio-based.eu/market_study/media/files/15-11-12-Bio-based-Building-Blocks-and-Polymers-in-the-World-short-version.pdf. Accessed July 2017

Alexandri M, Papapostolou H, Stragier L, Verstraete W, Papanikolaou S, Koutinas AA (2017) Succinic acid production by immobilized cultures using spent sulphite liquor as fermentation medium. Bioresour Technol 238:214–222

Asocaña (2015) Sector azucarero colombiano. Available at: http://www.observatoriovalle.org.co/wp-content/uploads/2014/12/presentaci%C3%B3n-sector-azucarero-colombiano-feb-15.pdf. Accessed July 2017

BCC Research (2014) Global market for bioproducts to reach $700.7 billion in 2018; non-energetic products moving at 14.9% CAGR. Available at: http://www.bccresearch.com/pressroom/egy/global-market-for-bioproducts-to-reach-$700.7-billion-2018. Accessed July 2017

Bozell JJ, Petersen GR (2010) Technology development for the production of biobased products from biorefinery carbohydrates: the US Department of Energy's "top 10". Green Chem 12:525–728

Braskem (2017) Green plastic. Available at: http://www.braskem.com/site.aspx/plastic-green. Accessed July 2017

BUN-CA (2002) Manuales sobre energía renovable: Biomasa. FOCER, San José. Available on: http://www2.congreso.gob.pe/sicr/cendocbib/con4_uibd.nsf/5EA2E564AF6F41D405257CC1005B2354/$FILE/Manuales_sobre_energ%C3%ADa_renovableBIOMASA.pdf. Accessed July 2017

Carvajal medios B2B (2017) Estados Unidos y China liderarán capacidad mundial de etileno para 2017. Available on: http://www.plastico.com/temas/Estados-Unidos-y-China-lideraran-capacidad-mundial-de-etileno-para-2017+102577. Accessed July 2017

Chen J (2014) Bioplastics. Available on: https://www.bccresearch.com/market-research/plastics/bioplastics-pls050c.html. Accessed July 2017

Choi S, Song CW, Shin JH, Lee SY (2015) Biorefineries for the production of top building block chemicals and their derivatives. Metab Eng 28:223–239

Díaz AB, Marzo C, Caro I, de Ory I, Blandino A (2017) Valorization of exhausted sugar beet cossettes by successive hydrolysis and two fermentations for the production of bio-products. Bioresour Technol 225:225–233

European Bioplastics (2017) Bioplastics. Available at: www.european-bioplastics.org. Accessed July 2017

Evans J (2011) Market research. Available on: https://www.bccresearch.com/market-research/biotechnology/commercial-amino-acids-bio007j.html. Accessed July 2017

Fedebiocombustibles (2017) Available on: www.fedebiocombustibles.com. Accessed July 2017

Food and Agriculture Organization of the United Nations (FAO) (2007) A review of the current state of bioenergy development in G8 + 5 countries. Available on: ftp://ftp.fao.org/docrep/fao/010/a1348e/a1348e00.pdf. Accessed July 2017

Food and Agriculture Organization of the United Nations (FAO) (2012) Estudio FAO: Riego y drenage. Available at: http://www.fao.org/3/a-i2800s.pdf. Accessed July 2017

Food and Agriculture Organization of the United Nations (FAO) (2014) Faostat. Available on: http://faostat3.fao.org/download/Q/QC/E. Accessed July 2017

Food and Agriculture Organization of the United Nations (FAO) (2017) Faostat. http://www.fao.org/faostat/en/#data/QC. Accessed July 2017

Gallezot P (2012) Conversion of biomass to selected chemical products. Chem Soc Rev 41:1538–1558

Gerssen-Gondelach SJ, Saygin D, Wicke B, Patel MK, Faaij APC (2014) Competing uses of biomass: assessment and comparison of the performance of bio-based heat, power, fuels and materials. Renew Sust Energ Rev 40:964–998

Gobina E (2014) Biorefinery products: Global market. Available on: https://www.bccresearch.com/market-research/energy-and-resources/biorefinery-products-market-egy117a.html. Accessed July 2017

Haro P, Ollero P, Trippe F (2013) Technoeconomic assessment of potential processes for bioethylene production. Fuel Process Technol 114:35–48

International Energy Agency (IEA) (2009) Biorefineries: adding value to the sustainable utilisation of biomass. Available on: http://www.ieabioenergy.com/publications/biorefineries-adding-value-to-the-sustainable-utilisation-of-biomass/. Accessed July 2017

Jiang M, Ma J, Wu M, Liu R, Liang L, Xin F, Zhang W, Jia H, Dong W (2017) Progress of succinic acid production from renewable resources: metabolic and fermentative strategies. Bioresour Technol. pii: S0960–8524(17)30884–2. https://doi.org/10.1016/j.biortech.2017.05.209

Jong E, Higson A, Walsh P, Wellisch M (2007) Bio-base chemicals. Value added products from biorefineries. Available on: http://www.qibebt.cas.cn/xscbw/yjbg/201202/P020120223415452622293.pdf. Accessed July 2017

Kumar V, Ashok S, Park S (2013) Recent advances in biological production of 3-hydroxypropionic acid. Biotechnol Adv 31:945–961

Moussa HI, Elkamel A, Young SB (2016) Assessing energy performance of bio-based succinic acid production using LCA. J Clean Prod 139:761–769

Murillo B, Zornoza B, de la Iglesia O, Téllez C, Coronas J (2016) Chemocatalysis of sugars to produce lactic acid derivatives on zeolitic imidazolate frameworks. J Catal 334:60–67

Mussatto SI, Dragone G, Guimarães PM, Silva JP, Carneiro LM, Roberto IC, Vicente A, Domingues L, Teixeira JA (2010) Technological trends, global market, and challenges of bio-ethanol production. Biotechnol Adv (6):817–830

PlasticsEurope (2014) Plastics—the facts 2014/2015. Available on: http://www.plasticseurope.org/documents/document/20150227150049-final_plastics_the_facts_2014_2015_260215.pdf. Accessed July 2017

Pratt CW, Cornely K (2012). Bioquimica (Spanish edition). Editorial El Manual Moderno, México

Procaña (2015) Revistas. Available on: https://issuu.com/procana.org/docs/procana111_baja. Accessed July 2017

Sanford K, Chotani G, Danielson N, Zhan JA (2016) Scaling up of renewable chemicals. Curr Opin Biotechnol 38:112–122

Solomon S (2011) Sugarcane by-products based industries in India. Sugar Technol 13:408–416

Statistica (2015) World sugar cane production from 1965 to 2014 (in million metric tons). Available on: https://www.statista.com/statistics/249604/sugar-cane-production-worldwide/. Accessed July 2017

Statistica (2016) Top sugar beet producers worldwide by volume (in 1,000 metric tons). Available on: https://www.statista.com/statistics/264670/top-sugar-beet-producers-worldwide-by-volume/. Accessed July 2017

U.S. Department of Agriculture (2017) Sugar. Available on: https://apps.fas.usda.gov/psdonline/circulars/Sugar.pdf. Accessed July 2017

Werpy T, Petersen G (eds) (2004) Top value added chemicals from biomass, vol 1. Results of screening for potential candidates from sugars and synthesis gas. Washington, DC: US-DOE

Xu Y, Hanna MA, Isom L (2008) Green chemicals from renewable agricultural biomass. Open Agric J 2:54–61

Chapter 3
Oleaginous Biomass for Biofuels, Biomaterials, and Chemicals

Simone P. Favaro, Cesar H.B. Miranda, Fabricio Machado, Itânia P. Soares, Alan T. Jensen, and Anderson M.M.S. Medeiros

Abstract Concerns about negative environmental impacts and questions of future availability surrounding the long-term use of fossil sources as a basis for production of fuels, and a plethora of derivatives, are matters of increasing importance. Consequently, plant biomass sources capable of efficiently replacing fossil fuel resources are gaining relevance as biofuels and in the oleochemical industry. The array of chemical compositions of vegetable oils and fats, the possibility of producing biomass in a sustainable way, and the development of routes for their transformation are the main drivers of this growing demand. This chapter covers topics of global production and consumption of the principal vegetable oil commodities, the comparative chemical composition of oils and fats, the potential use of the biological storage structures of oils and fats, the main processes of transforming oils into biofuels, and the production of bio-based polymers. Also, mechanisms of the functionalization of vegetable oils are stressed.

Keywords Oleochemistry • Fatty acids • Biodiesel • Polymers • Vegetable oils

3.1 Introduction

Lipids are important sources of food and renewable energy (Christie 2017; Cyberlipid Center 2017). Lipids encompass oils, fats, greases, steroids, cholesterols, lipid-soluble vitamins, and phospholipids and are the basis for a range of products, from personal care to the hardware industry. Conceptually, lipids are compounds that are insoluble in water but soluble in organic solvents such as ether, benzene, and alcohols. Oils and fats are the main commodities within lipids and are

---

S.P. Favaro (✉) • C.H. Miranda • I.P. Soares
Embrapa Agroenergia, Parque Estação Biológica, Brasília, DF, Brazil
e-mail: Simone.favaro@embrapa.br

F. Machado • A.T. Jensen • A.M.M.S. Medeiros
Instituto de Química, Universidade de Brasília, Brasília, DF, Brazil

S. Vaz Jr. (ed.), *Biomass and Green Chemistry*,
https://doi.org/10.1007/978-3-319-66736-2_3

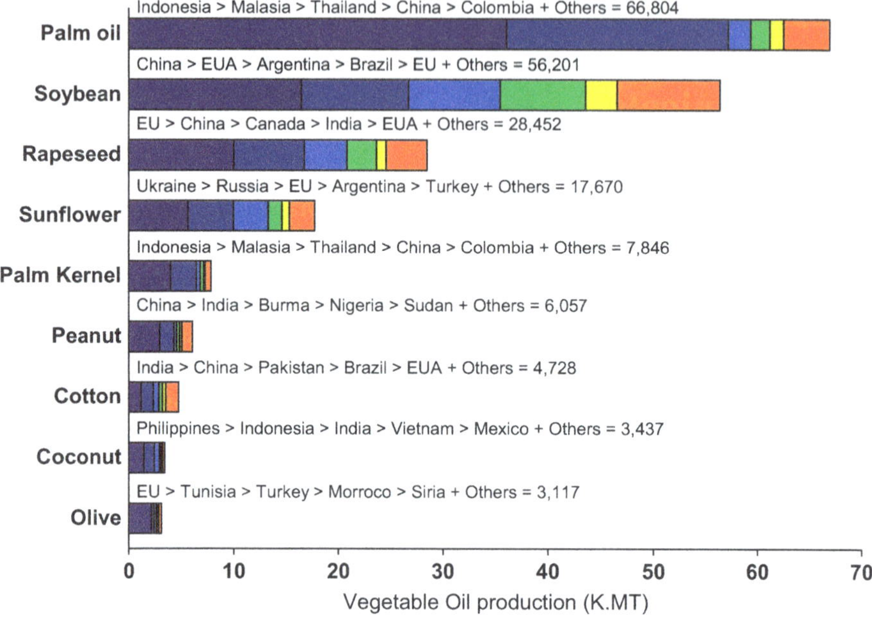

Fig. 3.1 Overview of the world's principal vegetable oil sources and production in 2017 (From USDA (2017))

the basis for the oleochemical industry. Their main market and their applications are in the food industry, but since 2001 a significant part of their increasing use is for biofuels production, especially biodiesel (Knothe 2010; Gunstone 2013).

The world production of fats and vegetable oils has increased steadily in the past 25 years, reaching 206 million tons in 2016 (USDA 2017). Although many promising oleaginous plant species are available in the wild, around 94% of the world's production comes from the seven best known sources (Fig. 3.1). Since 2004, palm oil (*Elaeis guineensis*) has become the main commodity produced and consumed, surpassing soybean (*Glycine max*), the long-standing leader. Considering its pulp and kernel oil, palm oil accounts for 38% of world production, followed by soybeans with 29% (Fig. 3.1).

The hydric and temperature needs of palm oil for productive growth and yield require its cultivation to be concentrated between parallels 10° north or south of the equator, whereas soybeans can be grown in a wider dispersion area. The main reasons for palm oil predominance as an industry base product are its desirable physicochemical characteristics and oil productivity. Averaging 3500 kg oil/ha (10,000 m^2), it can produce more than 6000 kg/ha, about tenfold the productivity of soybean, which supported the oleochemical industry for several decades and is still of great use. The soybean was domesticated in China more than 3000 years ago, and for centuries its oil provided light to cities and meal for animal feed. In the first quarter of the twentieth century, the soybean arrived in Europe and then the United States (USA), being then used mainly for industrial purposes to produce inks, varnishes, coatings, soap stocks, lubricants, and textiles. After soybean crushers were established in the USA, solvent extraction processes were developed and its oils replaced cotton and linseed oils, and, later, tallow. The need to feed the troops and the limitations on oil imports during the Second World War led to further development of soybean agricultural production systems and technological processes (Johnson and Myers 1995; Shurtleff and Aoyagi 2015). Nowadays soybean oil is mostly consumed as an edible oil. With the surge in biodiesel production it is filling a dual purpose, with strong environmental appeal: oil (food, fuel, oleochemical) and the world's leading source of high-quality vegetable protein.

The oleochemical industry also absorbs around 12×10^6 ton from several other oil sources, usually of local importance, which depends on their availability and the potential application of their added value. A good example is castor oil (*Ricinus communis*), which is used to produce specialized lubricants and coatings (McKeon 2016; Patel et al. 2016), and animal fats, used in biodiesel production (Bousba et al. 2013; Van Gerpen 2014). The usage of a given oil source is basically defined by the characteristics and proportions of its fatty acids profile. These fatty acids are the drivers of the technological routes that can be used and the final products to be obtained. As details on lipid structure and labeling are abundant in the literature (Lehninger et al. 2000; Scrimgeour 2005), this chapter focuses on the utilization of oils and fats in renewable chemistry.

3.1.1 Oils and Fats

The major components of oils and fats are glycerol-esterified fatty acids, which, as triols, can form mono-, di-, or triglycerides. In general, triglycerides (TAGs) predominate in the composition of oils and fats (Fig. 3.2).

Fatty acids are unbranched carbon chains, varying from 4 to 22 carbons, with a terminal carboxyl group. They are called saturated when all carbons are linked by single bonds and unsaturated when one or more carbons are linked with double bonds. With a single double bond, they are monounsaturated, and polyunsaturated when they contain two or more double bonds (Fig. 3.3).

The amphiphilic nature of fatty acid molecules causes oil to be stored intracellularly inside the organelles of oleaginous plants, called body oils (BOs) (Fig. 3.4), rather than storage as a dispersed continuous layer (Purkrtova et al. 2008). The BO structural model shows the TAG matrix encompassed within a phospholipid monolayer (Lin and Tzen 2004; Tzen et al. 1993), protected by a layer of structural proteins, predominantly oleosines (Huang 1992; Furse et al. 2013). The polar side is exposed to the cytosol, and the acyl group turns inside and interacts with TAGs (Beisson et al. 1996).

Figure 3.5 shows an example of body oil (BO) structures in the pulp and nut of the fruit of the macauba palm (*Acrocomia aculeata*), both of which contain high levels of oil (Lescano et al. 2015). The spherical fruit contains lipid bodies embedded in the cytoplasm. The oil bodies (OB) are scattered in the pulp, which is rich in

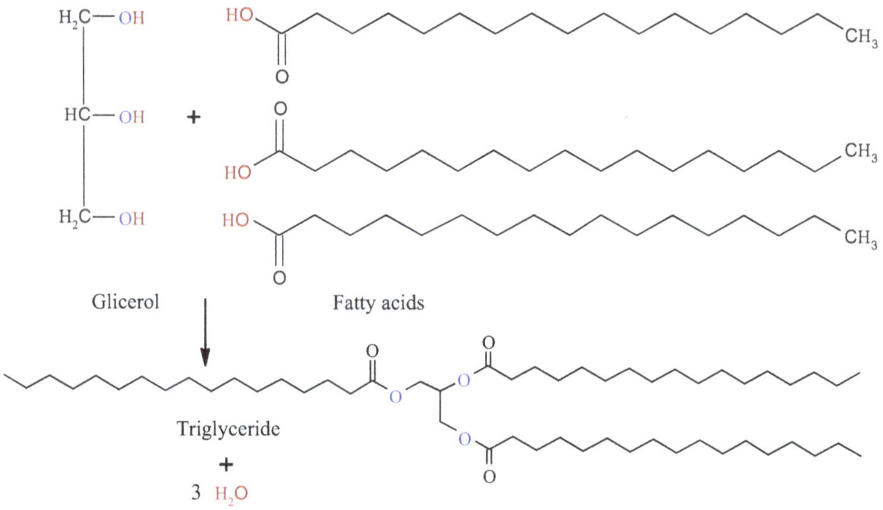

Glicerol

Fatty acids

Triglyceride + 3 H_2O

Fig. 3.2 Triglyceride formation

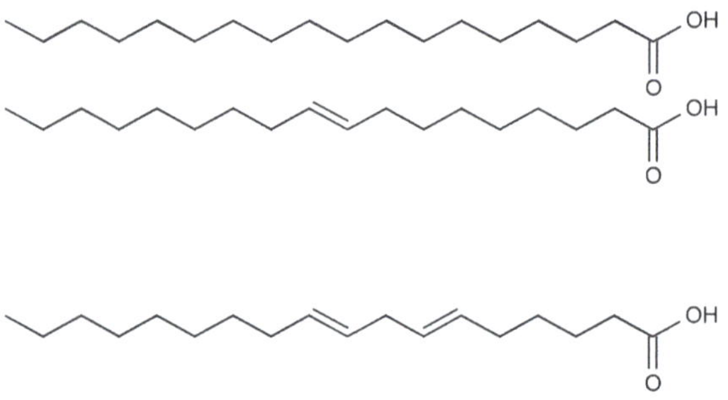

Saturated – Stearic acid (18:0)

Monounsaturated - Oleic acid (18:1)

Poly-unsaturated – Linoleic acid (18:2)

Fig. 3.3 Fatty acid structures

fibers (Fig. 3.5a). Nuts, in addition to their richness in oil, also have a significant amount of protein that is also stored as individualized structures (protein bodies, PB) (Fig. 3.5b).

The emulsifying character of BO allows the development of nano-emulsions capable of carrying hydrophobic molecules as functional components, antioxidant metabolites, vitamins, and drugs, among others (Zhao et al. 2016). Thus, they can be added to the formulation of foodstuffs (Nikiforidis et al. 2014), pharmaceuticals (Hou et al. 2003), and cosmetics (Marcoux et al. 2004) or utilized as biotechnological tools (Peng et al. 2004; Leng et al. 2016; Montesinos et al. 2016).

In fact, the development of oleaginous biomass-based products goes beyond the regular usage of extracted plant oils and fats. The industry of renewable derivatives is a fertile ground for the use of the original matrix of the vegetal tissue, or the products obtained by their modification.

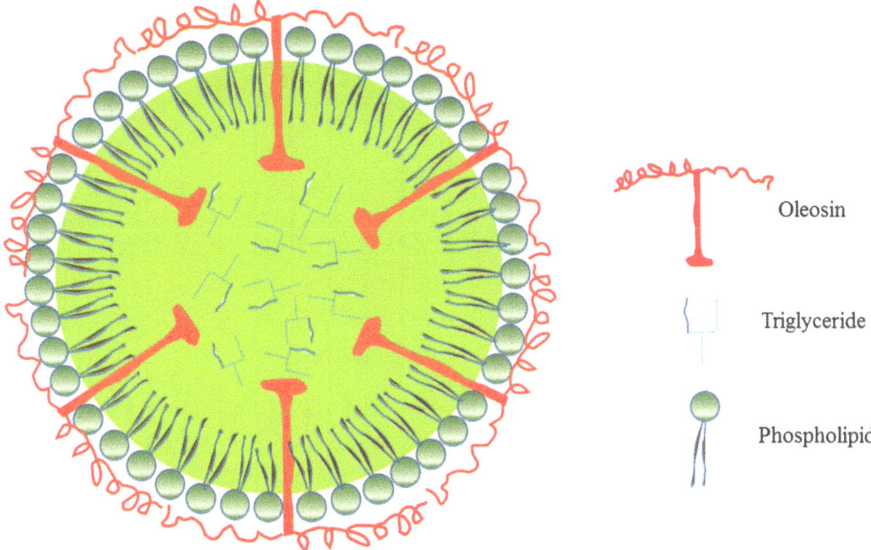

Fig. 3.4 Body oil structural model

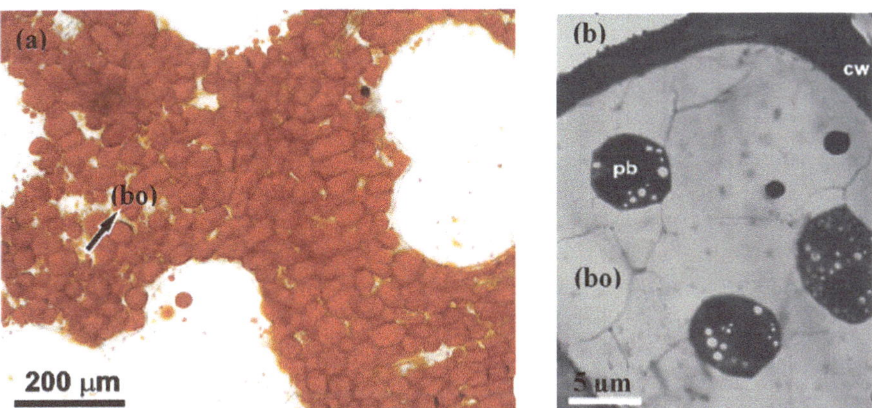

Fig. 3.5 Body oil (*bo*) and body protein (*pb*) in macauba (*Acrocomia aculeata*) pulp (**a**) and kernel (**b**). **a** Light microscopy of pulp body oil colored in *red* with Sudan III (Reis et al. 2012) (Reproduced with permission. Copyright © Rodriguesia). **b** Transmission electron microscopy of kernel (Moura et al. 2010) (Reproduced with permission. Copyright © Scientia Agricola)

3.1.2 Fatty Acids Composition

There is a broad variety of fatty acids composition in nature, varying as functions of oleaginous plant species, the storage organ, and intraspecies variability. Fatty acid composition also is influenced by environmental determinants of plant growth, such as soil characteristics and climate, as well as general post-harvest conditions. Table 3.1 was constructed to show a general view of fatty acids profiles of sources

Table 3.1 Fatty acid composition of different oleaginous biomasses

Sources	Caproic (C:6	Capryli c(C8:0)	Capric (C10:0	Lauric (C12:0)	Miristic (14:0)	Palmitic (C16:0)	Palmitoleic (C16:1)	Estearic (C18:0)	Oleic (C18:1)	Linoleic (C18:2)	Linolenic (C18:3 (9Z,12Z,15Z)	α-Eleostearic (18:3 9Z, 11E,13E)	β-Eleostearic (18:3 9E, 11E,13E)	Punic (18:3 9Z, 11E,13Z)	Arachidic (C20:0)	Gadoleic (20:1)	Behenic (C22:0)	Erucic (22:1)	Nervonic (24:0)	Ricinoleic (18:1 OH)	Iodine value (g of I/100 g of oil)
Indaiá (*Attalea dubia*)[1]	3.5	10.9	8.5	55.8	11.6	3.8		5.8													7,1
Coconut (*Cocus nucifera*)[2]	ND-0.7	4.6-10.0	5.0-8.0	45.1-53.2	16.8-21.0	7.5-10.2		2.0-4.0	5.0-10.0	1.0-2.5	ND-0.2				ND-0.2						6.3-10.6
Babassu (*Attalea speciosa*)[2]		2.6-7.3	1.2-7.6	40.0-55.0	5.2-11.0			1.8-7.4	9.0-20.0	1.4-6.6											10 - 18
Palm Kernel Oil[2]	ND-0.8	2.4-6.2	2.6-5.0	45.0-55.0	14.0-18.0	6.5-10.0		1.0-3.0	12.0-19.0	1.0-3.5											14.1-21.0
Bacuri (*Scheelea phalerata*)[3]		5.8	6.0	35.1	11.5			4.3	22.6	5.2											25,4
Tallow[4]					2.6	25.9	1.2	24.8	39.3	1.7	0.3				0.2						33.3
Macauba (*Acrocomia aculeata*) /kernel [5,6,7]		5.2	4.4	40.5	9.5	8.3	0,1	3,2	25.2	3.4	0,04				0,2						36
Palm Oil (*Elaeis guineenses*)[2]				ND-0.5	0.5-2.0	39.3-47.5	ND-0.6	3.5-6.0	36.0-44.0	9.0-12.0	ND-0.5				ND-1.0	ND-1,0					50.0-55.0
Macauba (*A aculeata*)/pulp [3,7,8,9,10,11]				0.8	0.6	14.8-24.4	1.32-4.6	2.1-4.2	47.05-72.6	1-13	1-4				0.2						66-74
Olive (*Olea europaea*)[12]						7.5-20.0	0.3-3.5	0.5-5.0	55.0-83.0	3.5-21.0											75-94
Jatropha curcas[13]				0.1	0.01	12.2	0.8	0.1	39.0	36.8	0.7				0.3						101
Crambe abyssinica[14,15]							1.7	0.5	18.4	10.8	2.3				0.9		1.6	59.1			88,7
Peanut (*Arachis* hypogaea)[2]				0.1		8.0-14.0		1.0-4.5	35.0-69.0	12.0-43.0	ND-0.3				1.0-2.0	0.7-1.7	1.5-4.5		0.5-2.5		86-107
Rapeseed (*Brassica napus*)[2]						1.5-6.0	ND-3.0	0.5-3.1	8.0-60.0	11.0-23.0	5.0-13.0				ND-3.0	3.0-15.0					94-120
Sunflower(Helianthus annuus)[2]						5.0-7.6		2.7-6.5	14.0-39.4	48.3-74					0.1-0.5		0.3-1.5				94-122
Cotton (*Gossypium spp*)[2]				ND-0.2	0.6-1.0	21.4-26.4	ND-1.2	2.1-3.3	14.7-21.7	46.7-58.2	ND-0.4				0.2-0.5						100-123
Soya bean (*Glycine* max)[2]						8.0-13.5		2.0-5.4	17-30	48.0-59.0	4.5-11.0				0.1-0.6						124-139
Castor (*Ricinus* communis)[16,17]						1.25		1.31	3.81	5.27	0.81					0.38				87,1	125-145
Tung (*Vernicia fordii*)[18,19]						2.3	2.2		6.4	7.23	0.28	75.3	6.3								161,3
Flaxseed (*Linum usitatissimum*) [20,21]						4.4-7.2		3.4-4.2	18.2-19.0	16.0	54.0- 57.1				0.20						170-196
Pomegranate (*Punica granatum*)[22]						2.2-4.2		1.7-2.8	7.6-9.1	7.5-8.8				71-77	0.6-0.8						214-227

[a]Ferreira et al. (2012)
[b]Codex Alimentarius (1999)

[c]Miyahira (2011)
[d]Yılmaz et al. (2010)
[e]Lescano et al. (2015)
[f]Silva et al. (2016)
[g]Oliveira et al. (2017)
[h]Nunes et al. (2015)
[i]da Conceição et al. (2015)
[j]Del Río et al. (2016)
[k]Ciconini (2012)
[l]Codex Alimentarius (1981)
[m]Souza et al. (2016)
[n]Silveira et al. (2016)
[o]Lalas et al. (2012)
[p]Ogunniyi (2006)
[q]Canoira et al. (2010)
[r]Chen et al. (2010)
[s]Zhang et al. (2014)
[t]Anwar et al. (2013)
[u]Khattab and Zeitoun (2013)
[v]Juhaimi et al. (2017)

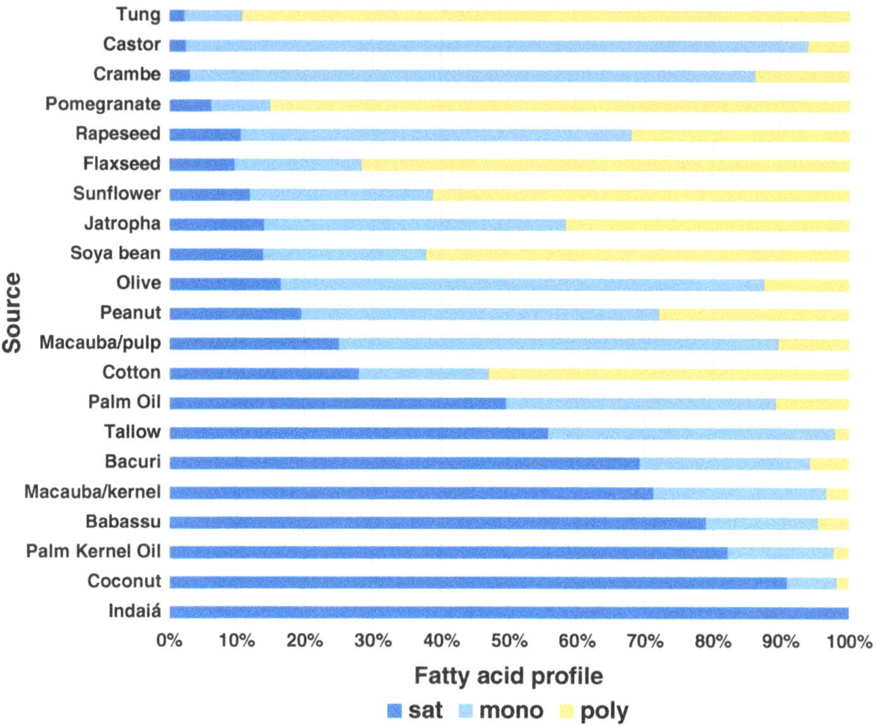

Fig. 3.6 Composition of saturated, monounsaturated, and polyunsaturated fatty acids in fats and oils

that predominantly produce saturated fatty acids up to those with highly unsaturated profiles. The iodine value (IV) is also presented. IV is a reference index for the identity of the producing source, and expresses the total unsaturation of a given oil or fat. IV is largely utilized by the industry as a referential to estimate the melting point and the oxidative stability, being also a quality control measurement of hydrogenation. More accurate methodologies have been developed to relate fatty acid composition to the physicochemical properties of the oils (Knothe 2002), but IV is still commonly used in processes with oleaginous plants and as determinant parameter in oils and fats trading. The mean values of fatty acids shown in Table 3.1 and Fig. 3.6 were assembled to illustrate the composition of the oils in saturated, mono-, and polyunsaturated fatty acids. The information contained in both table and figure contributes to the selection of an oil, or even to establishing mixtures of oils for specific applications.

Palm nuts are the primary sources of saturated fatty acids (Table 3.1 and Fig. 3.6). The indaia palm (*Attalea dubia*) has the lowest IV, with 100% saturated fatty acids, followed by coconut (*Cocos nucifera*), with IV up to 10.6. On the other hand, tallow is the most representative saturated animal fat in use, mainly for biodiesel production (Esteves et al. 2017). On the other extreme are oils from flaxseed (*Linum usita-*

tissimum, IV from 170 to 196) and pomegranate (*Punica granatum*, IV 212 to 212). Oils from crambe (*Crambe abyssinica*), tung (*Vernicia fordii*), and castor show less than 3% saturation. Tung, pomegranate, and flaxseed are rich polyunsaturated sources by the presence of three unsaturations. On the other hand, olive oil and macauba pulp oil show a high proportion of monounsaturation, being rich in oleic acid. Castor oil monounsaturation is the result of the high proportion of ricinoleic fatty acid, which is differentiated from others because it is a hydroxy acid. Erucic acid is responsible for the high proportion of monounsaturations in *Crambe*.

3.2 Vegetable Oils as Biofuels

Research in the usage of vegetable oils as fuel usually aims to replace the fossil fuels gasoline, kerosene, and diesel, derived from petrol. Gasoline is mainly a mixture of chained hydrocarbons, varying from 4 to 12 carbons, obtained by distillation of petrol in a temperature range of 30 °–220 °C; kerosene is a mixture of chained hydrocarbons, varying from 8 to 18 carbons, obtained by distillation within a 150 °–300 °C temperature range; and diesel is a combination of several classes of aliphatic, naphthenic, and aromatic hydrocarbons, and a lower concentration of some composts containing sulfur, nitrogen, and oxygen in a chain of 8–40 carbons (Haddad et al. 2012).

It is possible to regulate and optimize engines to work in accordance with the characteristics of a given fuel, bearing in mind that combustion is influenced by fuel density, viscosity, volatility, and oxidative stability. Apparently, at the beginning of the twentieth century, Rudolph Diesel, inventor of the Diesel cycle engine, successfully tested vegetable oils (Knothe 2001). However, their direct use in current engines may cause carbon deposition, injector blocking, and incomplete combustion because of their high viscosities, low volatilities, and polyunsaturated character, among others (Soares et al. 2008). On the other hand, small modifications in the injection system, or preheating of the fuel line or the fuel itself, can be an alternative solution to run stationary engines and gas turbine engines fully on vegetable oils alone, with some "green" advantages such as their low embodied fossil energy and renewable performance (Soo-Young 2017).

In this sequence, we present the main processes related to the usage of vegetable oils as fuels.

3.2.1 *Vegetable Oil Micro-Emulsions*

Micro-emulsions are small liquid drops of 100–1000 Å formed when two immiscible liquids and a surfactant are mixed, resulting in a system macroscopically homogeneous and thermodynamically stable (Attaphong and Sabatini 2012). Oil is used as a nonpolar phase, with the polar phase being an alcohol, plus a surfactant of

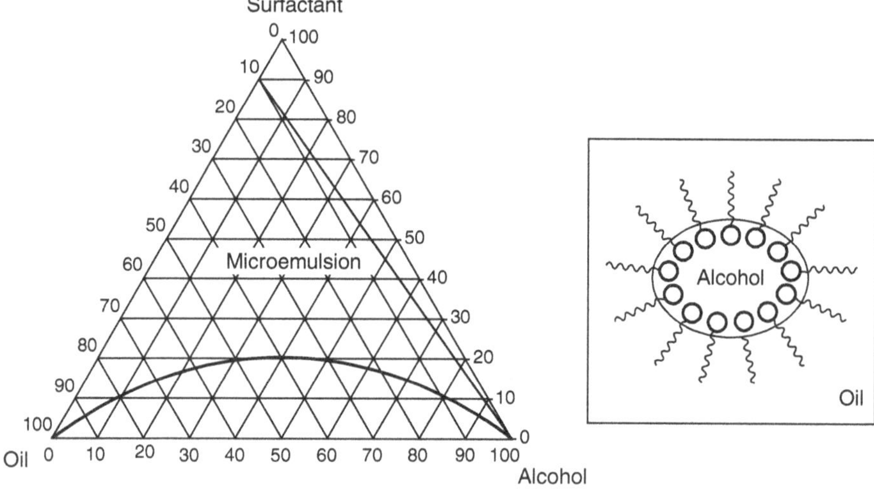

Fig. 3.7 Microemulsion of alcohol in oil

intermediary polarity (Fig. 3.7). The composition of these components can be rearranged in a phase diagram, forming an area of micro-emulsion that can reduce as much as ten times the viscosity of a vegetable oil, although it may give incomplete burning.

3.2.2 Vegetable Oil Cracking

Free fatty acids can be obtained from triglycerides of a vegetable oil after thermal decomposition in the temperature range 300 °–500 °C, releasing, after deoxygenation, chained hydrocarbons such as those present in gasoline, up to diesel (Speight 2008).

Cracking may be thermic or thermocatalytic. In the latter, a catalyst is added to redirect the catalytic route and generate a given product, with the advantage of processing at a lower temperature (Zhao et al. 2017). An example is the use of mesoporous matrixes of silica to generate hydrocarbons in the diesel range (Soltani et al. 2017). Several zeolites have also shown activity in cracking reactions, being useful as support to some metals (Emori et al. 2017). A scheme for a vegetable oil cracking is shown in Fig. 3.8.

3.2.3 Transesterification/Esterification

A transesterification reaction happens between an oil or a fat and an alcohol, usually of short length, in the presence of a catalyst and heating with the release of a monoalkyl ester (Fig. 3.9). If the reaction is between a free triglyceride and an alcohol, it is then called esterification, in which case it is necessary to use an acid catalyst to avoid neutralization reactions.

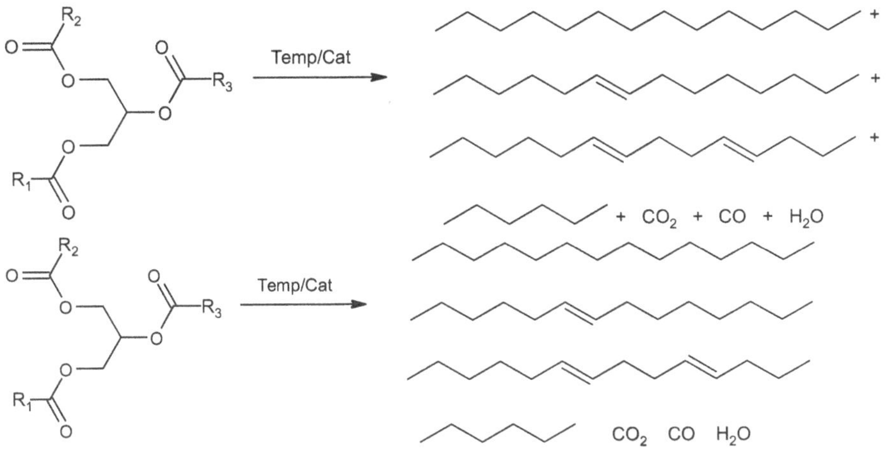

Fig. 3.8 Thermocatalytic cracking of a vegetable oil

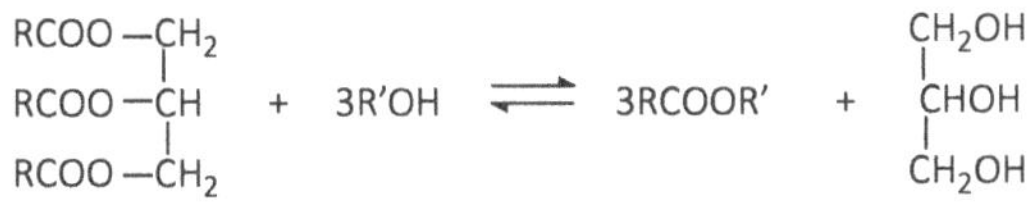

Fig. 3.9 Transesterification reaction scheme

$$\begin{matrix} RCOO-CH_2 \\ RCOO-CH \\ RCOO-CH_2 \end{matrix} + 3R'OH \rightleftharpoons 3RCOOR' + \begin{matrix} CH_2OH \\ CHOH \\ CH_2OH \end{matrix}$$

Table 3.2 Characteristics of diesel derived from petrol and of biodiesel

Property	Diesel oil	Biodiesel
Specific mass (kg l^{-1})	0.883	0.880
Calorific value (MJ^{-}/l^{-1})	38.3	33.3
Viscosity (mm^2/s at 40 °C)	3.86	4.70

The resulting monoalkyl ester shows characteristics similar to those of diesel derived from petrol (Table 3.2), being then denominated as biodiesel (Knothe 2016). Biodiesel may be used as a fuel of its own or in blends with diesel oil. Both their specific mass and viscosity are close, but calorific value of biodiesel is lesser, because it contains more oxygenated molecules.

In the transesterification/esterification reactions, both homogeneous and heterogeneous (acids and bases) catalysts may be used, either chemical or enzymatic (Akoh et al. 2007). Nowadays, industrial processes of transesterification mostly use homogeneous base catalysts, whereas esterification processes are mostly based on homogeneous acid catalysts.

An alternative process is hydroesterification, by which a triglyceride is hydrolyzed to glycerol and acid, and then the acid is converted to ester by esterification. Such a process is preferred when the fatty raw material shows high acidity, and the excess free fatty acids may be an inconvenient path to transesterification (Pourzolfaghar et al. 2016).

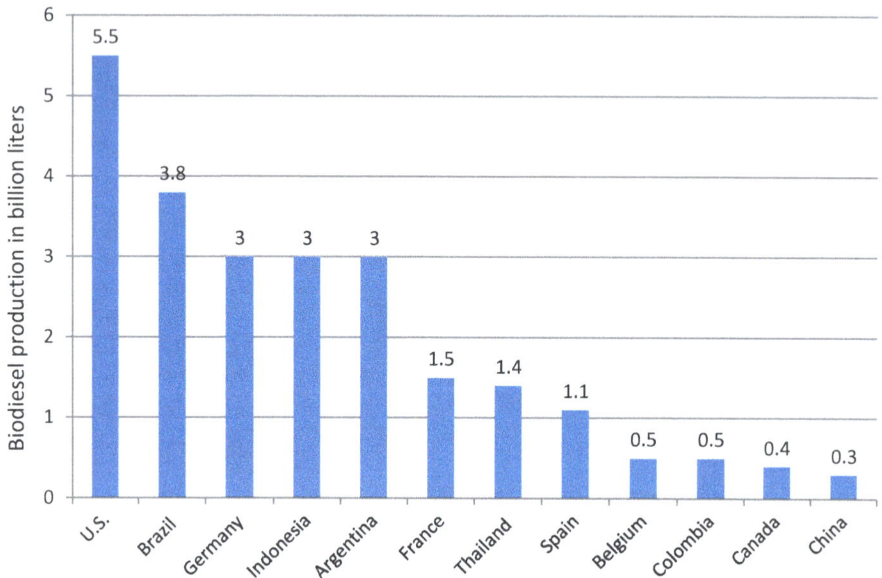

Fig. 3.10 Global biodiesel production by country in 2016

3.2.4 Biodiesel Production

In the past decade, biodiesel production has gained worldwide projection. Nowadays, 64 countries around the world have either targets or mandates to use biofuels, a broad term that includes biodiesel or bioethanol (Biofuelsdigest 2016). Also, the biodiesel production technologies acquired a maturity stage during this time.

The biodiesel industry is based on a few raw material sources, resulting from adjusting well-established crop production systems around the world. Thus, soybean, canola oil, palm oil, cotton seed, and sunflower seed are the main sources in use. In each case, attention is necessary for their saturated and polyunsaturated fatty acids contents, because they may affect the cold flow and stability of the resulting biodiesel. Independently of the oleaginous source used, the use of biodiesel as an alternative fuel has resulted in important environmental, social, and economic gains. The USA (5.5 billion l) and Brazil (3.8 billion l) are the world's largest producers so far, but Argentina, Germany, and Indonesia are also important (Fig. 3.10) (The Statistics Portal, 2017).

3.2.4.1 Biodiesel Quality

Despite its physicochemical similarity to diesel oil, biodiesel has some particularities that need to be looked after when blending it (Rodrigues et al. 2017). It is less stable than diesel, because of the unsaturated chains, caused by oxygen; absorbs

Table 3.3 Biodiesel standards in Europe, USA, and Brazil

Biodiesel standards		Europe	USA	Brazil
Specification		EN 14214:2012	ASTM D 6751-15	Res. 45/2014
Density 15 °C	g $(cm^3)^{-1}$	0.86–0.90		0.85–0.9 (20 °C)
Viscosity 40 °C	mm^2/s	3.5–5.0	1.9–6.0	3.0–6.0
Distillation	% / °C		90%, 360 °C	
Flashpoint (Fp)	°C	101 min	93 min	100 min
CFPP	°C	[a]Country specific		[a]Per region
Cloud point	°C		[a]report	
Sulfur	mg kg^{-1}	10 max	15 max	10 max
Carbon residue	%mass		0.05 max	
Sulfated ash	%mass	0.02 max	0.02 max	0.02 max
Water	mg kg^{-1}	500 max		200 max
Total contamination	mg kg^{-1}	24 max		
Cu corrosion max	3 h/50 °C	1	3	1
Oxidation stability	hrs;110 °C	8 h min	3 h min	8 h min
Cetane number		51 min	47 min	[a]Report
Acid value	mg KOH g^{-1}	0.5 max	0.5 max	0.5 max
Methanol	%mass	0.20 max	0.2 max or Fp < 130 °C	0.20 max
Ester content	%mass	96.5 min		96.5 min
Monoglyceride	%mass	0.7 max	0.4 max	0.7 max
Diglyceride	%mass	0.2 max		0.2 max
Triglyceride	%mass	0.2 max		0.2 max
Free glycerol	%mass	0.02 max	0.02 max	0.02 max
Total glycerol	%mass	0.25 max	0.24 max	0.25 max
Iodine value		120 max	[a]Report	
Linolenic acid ME	%mass	12 max		
Phosphorus	mg kg^{-1}	4 max	10 max	10 max
Na, K	mg kg^{-1}	5 max	5 max	5 max
Ca, Mg	mg kg^{-1}	5 max	5 max	5 max

[a]*CFPP*, cold filter plugging point

water in an easier pattern; and may be obtained from several oleaginous sources, with different chemical and fatty acids profiles. Because of these characteristics, most countries have devised strict policies for quality control of biodiesel use and its blend with diesel (seen in Table 3.3).

Most countries have applied their own legislation regarding biodiesel quality, but for some parameters there is not a given mandatory specific range. In this case, results are monitored and checked for more detailed evaluation if values are far outliers.

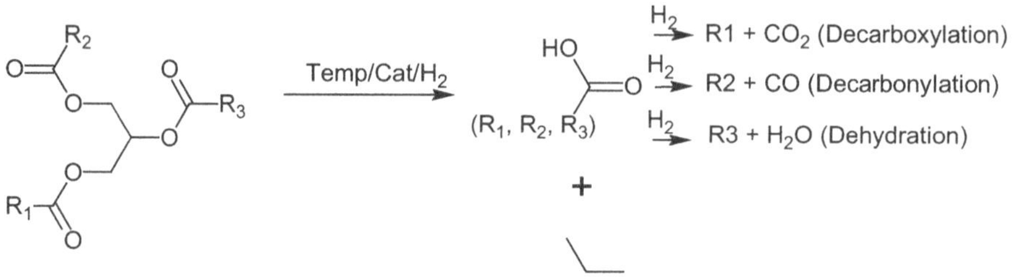

Fig. 3.11 Hydrogenation of a triglyceride

3.2.4.2 Biodiesel Glycerin

Glycerin is an important subproduct of biodiesel production. There is a large market for glycerin, which finds use in the food industry as well as cosmetics and pharmaceuticals. One kilogram (kg) of raw glycerin is generated for every 10 kg of biodiesel produced. As it still contains some residual fatty acids, methanol, and ashes, the raw glycerin must be purified for further use, which, depending on the desired purity, and hence the required purification process, may result in elevated costs. Considering such costs and the large volume produced, several studies are being conducted to find other added-value bioproducts from raw glycerin. Most promising is the use of microorganisms that produce metabolites such as succinic, citric, propionic, lactic, and glyceric acids (Vivek et al. 2017).

3.2.5 Hydrotreating

Hydrotreating is the elimination of heteroatoms from organic compounds by reactions with oxygen, under controlled temperature and pressure, and in the presence of a catalyst such as nickel and molybdenum in a high specific area material. When applied to vegetable oils, a hydroprocessed vegetable oil (HVO) results (Vrtiska and Simacek 2016). To convert vegetable oils to hydrocarbons with the physicochemical characteristics of a diesel oil, which would be then called green diesel, the reaction procedure is of hydrodeoxygenation (HDO) (Pattanaik and Misra 2017; Sugami et al. 2017) (Fig. 3.11).

Last, but not the least, there are other relevant issues to be taken into consideration besides the physicochemical characteristics of the biodiesel when planning to substitute a petrol-derived fuel, as pointed out by Refaat (2009). Perhaps the most important characteristic is the scale of the necessary substitution, as it is a function of raw material availability and affordability: this may be the driving force behind investors' evaluations on the technological route and the raw material.

3.3 Bio-Based Polymers

3.3.1 Functionalization of Vegetable Oils to Produce Bio-Based Polymers

The use of vegetable oils to produce biomaterials such as bio-based polymers has gained great attention for its versatility and abundant availability as well as eco-friendly global initiatives (Lligadas et al. 2013; Fernandes et al. 2017; Mucci et al. 2017). Plant oils-based polymers have shown a powerful capacity of application in varied technological fields to produce coatings, resins, paint, inks, and lubricants (Mucci et al. 2017).

Vegetable oils had been employed in paints and coatings since a long time ago once the unsaturated bonds are able to polymerize when exposed to the air (Van De Mark and Sandefur 2005). Also, biorenewable polymers have been developed by using unmodified vegetable oils once the carbon–carbon double bonds are capable to react by thermal or cationic polymerization as described by Larock's research group (Li and Larock 2003, 2005).

However, depending on the final application or the polymerization route, it is more practical to functionalize its chemical structure. Recently, new strategies to achieve vegetable oil-based polymers also involve functionalized vegetable oils. Use of modified vegetable oils in free radical polymerization, step-growth polymerization, acyclic diene metathesis polymerization (ADMET), and ring-opening metathesis polymerization (ROMP) has been reported in the scientific literature, as well as the available commercial materials based on these biosources (e.g., Vikoflex 7190, Ebecryl 860, Drapex 6.8).

The functionalization of vegetable oils has a fundamental role in polymer chemistry because it is possible to modify some final properties of polymeric matrices, for example, to impart stiffness, or improve the characteristics of commercial polymers to increase the potential application as resin or coating (Wool and Sun 2005).

Basically, the major component of vegetable oils corresponds to triglycerides, whose structure is shown in Fig. 3.12.

The potential sites to functionalize the triglycerides are (i) double bonds, (ii) allylic carbons, (iii) ester group, and (iv) carbons alpha to the ester group (Bonnaillie and Wool 2007). According to the open literature, the most important pathways to modify the vegetable oils for production of bio-based polymers can be organized in these three methods (Khot et al. 2001):

i. Functionalization of double bonds of triglycerides by epoxidation or maleinization followed by attaching of vinyl functionalities to the triglyceride chains
ii. Conversion of triglycerides to monoglycerides through glycerolysis or amidation reaction (ester/glycerol linkage)
iii. Synthesis of monoglycerides or diglycerides through glycerolysis or amidation reaction and functionalization of unsaturations by hydroxylation/acrylation (combination of methods i and ii).

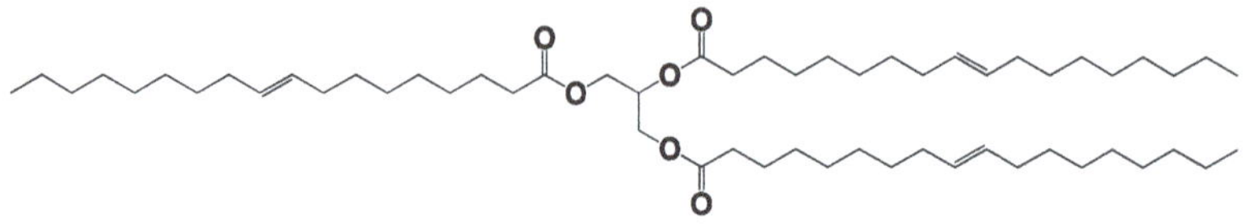

Fig. 3.12 Generic structure of triglycerides

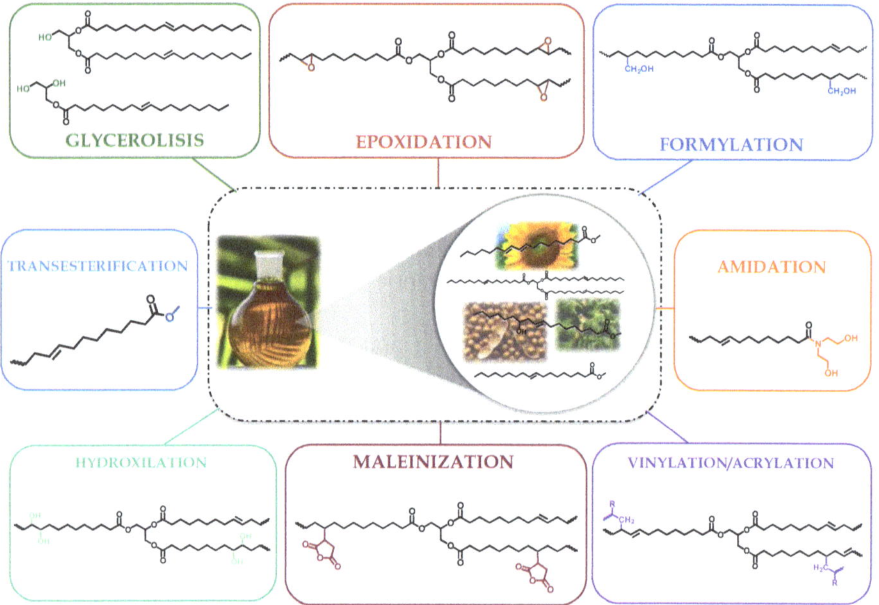

Fig. 3.13 Illustrative scheme with some possibilities of modifying the structure of vegetable oils

Figure 3.13 reviews schematically some important routes of vegetable oil functionalization.

Depending on the modification strategy, the modified monomers are able to polymerize through free radical polymerization or step-growth polymerization. It is noteworthy that natural epoxy or hydroxyl functional triglycerides such as castor oil, *Sterculia striata* (chicha oil), and *Exocarpos cupressiformis* are also easily modified by these cited pathways (Mangas et al. 2012).

Functionalized monomers from vegetable oils as soybean, sunflower, and castor oil are extensively employed to produce bio-based monomers through free radical polymerization (Bonnaillie and Wool 2007; Scala and Wool 2002; Campanella et al. 2010; Jensen et al. 2014; Medeiros et al. 2015). In this scenario, the work developed by Wool (Bonnaillie and Wool 2007), La Scala (Scala and Wool 2002), Campanella (Campanella et al. 2010), Jensen (Jensen et al. 2014), and Medeiros et al. (2015) are examples of functionalization of triglycerides/monoglycerides with in situ generated organic peracid followed by a ring-opening reaction with acrylic acid (Fig. 3.14).

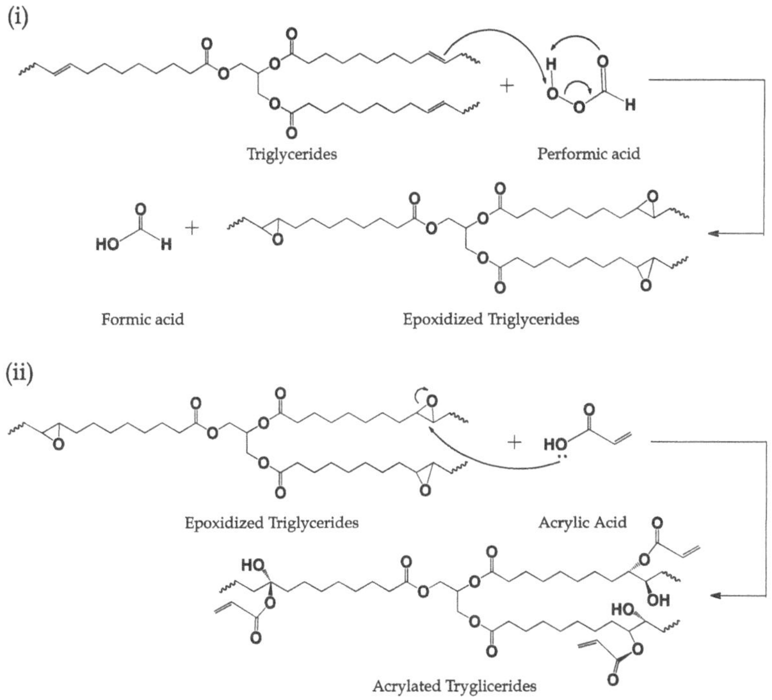

Fig. 3.14 Epoxidation reaction of triglycerides from vegetable oil (i) and ring-opening reaction with acrylic acid (ii) (Medeiros et al. 2015) (Adapted and reproduced with permission. Copyright © 2015 Elsevier Ltd.)

Can et al. (2001, 2002) have reported studies on the functionalization of vegetable oil to produce bio-based thermosettings. In accordance with these studies, maleate half-ester monoglycerides from soybean oil are obtained by two steps. First, the glycerolysis process is carried out to obtain monoglycerides (SOMGs), followed by reaction with maleic anhydride (Fig. 3.15).

In the same way, triglycerides can be functionalized by dicarboxylic acids such as maleic acid and cyclohexane dicarboxylic anhydride to generate oligomers (Fig. 3.16) (Khot et al. 2001). The introduction of cyclic rings into the structure provokes the increase of the entanglement density as well as the stiffness of the polymeric material. According to the authors, the oligomers can be blended with styrene and cured in the same manner as an unmodified AESO resin (Khot et al. 2001).

The use of unsaturated vegetable oils in the polymer field intended to produce bio-based polymeric materials has several advantages, as, for instance, low cost associated with the production process, a large range of structural changes of the vegetable oil and/or polymers, and reduction of the environmental impact by using renewable resources.

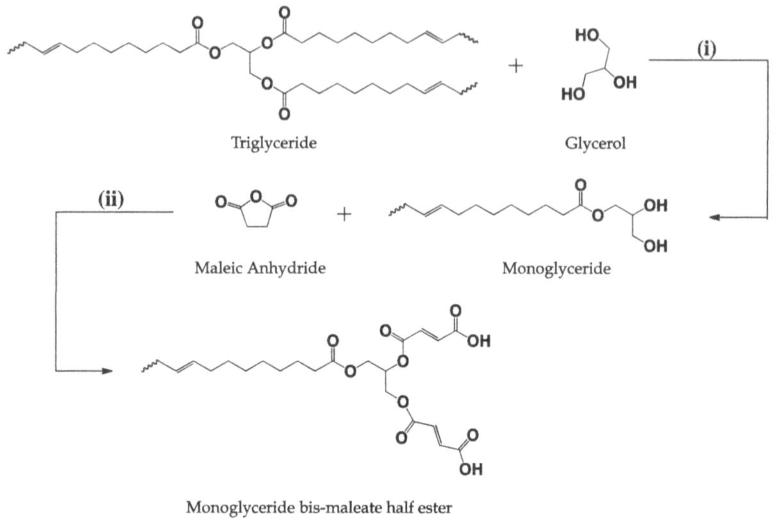

(i)

Triglyceride + Glycerol

(ii)

Maleic Anhydride + Monoglyceride

Monoglyceride bis-maleate half ester

Fig. 3.15 Synthesis of (i) SOMGs and (ii) modified maleinated SOMGs (Can et al. 2002) (Adapted and reproduced with permission. Copyright © 2002 Wiley Periodicals)

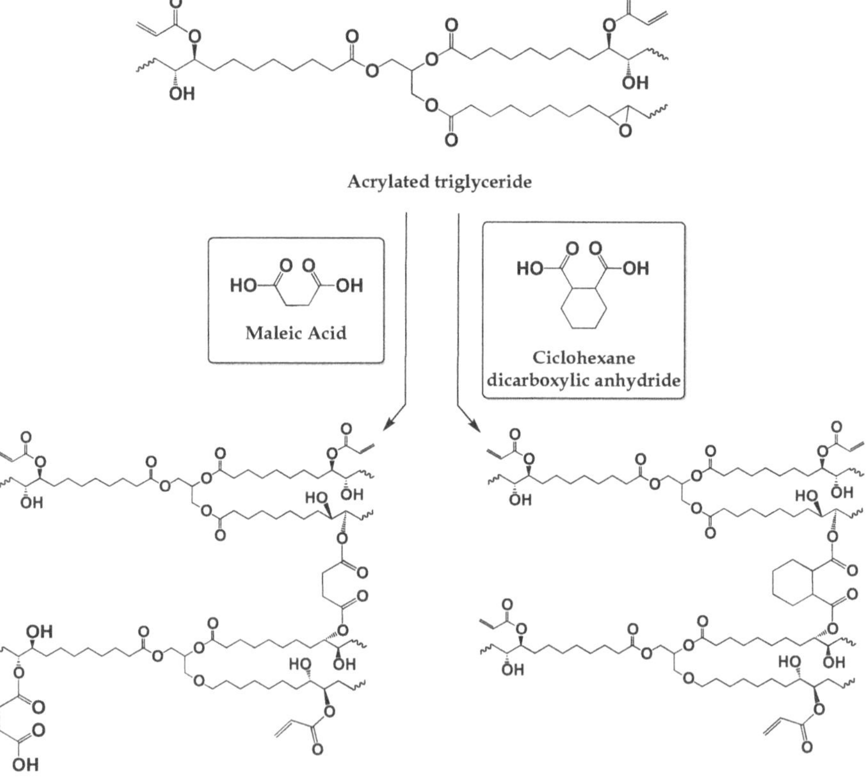

Fig. 3.16 Oligomers from AESO obtained with maleic acid and cyclohexane dicarboxylic anhydride (Khot et al. 2001) (Adapted and reproduced with permission. Copyright © 2001 Wiley Periodicals, Inc.)

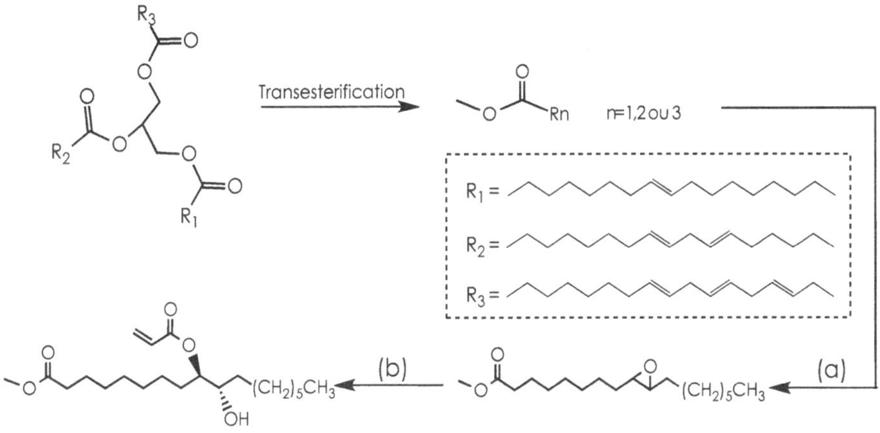

Fig. 3.17 Epoxidation of fatty acid methyl ester (FAME) (a) and ring-opening reaction of epoxidized fatty acid methyl ester (EFAME) (b) (Medeiros et al. 2017) (Reproduced with permission. Copyright © 2017 Wiley Periodicals, Inc.)

In this field, the combination of modified vegetable oils with traditional vinylic monomers is very attractive as new polymeric materials can be successfully synthesized with tailor-made final properties by using the classical industrial polymerization processes, such as, for instance, mass, solution, suspension, emulsion, or mini-emulsion.

Medeiros et al. (2017) have described the synthesis of a bio-based monomer, acrylated fatty acid methyl ester (AFAME), from soybean oil by the epoxidation reaction followed by ring-opening using acrylic acid (see Fig. 3.17). It was demonstrated that the synthesis of poly(styrene-*co*-AFAME) is easily accomplished by free radical copolymerization of styrene and AFAME in a mini-emulsion polymerization process.

Research has shown that increase of AFAME content in the polymeric structure led to a significant decrease in the glass transition temperature of the poly(styrene-co-*AFAME*). As stated by the authors, as the glass transition temperature decreased with increase of AFAME content, the synthesized polymers exhibited improved softness and malleability features (Fig. 3.18).

Ferreira et al. (2015) have evaluated the copolymerization reaction between epoxy-acrylated fatty acids from soybean oil and methyl methacrylate. Figure 3.19 illustrates the experimental steps used for the synthesis of bio-based polymers. It was demonstrated that copolymerization reactions exhibiting high reaction rates can be performed with excellent colloidal stability. In addition, the polymer particles obtained showed a very narrow particle-size distribution, and both average molar mass and glass transition temperature were very dependent on the epoxy-acrylated fatty acid mixture composition in the reaction medium.

In addition, the synthesis of bio-based polymers derived from plant oils has been extensively studied for pressure-sensitive adhesive (PSA) purposes. Numerous

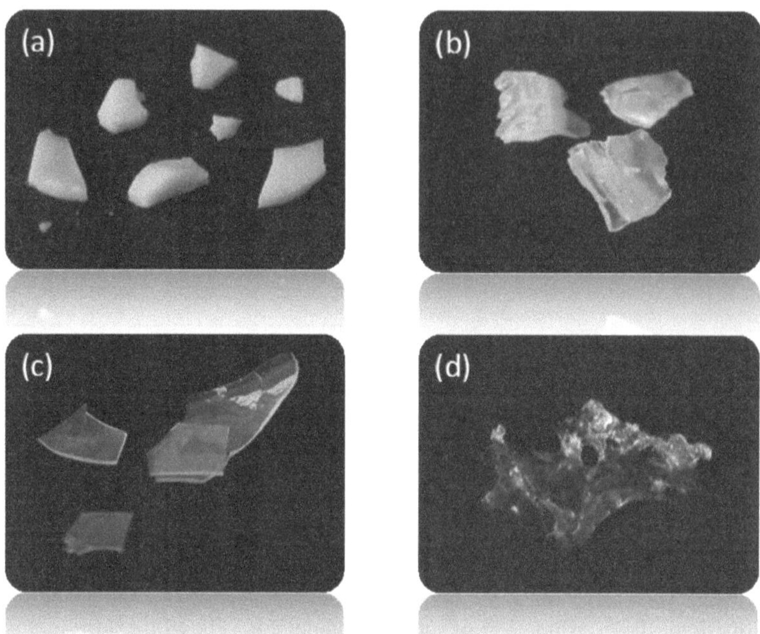

Fig. 3.18 Images of poly(styrene-*co*-AFAME): 100/0 (**a**), 95/5 (**b**), 75/25 e (**c**), 50/50 (**d**) (Medeiros et al. 2017) (Reproduced with permission. Copyright © 2017 Wiley Periodicals, Inc.)

studies have described the synthesis of PSAs from biorenewable feedstock with compatible or improved properties compared to commercial PSAs (Meier et al. 2007; Sharmin et al. 2015; Li and Sun 2015; Maassen et al. 2016; Peykova et al. 2012).

Meier and co-authors have demonstrated the excellent adhesion properties of synthesized bio-based polymers from plant oils. Bio-based polymers were obtained by polymerization of monomers derived from acrylated methyl oleate (AMO). The researchers proposed the modification of bio-based monomer precursors via a one-step, two-step, or three-step route (Fig. 3.20) (Maaßen et al. 2015).

As illustrated in Fig. 3.21, the synthesized AMO homopolymer p(AMO) presented cohesive forces (*left* in the figure) in relation to cure time. The authors affirmed that p(AMO) exhibited pronounced maximum peel strength and tack in about 5 h of curing time because of spanning network formation.

In addition, it was also shown in the study by Meier et al. (Maassen et al. 2016) that better performance, such as adhesive force and peel strength (Fig. 3.22a), and in presence of water (Fig. 3.21b), occurred on low-energy substrates of plant oils-based PSAs as compared with commercial PSAs.

As displayed in Fig. 3.22, it is reasonable to affirm that the homopolymer based on plant oil feedstock (pAMO and pAMO/MMA) achieved better results when compared to commercial PSAs, such as Acronal V212. The graphs indicate that pAMO and pAMO/MMA showed higher adhesive capacity and retained peel strength in water immersion.

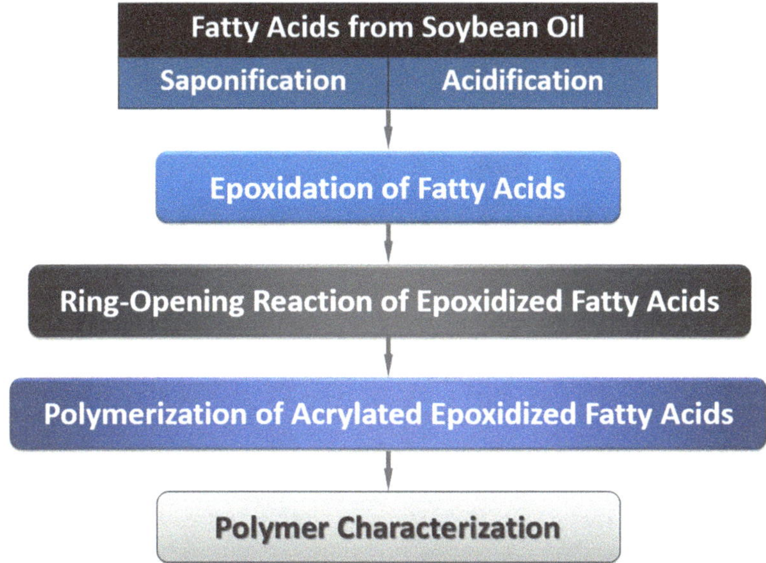

Fig. 3.19 Main experimental steps employed for the synthesis of soybean-based polymeric compounds (Ferreira et al. 2015) (Reproduced with permission. Copyright 2014 © Elsevier)

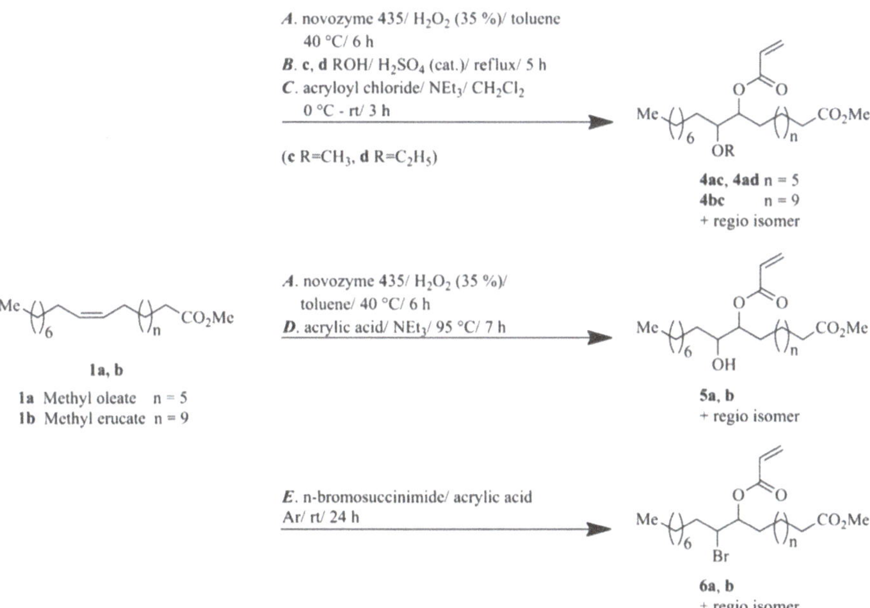

Fig. 3.20 Schematic route related to the synthesis pathways to oleate and erucate derivatives: 4AC (AMO), 4AD, 4BC, 5AB, and 6AB (Maaßen et al. 2015) (Reproduced with permission. Copyright © 2015 Wiley-VCH Verlag GmbH & Co. KGaA, Weinheim)

Fig. 3.21 Photographs of cohesive (*left*) and adhesive (*right*) failure in tack and peel measurements of cured p(AMO) (Maassen et al. 2016) (Reproduced with permission. Copyright © 2015 Elsevier Ltd.)

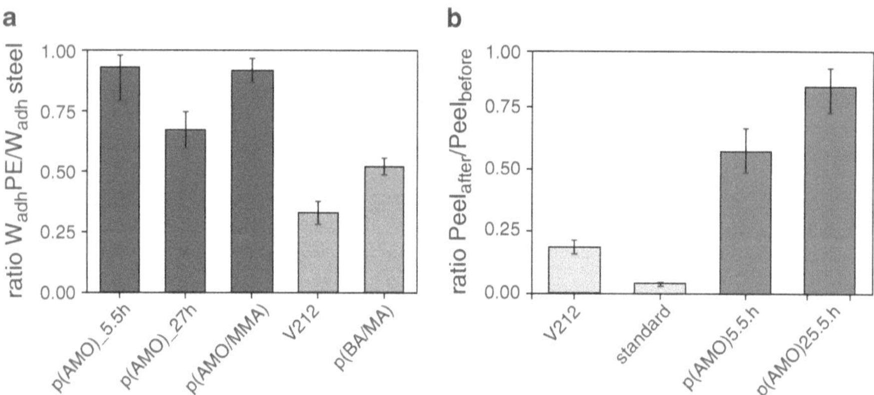

Fig. 3.22 (**a**) Ratio of W_{adh} (adhesion force) measured on polyethylene (PE) and on steel of a cured homopolymer p(AMO) at 5.5 h and of p(AMO) at 27 h compared to acrylate copolymer dispersion Acronal V212 and model solution-based copolymer p(BA/MA). (**b**) Remaining peel strength after 24 h water immersion of acrylate copolymer Acronal V212, a standard office tape (tesa SE product), and cured p(AMO) after 5.5 h and 25.5 h of curing time, respectively (Maassen et al. 2016) (Reproduced with permission. Copyright © 2015 Elsevier Ltd.)

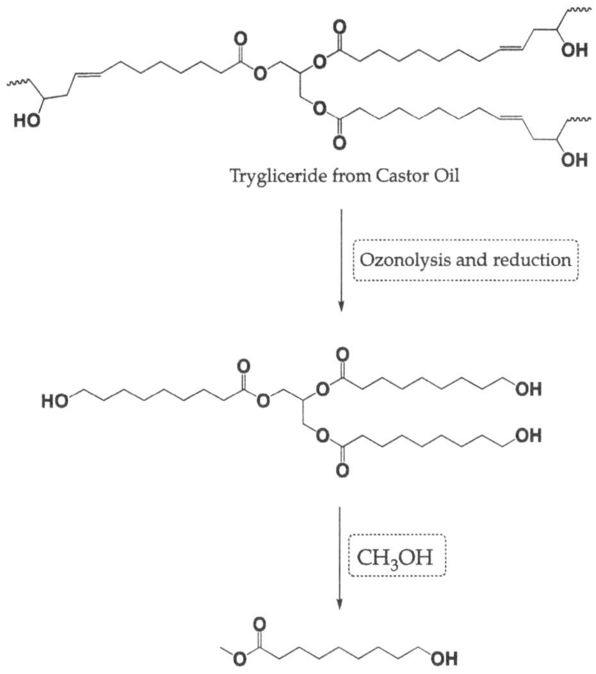

Fig. 3.23 Synthesis of high molecular weight linear polyester (HNME) from castor oil (Petrovic et al. 2010) (Adapted and reproduced with permission. Copyright © 2010, American Chemical Society)

Trygliceride from Castor Oil

Ozonolysis and reduction

CH_3OH

Modified bio-based monomers have been also used to synthesize polymers from a condensation process (Das et al. 2013; Ng et al. 2017; Fan et al. 1999; Deng et al. 1999; Koch 1977; Nayak 2000). Polyurethanes, polyesters, and polyamides obtained from modified bio-based monomers, mainly those derived from castor, soybean, and rapeseed oil, have exhibited suitable properties compared to the commercial products, as, for instance, nylon 22, which has been developed by Noordover and co-authors (Noordover 2011). According to the authors, they have developed a bio-based polyamide from castor oil, one of the most important bio-based polymers, which is used in the manufacture of automotive and engineering components.

Bio-based polyesters from castor oil that have the potential to replace industrial polyester resins have been reported. Slivniak and co-authors (Slivniak et al. 2005, 2006; Slivniak and Domb 2005) demonstrated the synthesis of copolyester produced by different ratios of ricinoleic acid (RA) and lactic acid (LA). The authors achieved liquid polyester at room temperature by random polymerization, using 15% or more than 50% RA, with potential application as a sealant or an injectable drug carrier.

Petrovic et al. (2010) have demonstrated the synthesis of thermoplastic with high molecular weight linear polyester (HNME) by ozonolysis followed by methanolysis of castor oil according to the route depicted in Fig. 3.23. The authors affirmed that the bio-based polyester obtained by self-transesterification of HNME leads to formation of high molecular weight polymeric chains with a structure similar to polycaprolactone (PCL). The presence of long hydrocarbon chains between ester groups allowed it to display better thermal stability (~250 °C), higher melting point (70 °C), higher glass transition temperature (−31 °C), and lower solubility in chlorinated solvents than PCL.

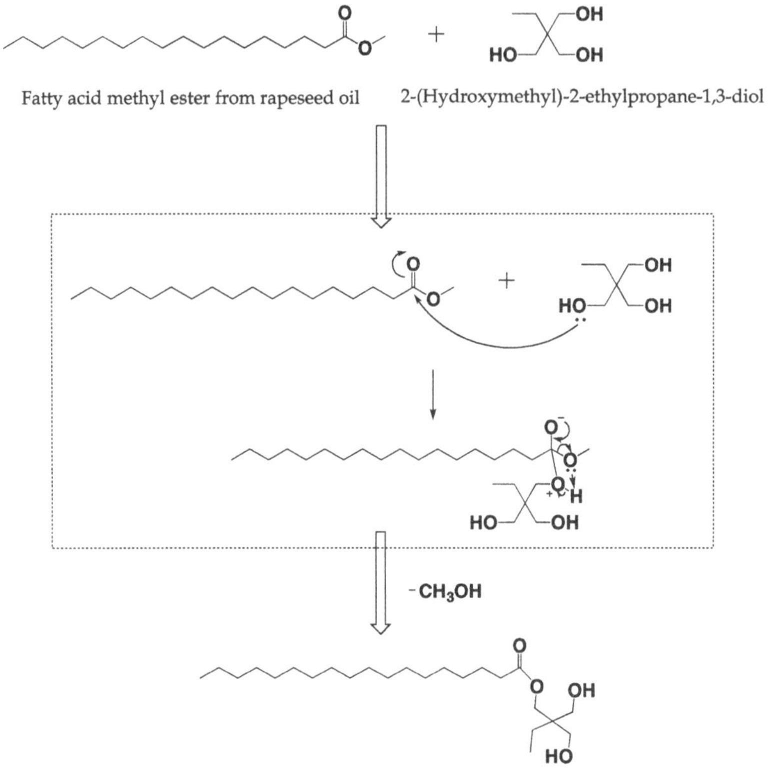

Fatty acid methyl ester from rapeseed oil

2-(Hydroxymethyl)-2-ethylpropane-1,3-diol

$-CH_3OH$

Fig. 3.24 Equimolar use of fatty acid methyl esters and 2-(Hydroxymethyl)-2-ethylpropane-1,3-diol (TMP) leads statistically to a fictive diol (Philipp and Eschig 2012) (Adapted and reproduced with permission. Copyright © 2011 Elsevier B.V.)

Waterborne polyurethanes based on soybean, castor, and rapeseed oil have been also synthesized for use as coating materials (Akram et al. 2017; Das et al. 2013; Ng et al. 2017; Philipp and Eschig 2012; Xia and Larock 2011). Philipp and co-authors (Philipp and Eschig 2012) have evaluated the use of fatty acid methyl esters as alternatives to technical fatty acids and vegetable oils in the synthesis of polyester polyurethane coatings (see Fig. 3.24) and converted to polyurethane dispersions. According to the authors, the use of fatty acid methyl esters leads to a significant reduction of the reaction time during polycondensation. However, as stated by Philipp and collaborators, the use of fatty acids is more favored in coating applications when compared to fatty acid methyl esters because of the unsaturation(s) of some fatty acids (for example, when esters are derived from oleic, linoleic, linolenic, ricinoleic, and other unsaturated compounds).

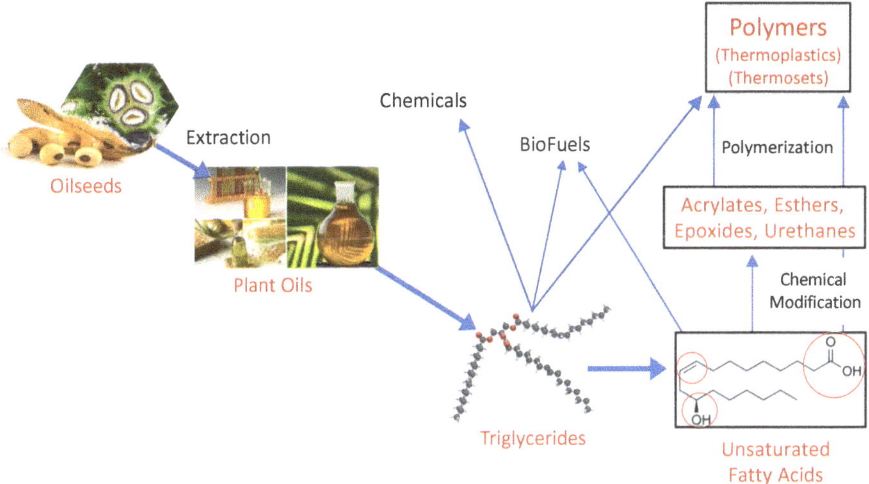

Fig. 3.25 Flowchart for the production, procurement, modification, and use of materials from renewable sources to produce polymers

3.3.2 Vegetable Oil-Based Polymers

Currently, several lines of research are being focused on the potential of vegetable oils and their derivatives in the polymer materials industry. These studies have been aimed mainly at obtaining thermoplastic or thermosetting polymers with different or superior properties in relation to commercial products, as well as the reduction or elimination of the use of petroleum raw material and the possibility of obtaining biodegradable materials that lead to a better response to current principles of sustainability.

In this sense, this section discusses the main polymerization reactions and modifications of synthetic routes of the various plant oils, derivatives, and constituents that are being studied and have potential use in the chemical and polymer industries (Fig. 3.25).

The industrial production of thermosetting polymers achieved approximately 35 million tons in 2015 (MordorIntelligence 2017), which corresponds to 13% of the world production of plastic materials in the same year (about 269 million tons: related to thermoplastics, thermosets, adhesives, coatings, sealants, fibers, and biopolymers, among others), according to *PlasticsEurope* (the Association of Plastics Manufacturers in Europe) (PlasticsEurope 2016). The value of 13% of the total of a market is impressive and justifies that financial investments be made to incorporate and produce biopolymers from renewable matrices.

Thermosetting polymeric materials generally have good chemical resistance, high stiffness, and excellent thermomechanical properties, and they decompose at high temperatures. These properties are closely related to the cross-linking ability of the multifunctional monomers to form cross-linked materials during the polymerization reaction. The main types of thermoset polymer materials are epoxy resins, polyurethane networks, polybenzoxazines, and unsaturated polyesters. Production is mainly focused on the market for adhesives, coatings, molding parts, automotive parts, flooring, electrical insulation, mortars, and other items (Llevot 2017).

Thermoplastic materials have in common the fact that they are formed by an arrangement of linear chains that present as their main characteristic the possibility of being moldable several times when submitted to temperature action. These materials are generally obtained through a chain-growth polymerization mechanism, and the use of monomers from renewable sources has contributed to the development of bio-based and biodegradable thermoplastic elastomers (Maisonneuve et al. 2013).

3.3.3 Analytical Techniques

Regardless of the classification of the polymer material, the final, physical, and chemical properties of polymers define the commercial and industrial applications, as well as the route of manufacture. Properties such as average molar mass, glass transition temperature, thermal stability, tensile and impact strength, stiffness, cohesion, and adhesion are extremely important and require different analytical techniques for characterization.

To produce biomaterials that may replace some commercial polymers obtained from nonrenewable feedstocks, rheological studies essential to characterize the mechanical properties have been carried out by several authors (Li et al. 2017; Hu et al. 2015; Garrison et al. 2014; Lu and Larock 2008; Tüzün et al. 2016). Lu and Larock (2008) and Garrison and collaborators (Garrison et al. 2014) have studied the effect of the fraction of unsaturated compounds (carbon–carbon double bonds), degree of hydroxylation, and ring-opening method on the mechanical properties of polymer films obtained from polyurethane dispersions that were synthesized from polyols derived from vegetable oils (peanuts, soybeans, etc.). Tüzün and coworkers (Tüzün et al. 2016) have obtained films of thermoset polymers based on benzoxazine with different degrees of hardness and flexibility through thermal curing of bis-benzoxazine monomers via the fatty acid metathesis reaction.

Dynamic mechanical analyses that evaluate such factors as tension, hardness, tensile strength, impact and compression, adhesion, cohesion, and tack (Fig. 3.26) are of extreme importance to the characterization of physical properties. For example, the experimental results obtained by Lu and Larock (2008) showed that the mechanical properties (stress–strain) of oil-based polymeric materials can be modulated, varying from elastomers to rigid plastics (Fig. 3.26a), and that the increase of the residual double content raises the glass transition temperature values, toughness, rupture resistance, modulus, and reductions in breakdown stress values. However, enhancement in the degree of hydroxylation and rigid sequencing lead to a rise in the degree of cross-linking and an effective improvement of the hydrogen bonds between the chains.

Stress–strain measurements performed by Tüzün et al. (2016) indicate that materials exhibiting different mechanical properties can be tailor made (Fig. 3.26b) from the ideal choice of a monomer precursor (benzoxazine) containing stiffened ester groups in its main chain.

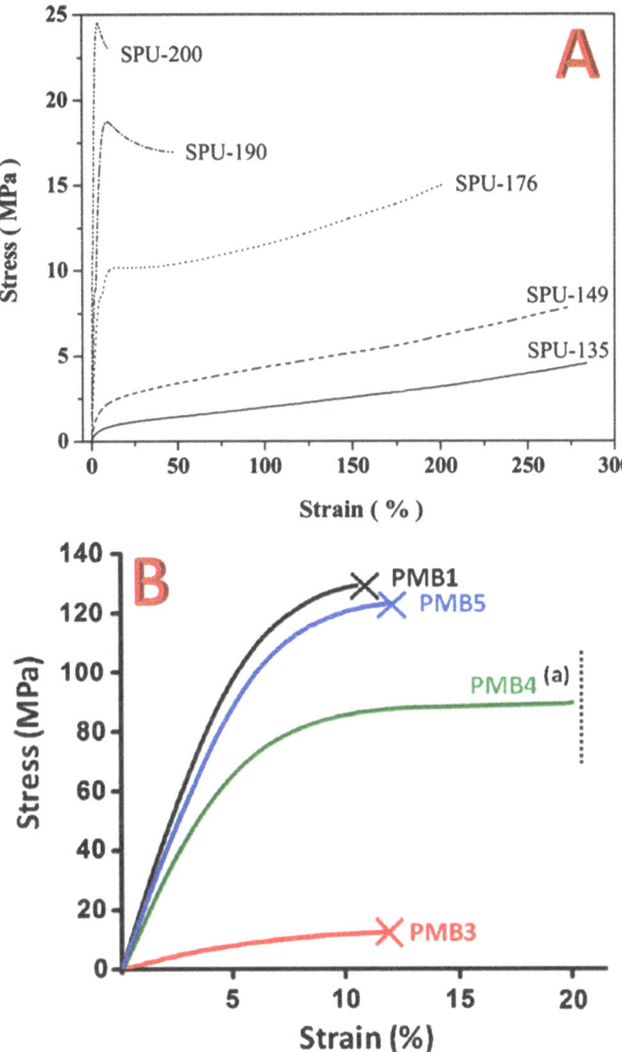

Fig. 3.26 (**A**) Soybean-oil-based waterborne polyurethane (SPU) films from methoxylated soybean oil polyols(MSOLs) with different OH numbers; (**B**) Dynamic mechanical thermal analysis (DMTA) measurements of bis-NPhenybenzoxazine derivatives (MB1, MB4, and MB5) and N-Propyl benzoxazine derivative (MB3) samples [N. B. In Figure (**B**) the letter (a) corresponds to the curve ends at the DMTA measurement limit (no sample break)]

Nuclear magnetic resonance (NMR) appears to be one of the main analytical techniques for monitoring reactions of modification of the reagents/monomers (Bunker and Wool 2002; Sehlinger et al. 2015; Medeiros et al. 2015), being fundamental for the characterization of bio-based materials obtained through the polymerization processes (Medeiros et al. 2015; Lluch et al. 2015; Gratia et al. 2015). NMR allows the analysis of several properties of a polymer (global conversion, microstructure, composition, sequences, etc.) and can also be used in the monitor-

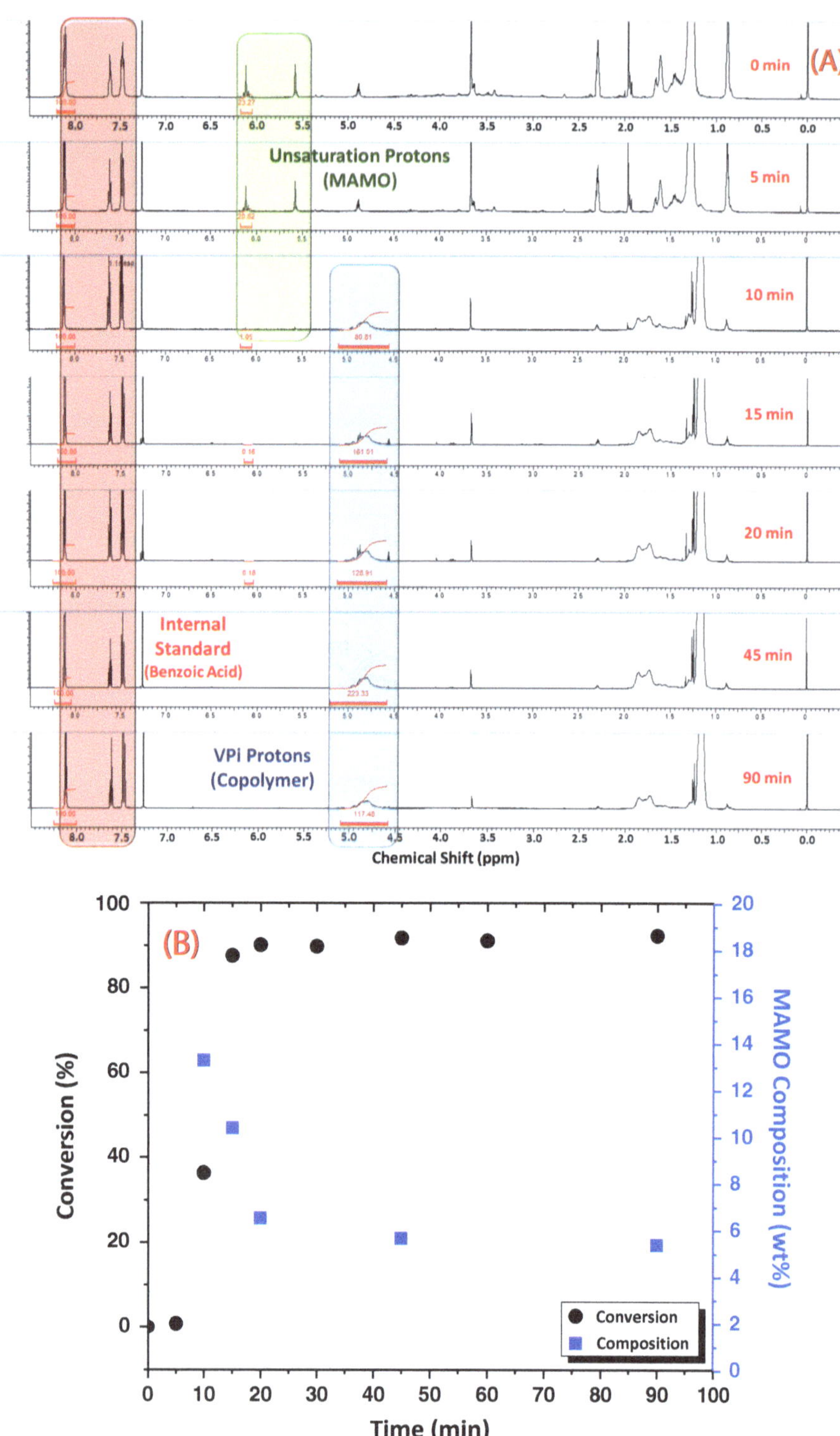

Fig. 3.27 (**a**) Off-line monitoring of the polymerization reaction via quantitative nuclear magnetic resonance (NMR) and the composition profile of methacrylated methyl oleate (MAMO) (**b**) with polymerization reaction (Jensen et al. 2016) (Reproduced with permission. Copyright © 2016 Wiley Periodicals, Inc.)

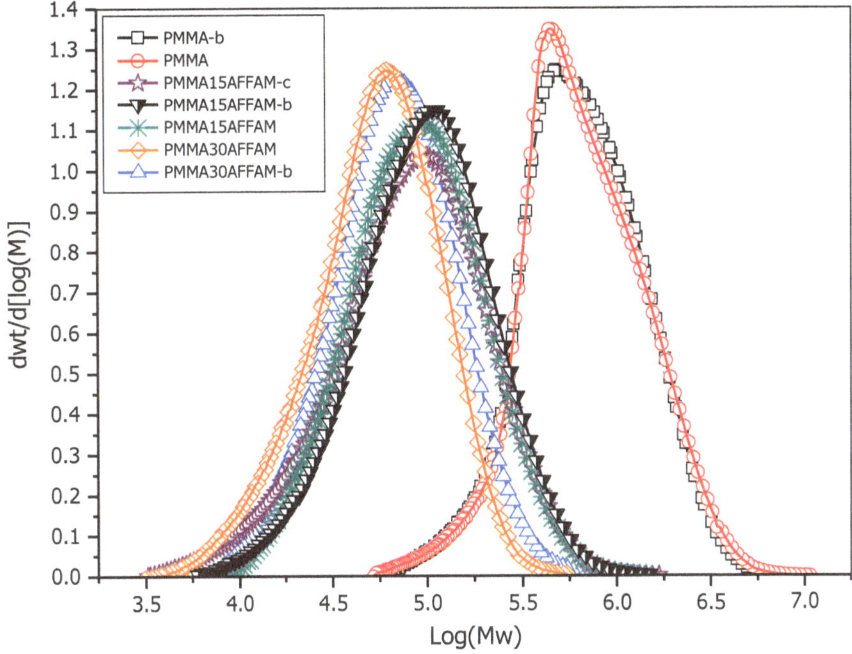

Fig. 3.28 Molar mass distributions of copolymers characterized by gel permeation chromatography (GPC) (Ferreira et al. 2015) (Reproduced with permission. Copyright 2014 © Elsevier)

ing and characterization of the molar fraction and reactivity of the monomers in copolymerization reactions. Figure 3.27 portrays experimental results obtained by Jensen et al. (2016) during classic emulsion copolymerization of a modified fatty acid (methacrylated methyl oleate, MAMO) with a vinyl monomer, vinyl pivalate (VPi), where the copolymer composition was monitored off line via quantitative NMR. Based on the NMR measurements, it was verified that MAMO was effectively incorporated into the polymer chains and that this vegetable oil-based monomer presents significant reactivity compared to VPi, which means that oleic acid-derived monomers can be successfully used in dispersed medium polymerizations, such as suspension, emulsion, or mini-emulsion.

The final properties of a polymer are directly influenced by both the average molar masses and molar mass dispersity. In spite of the existence of different forms of characterization [for instance, dynamic light scattering and NMR (Türünç et al. 2011)], gel permeation chromatography (GPC) (Türünç and Meier 2010; Kolb et al. 2014) is currently the most widely used characterization technique despite the need for particular solvent solubility and/or external calibration standards, depending on the type of detector. The molar mass of a polymer can be controlled by several methods and can also vary with the polymerization process being employed. Thus, Ferreira et al. (2015) evaluated the emulsion copolymerization of an acrylated monomers mixture (AFFAM) from soybean oil (see Fig. 3.18). The molar mass distributions obtained via GPC (Fig. 3.28) indicate that a reduction of the mass-average molar mass occurs as a result of the increase in the AFFAM fraction in the copolymer chains.

Thermal analyses such as thermogravimetry (TG) and differential scanning calorimetry (DSC) are important for the characterization of some properties of the polymeric materials, such as glass transition temperature of amorphous or semi-crystalline polymers and melting and crystallization temperatures. In this scenario, glass transition temperature appears as one of the most important features of copolymeric materials, because these materials may have distinct characteristics that depend on the organization of the polymer chains. Thus, studies of substances from renewable sources as monomers in polymerization reactions use DSC to evaluate different thermal events, as well as to establish a relationship between the renewable monomer content and the physical and chemical properties of the bio-based material.

Miao et al. (2013) have synthesized polymers based on soybean oil that present structural memory properties and which may have their shape altered at temperatures that vary according to the glass transition temperature (Tg) of the materials. Caillol et al. (2012) and Miao et al. (2010) have observed an increase in the Tg of thermosetting materials because there is an increase in the fraction of hydroxyl groups in the chains of polymerizable vegetable oils, whereas several studies (Liu et al. 2015; Ferreira et al. 2015; Jensen et al. 2014, 2016) have shown that reduction in the Tg of the polymer may reflect a primary effect of the increase in the fraction of modified vegetable oils incorporated into the polymer chains (Fig. 3.29) (Jensen et al. 2016).

Péres and collaborators (2014) have developed a superparamagnetic biopolyester based on ricinoleic acid and magnetite nanoparticles. It was observed that the polymerization of ricinoleic acid with surface-modified iron oxide magnetic nanoparticles presented a high reaction rate, indicating a catalytic effect attributed to the presence of the magnetite nanoparticles. As additional information, the magnetic nano-composites exhibited good magnetic response and superparamagnetic behavior.

More recently, Péres and coworkers (2017) have evaluated the synthesis of a new bio-based magnetic poly(urethane ester) from ricinoleic acid, 1,6-diisocyanatehexane, and glycerol (Fig. 3.30). According to the authors, the observed increase in the polymer chains enhances the thermal stability of the final material. It was also observed that the glass transition temperature of the superparamagnetic bio-based poly(urethane ester) was significantly increased in comparison to that observed in poly(urethane ester).

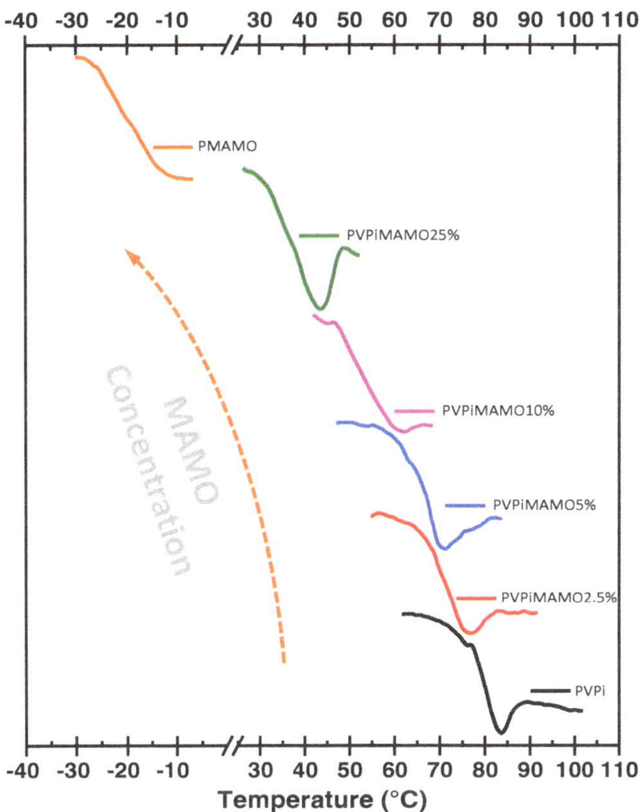

Fig. 3.29 Glass transition temperatures (Tg) and the effect of increasing the MAMO fraction on the copolymer chains obtained with vinyl pivalate (VPi) (Jensen et al. 2016) (Reproduced with permission. Copyright © 2016 Wiley Periodicals, Inc.)

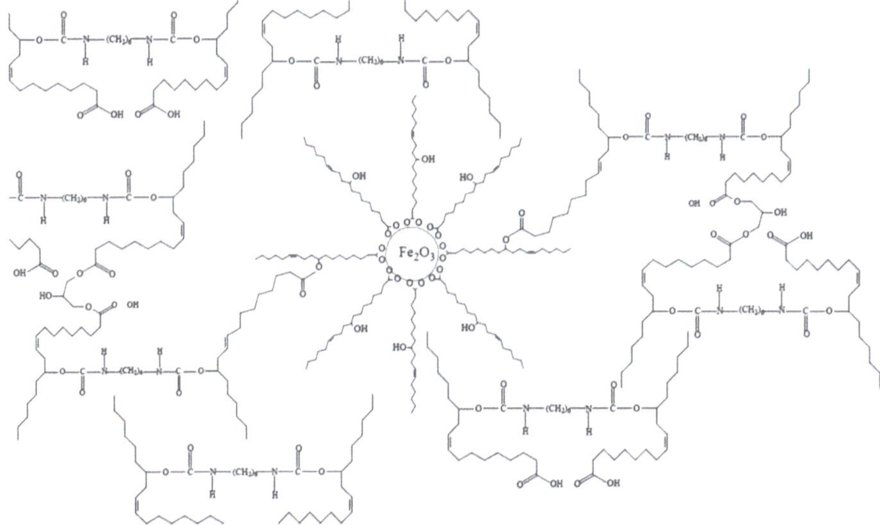

Fig. 3.30 Representation of the magnetic poly(urethane ester) based on ricinoleic acid, 1,6-diisocyanatehexane, and glycerol (Péres et al. 2017) (Reproduced with permission. Copyright © 2017 Wiley-VCH Verlag GmbH & Co. KGaA, Weinheim)

References

Akoh CC, Chang SW, Lee GC, Shaw JF (2007) Enzymatic approach to biodiesel production. J Agric Food Chem 55:8995–9005. https://doi.org/10.1021/jf071724y

Akram D, Hakami O, Sharmin E, Ahmad S (2017) Castor and Linseed oil polyurethane/TEOS hybrids as protective coatings: a synergistic approach utilising plant oil polyols, a sustainable resource. Prog Org Coat 108:1–14. https://doi.org/10.1016/j.porgcoat.2017.03.012

Anwar F, Zreen Z, Sultana B, Jamil A (2013) Enzyme-aided cold pressing of flaxseed (*Linum usitatissimum* L.): enhancement in yield, quality and phenolics of the oil. Grasas y Aceites 64:463–471. https://doi.org/10.3989/gya.132212

Attaphong C, Sabatini DA (2012) Vegetable oil-based microemulsions using carboxylate-based extended surfactants and their potential as an alternative renewable biofuel. Fuel 94:606–613

Beisson F, Ferte N, Noat G (1996) Oil bodies from sunflower (*Heliantus annuus* L.) seeds. Biochem J 317:955–958

Biofuelsdigest (2016) Biofuels mandates around the world 2017. http://www.biofuelsdigest.com/bdigest/2016/12/28/biofuels-mandates-around-the-world-2017/. Accessed 08 Jun 2017

Bonnaillie LM, Wool RP (2007) Thermosetting foam with a high bio-based content from acrylated epoxidized soybean oil and carbon dioxide. J Appl Polym Sci 105(3):1042–1052. https://doi.org/10.1002/app.26182

Bousba H, Liazid A, Sary A, Tazerout M (2013) Numerical investigations on the use of waste animal fat as fuel on DI diesel engine. J Pet Technol Altern Fuels 47:143–154

Bunker SP, Wool RP (2002) Synthesis and characterization of monomers and polymers for adhesives from methyl oleate. J Polym Sci A Polym Chem 40(4):451–458. https://doi.org/10.1002/pola.10130

Caillol S, Desroches M, Boutevin G, Loubat C, Auvergne R, Boutevin B (2012) Synthesis of new polyester polyols from epoxidized vegetable oils and biobased acids. Eur J Lipid Sci Technol 114(12):1447–1459. https://doi.org/10.1002/ejlt.201200199

Campanella A, Rustoy E, Baldessari A, Baltanás MA (2010) Lubricants from chemically modified vegetable oils. Bioresour Technol 101(1):245–254. https://doi.org/10.1016/j.biortech.2009.08.035

Can E, Küsefoğlu S, Wool RP (2001) Rigid, thermosetting liquid molding resins from renewable resources. I. Synthesis and polymerization of soy oil monoglyceride maleates. J Appl Polym Sci 81(1):69–77. https://doi.org/10.1002/app.1414

Can E, Küsefoğlu S, Wool RP (2002) Rigid thermosetting liquid molding resins from renewable resources. II. Copolymers of soybean oil monoglyceride maleates with neopentyl glycol and bisphenol A maleates. J Appl Polym Sci 83(5):972–980. https://doi.org/10.1002/app.2277

Canoira L, Galean JG, Alcantara R, Lapuerta M, Garcia-Contreras R (2010) Fatty acid methyl esters (FAMEs) from castor oil: production process assessment and synergistic effects in its properties. Renew Energy 35:208–217

Chen YH, Chen JH, Chang CY, Chang CC (2010) Biodiesel production from tung (*Vernicia montana*) oil and its blending properties in different fatty acid composition. Bioresour Technol 101:9521–9526

Christie WW (2017) What is a lipid? AOCS Library. http://lipidlibrary.aocs.org. Accessed 20 May 2017

Ciconini G (2012) Characterization of fruits and pulp oil of macauba palm in the Cerrado and Pantanal of Mato Grosso do Sul State, Brazil (Caracterização de frutos e óleo de polpa de macaúba dos biomas Cerrado e Pantanal do estado de Mato Grosso do Sul, Brasil). Dissertation, Universidade Católica Dom Bosco

Codex Alimentarius (1981) Codex alimentarius standards for named vegetable oils codex stan 33. www.fao.org/input/download/standards/88/CXS_033e_2015.pdf. Accessed 20 May 2017

Codex Alimentarius (1999) Standards for fats and oils from vegetable sources. Codex Stan 210. www.fao.org/input/download/standards/.../CXS_210e_2015.pdf. Accessed 20 May 2017

Conceição LDHCS, Antoniassi R, Junqueira NTV, Braga MF, Machado AFF, Rogério JB, Durarte ID, Bizzo HR (2015) Genetic diversity of macauba from natural populations of Brazil. BMC Res Notes 8:406–414

Cyber Lipid Center (2017) http://cyberlipid.org. Accessed 20 May 2017

Das B, Konwar U, Mandal M, Karak N (2013) Sunflower oil based biodegradable hyperbranched polyurethane as a thin film material. Ind Crop Prod 44:396–404. https://doi.org/10.1016/j.indcrop.2012.11.028

Del Río JC, Evaristo AB, Marques G, Martín-Ramos P, Martín-Gil J, Gutiérrez A (2016) Chemical composition and thermal behavior of the pulp and kernel oils from macauba palm (*Acrocomia aculeata*) fruit. Ind Crop Prod 84:294–304

Deng YL, Fan XD, Waterhouse J (1999) Synthesis and characterization of soy-based copolyamides with different α-amino acids. J Appl Polym Sci 73(6):1081–1088. https://doi.org/10.1002/(SICI)1097-4628(19990808)73:6<1081::AID-APP27>3.0.CO;2-J

Emori EY, Hirashima FH, Zandonai CH, Ortiz-Bravo CA, Fernandes-Machado NRC, Olsen-Scaliante MHN (2017) Catalytic cracking of soybean oil using ZSM5 zeolite. Catal Today 279:168–176. https://doi.org/10.1016/j.cattod.2016.05.052

Esteves VPP, Esteves EMM, Bungenstab DJ, Feijó GLD, Araújo OQF, Morgado CRV (2017) Assessment of greenhouse gases (GHG) emissions from the tallow biodiesel production chain including land use change (LUC). J Clean Prod 151:578–591. https://doi.org/10.1016/j.jclepro.2017.03.063

Fan XD, Deng YL, Waterhouse J, Pfromm P, Carr WW (1999) Development of an easily deinkable copy toner using soy-based copolyamides. J Appl Polym Sci 74(6):1563–1570. https://doi.org/10.1002/(SICI)1097-4628(19991107)74:6<1563::AID-APP31>3.0.CO;2-0

Fernandes FC, Kirwan K, Lehane D, Coles SR (2017) Epoxy resin blends and composites from waste vegetable oil. Eur Polym J 89:449–460. https://doi.org/10.1016/j.eurpolymj.2017.02.005

Ferreira BS, Faza LP, Hyaric MA (2012) Comparison of the physicochemical properties and fatty acid composition of Indaiá (*Attalea dubia*) and Babassu (*Orbignya phalerata*) oils. Sci World J 2012:1–4

Ferreira GR, Braquehais JR, da Silva WN, Machado F (2015) Synthesis of soybean oil-based polymer lattices via emulsion polymerization process. Ind Crop Prod 65:14–20. https://doi.org/10.1016/j.indcrop.2014.11.042

Furse S, Liddell S, Ortore CA, Williams H, Neylon DC, Scott DJ, Barret DJ, Gray DA (2013) The lipidome and proteome of oil bodies from *Helianthus annuus* (common sunflower). J Chem Biol 6:63–76. https://doi.org/10.1007/s12154-012-0090-1

Garrison TF, Kessler MR, Larock RC (2014) Effects of unsaturation and different ring-opening methods on the properties of vegetable oil-based polyurethane coatings. Polymer 55(4):1004–1011. https://doi.org/10.1016/j.polymer.2014.01.014

Gratia A, Merlet D, Ducruet V, Lyathaud C (2015) A comprehensive NMR methodology to assess the composition of biobased and biodegradable polymers in contact with food. Anal Chim Acta 853:477–485. https://doi.org/10.1016/j.aca.2014.09.046

Gunstone FD (2013) Biodiesel. Oils and fats in the marketplace. AOCS Library. http://lipidlibrary.aocs.org. Accessed 20 May 2017

Haddad R, Regiani T, Klitze CF, Sanvido GB, Corilo YE, Augusti DV, Pasa VMD, Pereira RCC, Eomao W, Vaz BG, Augusti R, Eberlin MN (2012) Gasoline, kerosene, and diesel fingerprinting via polar markers. Energy Fuel 26:3542–3547. https://doi.org/10.1021/ef300277c

Hou RC, Lin MY, Wang MM, Tzen JT (2003) Increase of viability of entrapped cells of *Lactobacillus delbrueckii* ssp. *bulgaricus* in artificial sesame oil emulsions. J Dairy Sci 86:424–428. https://doi.org/10.3168/jds.S0022-0302(03)73620-0

Hu F, Yadav SK, La Scala JJ, Sadler JM, Palmese GR (2015) Preparation and characterization of fully furan-based renewable thermosetting epoxy-amine systems. Macromol Chem Phys 216(13):1441–1446. https://doi.org/10.1002/macp.201500142

Huang AHC (1992) Oil bodies and oleosins in seeds. Annu Rev Plant Physiol Plant Mol Biol 43:177–200. https://doi.org/10.1146/annurev.pp.43.060192.001141

Jensen AT, Sayer C, Araújo PHH, Machado F (2014) Emulsion copolymerization of styrene and acrylated methyl oleate. Eur J Lipid Sci Technol 116(1):37–43. https://doi.org/10.1002/ejlt.201300212

Jensen AT, de Oliveira ACC, Gonçalves SB, Gambetta R, Machado F (2016) Evaluation of the emulsion copolymerization of vinyl pivalate and methacrylated methyl oleate. J Appl Polym Sci 133(45):44129. https://doi.org/10.1002/app.44129

Johnson LA, Myers DJ (1995) Industrial uses for soybeans. In: Erickson DR (ed) Practical handbook of soybean processing and utilization. AOCS Press/United Soybean Board, Urbana, pp 380–427

Juhaimi FA, Özcan MM, Ghafoor K (2017) Characterization of pomegranate (*Punica granatum* L.) seed and oils. Eur J Lip Sci Tech. https://doi.org/10.1002/ejlt.207700124

Khattab RY, Zeitoun MA (2013) Quality evaluation of flaxseed oil obtained by different extraction techniques. LWT Food Sci Technol 53:338–345. https://doi.org/10.1016/j.lwt.2013.01.004

Khot SN, Lascala JJ, Can E, Morye SS, Williams GI, Palmese GR, Kusefoglu SH, Wool RP (2001) Development and application of triglyceride-based polymers and composites. J Appl Polym Sci 82(3):703–723. https://doi.org/10.1002/app.1897

Knothe G (2001) Historical perspectives on vegetable oil-based diesel fuels. AOCS Inform 12:1003–1007

Knothe G (2002) Structure indices in FA chemistry. How relevant is the iodine value? J Am Oil Chem Soc 79:847–854

Knothe G (2010) Biodiesel: current trends and properties. Top Catal 53:714–720. https://doi.org/10.1007/s11244-010-9457-0

Knothe G (2016) Biodiesel and its properties. In: McKeon T, Hayes T, Hildebrand D, et al (eds) Industrial oil crops. Academic Press, New York, p 474; Chapter 2, pp 15–42

Koch R (1977) Nylon 11. Polym News 3:302–307

Kolb N, Winkler M, Syldatk C, Meier MAR (2014) Long-chain polyesters and polyamides from biochemically derived fatty acids. Eur Polym J 51:159–166. https://doi.org/10.1016/j.eurpolymj.2013.11.007

Lalas S, Gortzi O, Athanasiadis V, Dourgtoglou E, Dortoglou V (2012) Full characterization of *Crambe abyssinica* Hochst. seed oil. J Am Chem Soc 89:2253–2258. https://doi.org/10.1007/s11746-012-2122-y

Lehninger AL, Nelson DL, Cox MM (2000) Lehninger principles of biochemistry. Worth Publishers, New York

Leng SH, Yang CE, Tsai SL (2016) Designer oleosomes as efficient biocatalysts for enhanced degradation of organophosphate nerve agents. Chem Eng J 287:568–557

Lescano CH, Oliveira IP, Silva LR, Baldivia DS, Sanjinez-Argandoña EJ, Arruda EJ, Moraes ICF, Lima FF (2015) Nutrients content, characterization and oil extraction from *Acrocomia aculeata* (Jacq.) Lodd. fruits. Afr J Food Sci 9:113–119. https://doi.org/10.5897/AJFS2014.1212

Li FK, Larock RC (2003) Synthesis, structure and properties of new tung oil-styrene-divinylbenzene copolymers prepared by thermal polymerization. Biomacromolecules 4:1018–1025. https://doi.org/10.1021/bm034049j

Li FK, Larock RC (2005) Natural fibers, biopolymers and biocomposites. CRC Press, Baton Rouge

Li Y, Sun XS (2015) Camelina oil derivatives and adhesion properties. Ind Crop Prod 73:73–80. https://doi.org/10.1016/j.indcrop.2015.04.015

Li Y, Yang L, Zhang H, Tang Z (2017) Synthesis and curing performance of a novel bio-based epoxy monomer from soybean oil. Eur J Lipid Sci Technol 119:1600429. https://doi.org/10.1002/ejlt.201600429

Lin LJ, Tzen JT (2004) Two distinct steroleosins are present in seed oil bodies. Plant Physiol Biochem 42:601–608. https://doi.org/10.1016/j.plaphy.2004.06.006

Liu K, Madbouly SA, Kessler MR (2015) Biorenewable thermosetting copolymer based on soybean oil and eugenol. Eur Polym J 69:16–28. https://doi.org/10.1016/j.eurpolymj.2015.05.021

Llevot A (2017) Sustainable synthetic approaches for the preparation of plant oil-based thermosets. J Am Oil Chem Soc 94(2):169–186. https://doi.org/10.1007/s11746-016-2932-4

Lligadas G, Ronda JC, Galià M, Cádiz V (2013) Monomers and polymers from plant oils via click chemistry reactions. J Polym Sci A Polym Chem 51(10):2111–2124. https://doi.org/10.1002/pola.26620

Lluch C, Calle M, Lligadas G, Ronda JC, Galià M, Cádiz V (2015) Versatile post-polymerization modifications of a functional polyester from castor oil. Eur Polym J 72:64–71. https://doi.org/10.1016/j.eurpolymj.2015.09.005

Lu Y, Larock RC (2008) Soybean-oil-based waterborne polyurethane dispersions: effects of polyol functionality and hard segment content on properties. Biomacromolecules 9(11):3332–3340. https://doi.org/10.1021/bm801030g

Maaßen W, Oelmann S, Peter D, Oswald W, Willenbacher N, Meier MAR (2015) Novel insights into pressure-sensitive adhesives based on plant oils. Macromol Chem Phys 216(15):1609–1618. https://doi.org/10.1002/macp.201500136

Maassen W, Meier MAR, Willenbacher N (2016) Unique adhesive properties of pressure sensitive adhesives from plant oils. Int J Adhes Adhes 64:65–71. https://doi.org/10.1016/j.ijadhadh.2015.10.004

Maisonneuve L, Lebarbe T, Grau E, Cramail H (2013) Structure–properties relationship of fatty acid-based thermoplastics as synthetic polymer mimics. Polym Chem 4(22):5472–5517. https://doi.org/10.1039/C3PY00791J

Mangas MBP, Rocha FN, Suarez PAZ, Meneghetti SMP, Barbosa DC, dos Santos RB, Carvalho SHV, Soletti JI (2012) Characterization of biodiesel and bio-oil from *Sterculia sriata* (chicha) oil. Ind Crop Prod 36(1):349–354. https://doi.org/10.1016/j.indcrop.2011.10.021

Marcoux D, Gorkiewicz-Petkow A, Hirsch R, Korting H-C (2004) Dry skin improvement by an oleosome emulsion as a carrier for sphingolipid. J Am Acad Dermatol 50(suppl 3)):77. https://doi.org/10.1016/j.jaad.2003.10.631

McKeon (2016) Castor oil. In: McKeon T, Hayes T, Hildebrand D, Weselake RJ (eds) Industrial oil crops. Academic Press, New York, pp 75–112

Medeiros AMMS, Machado F, Rubim JC (2015) Synthesis and characterization of a magnetic bio-nanocomposite based on magnetic nanoparticles modified by acrylated fatty acids derived from castor oil. Eur Polym J 71:152–163. https://doi.org/10.1016/j.eurpolymj.2015.07.023

Medeiros AMMS, Machado F, Rubim JC, McKenna TFL (2017) Bio-based copolymers obtained through miniemulsion copolymerization of methyl esters of acrylated fatty acids and styrene. J Polym Sci A Polym Chem 55(8):1422–1432. https://doi.org/10.1002/pola.28511

Meier MAR, Metzger JO, Schubert US (2007) Plant oil renewable resources as green alternatives in polymer science. Chem Soc Rev 36(11):1788–1802. https://doi.org/10.1039/B703294C

Miao S, Zhang S, Su Z, Wang P (2010) A novel vegetable oil–lactate hybrid monomer for synthesis of high-Tg polyurethanes. J Polym Sci A Polym Chem 48(1):243–250. https://doi.org/10.1002/pola.23759

Miao S, Callow N, Wang P, Liu Y, Su Z, Zhang S (2013) Soybean oil-based polyurethane networks: shape-memory effects and surface morphologies. J Am Oil Chem Soc 90(9):1415–1421. https://doi.org/10.1007/s11746-013-2273-5

Miyahira MAM (2011) Characterization of fruits and nuts of Bacuri (*Scheelea phalerata* Mart) e Baru (*Dipteryx alata* Vog.) [Caracterização dos frutos e amêndoas de Bacuri (*Scheelea phalerata* Mart) e Baru (*Dipteryx alata* Vog.)]. Dissertation, Universidade Católica Dom Bosco

Montesinos L, Bundo M, Izquierdo E (2016) Production of biologically active Cecropin a peptide in rice seed oil bodies. PLoS One 11. https://doi.org/10.1371/journal.pone.0146919

MordorIntelligence (2017) Thermosetting plastics market—segmented by type, industry and geography: trends and forecasts (2015–2020). https://www.mordorintelligence.com/industry-reports/global-thermosetting-plastics-market-industry. Accessed 27 May 2017

Moura EF, Ventrella MC, Motoike SY (2010) Anatomy, histochemistry and ultrastructure of seed and somatic embryo of *Acrocomia aculeata* (Arecaceae). Sci Agric 67:399–407. https://doi.org/10.1590/S0103-90162010000400004

Mucci VL, Ivdre A, Buffa JM, Cabulis U, Stefani PM, Aranguren MI (2017) Composites made from a soybean oil biopolyurethane and cellulose nanocrystals. Polym Eng Sci:1–8. https://doi.org/10.1002/pen.24539

Nayak PL (2000) Natural oil-based polymers: opportunities and challenges. J Macromol Sci Rev Macromol Chem Phys 40:1–21. https://doi.org/10.1081/MC-100100576
Ng WS, Lee CS, Chuah CH, Cheng S-F (2017) Preparation and modification of water-blown porous biodegradable polyurethane foams with palm oil-based polyester polyol. Ind Crop Prod 97:65–78. https://doi.org/10.1016/j.indcrop.2016.11.066
Nikiforidis CV, Matsakidou A, Kiosseoglou V (2014) Composition, properties and potential food applications of natural emulsions and cream materials based on oil bodies. RSC Adv 4:25067–25078
Noordover BAJ (2011) Polyesters, polycarbonates and polyamides based on renewable resources. In: Mittal V (ed) Renewable polymers: synthesis, processing, and technology. Wiley, Hoboken, pp 305–354. https://doi.org/10.1002/9781118217689.ch7
Nunes AA, Favaro SP, Galvani F, Miranda CHB (2015) Good practices of harvest and processing provide high quality Macauba pulp oil. Eur J Lipid Sci Technol 117:2036–2043. https://doi.org/10.1002/ejlt.201400577
Ogunniyi DS (2006) Castor oil: a vital industrial raw material. Bioresour Technol 97:1086–1091. https://doi.org/10.1016/j.biortech.2005.03.028
Oliveira IP, Correa WA, Neves PVR, Silva PVB, Lescano CH, Michels FS, Passos WE, Muzzi RM, Oliveira SL, Caires ARL (2017) Optical analysis of the oils obtained from *Acrocomia aculeata* (Jacq.) Lodd: mapping absorption-emission profiles in an induced oxidation process. Photo-Dermatology 4:3. https://doi.org/10.3390/photonics4010003
Patel VR, Dumancas GG, Subong JJ (2016) Castor oil: properties, uses, and optimization of processing parameters in commercial production. Lipid Insights 9:1–12. https://doi.org/10.4137/LPI.S40233
Pattanaik P, Misra RD (2017) Effect of reaction pathway and operating parameters on the deoxygenation of vegetable oils to produce diesel range hydrocarbon fuels: a review. Renew Sustain Energy Rev 73:545–557
Peng CC, Shyu DJ, Chou WM, Chen MJ, Tzen JT (2004) Method for bacterial expression and purification of sesame cystatin via artificial oil bodies. J Agric Food Chem 52:3115–3119. https://doi.org/10.1021/jf049849f
Péres EUX, de Souza FG Jr, Silva FM, Chaker JA, Suarez PAZ (2014) Biopolyester from ricinoleic acid: synthesis, characterization and its use as biopolymeric matrix for magnetic nanocomposites. Ind Crop Prod 59:260–267. https://doi.org/10.1016/j.indcrop.2014.05.031
Péres EUX, Sousa MH, Gomes de Souza F, Machado F, Suarez PAZ (2017) Synthesis and characterization of a new biobased poly(urethane-ester) from ricinoleic acid and its use as biopolymeric matrix for magnetic nanocomposites. Eur J Lipid Sci Technol 119:1600451. https://doi.org/10.1002/ejlt.201600451
Petrovic ZS, Milic J, YJ X, Cvetkovic I (2010) A chemical route to high molecular weight vegetable oil-based polyhydroxyalkanoate. Macromolecules 43:4120–4125. https://doi.org/10.1021/ma100294r
Peykova Y, Lebedeva OV, Diethert A, Müller-Buschbaum P, Willenbacher N (2012) Adhesive properties of acrylate copolymers: effect of the nature of the substrate and copolymer functionality. Int J Adhes Adhes 34:107–116. https://doi.org/10.1016/j.ijadhadh.2011.12.001
Philipp C, Eschig S (2012) Waterborne polyurethane wood coatings based on rapeseed fatty acid methyl esters. Prog Org Coat 74(4):705–711. https://doi.org/10.1016/j.porgcoat.2011.09.028
PlasticsEurope (2016) Plastics—the facts 2016: an analysis of European plastics production, demand and waste data. http://www.plasticseurope.org/cust/documentrequest.aspx?DocID=67651. Accessed 27 May 2017
Pourzolfaghar H, Abnisa F, Daud W, Aroua MK (2016) A review of the enzymatic hydroesterification process for biodiesel production. Renew Sustain Energy Rev 61:245–257. https://doi.org/10.1016/j.rser.2016.03.048
Purkrtova Z, Jolivet P, Miguel M, Chardot T (2008) Structure and function of seed lipid body-associated proteins. C R Biol 331:746–754. https://doi.org/10.1016/j.crvi.2008.07.016
Reis SB, Mercadante-Simões MO, Ribeiro LM (2012) Pericarp development in the macaw palm *Acrocomia aculeata* (Arecaceae). Rodriguésia 63:541–550

Refaat AA (2009) Correlation between the chemical structure of biodiesel and its physical properties. Int J Environ Sci Technol 6:677–694

Rodrigues JS, do Valle CP, Guerra PAGP, Rios MAS, Malveira JQ, Ricardo NMPS (2017) Study of kinetics and thermodynamic parameters of the degradation process of biodiesel produced from fish viscera oil. Fuel Process Technol 161:95–100. https://doi.org/10.1016/j.fuproc.2017.03.013

Scala JL, Wool RP (2002) The effect of fatty acid composition on the acrylation kinetics of epoxidized triacylglycerols. J Am Oil Chem Soc 79(1):59–63. https://doi.org/10.1007/s11746-002-0435-4

Scrimgeour C (2005) Chemistry of fatty acids. In: Shahidi F (ed) Bailey's industrial oil and fat products, 6th edn. Wiley-Interscience, New York, pp 1–43

Sehlinger A, Ochsenreither K, Bartnick N, Meier MAR (2015) Potentially biocompatible polyacrylamides derived by the Ugi four-component reaction. Eur Polym J 65:313–324. https://doi.org/10.1016/j.eurpolymj.2015.01.032

Sharmin E, Zafar F, Akram D, Alam M, Ahmad S (2015) Recent advances in vegetable oils based environment friendly coatings: a review. Ind Crop Prod 76:215–229. https://doi.org/10.1016/j.indcrop.2015.06.022

Shurtleff W, Ayoagy A (2015) History of soybean crusching—soy oil and soybeans meal (980-2016): extensively annotated bibliography and sourcebook. Soyinfo Center, Lafayette, 3665 p. ISBN 9781928914891. Available at http://www.soyinfocenter.com/pdf/196/Crus.pdf

Silva LN, Fortes ICP, de Sousa FP, Pasa VMD (2016) Biokerosene and green diesel from macauba oils via catalytic deoxygenation over Pd/C. Fuel 164:329–338. https://doi.org/10.1016/j.fuel.2015.09.081

Silveira EV, Vilela LS, Castro CFS, Liao LM, Gambara Neto FF, Oliveira PMS (2017) Chromatographic characterization of the crambe (*Crambe abyssinica* Hochst) oil and modeling of some parameters for its conversion in biodiesel. Ind Crop Prod 97:545–551. https://doi.org/10.1016/j.indcrop.2016.12.033

Slivniak R, Domb AJ (2005) Lactic acid and ricinoleic acid based copolyesters. Macromolecules 38(13):5545–5553. https://doi.org/10.1021/ma0503918

Slivniak R, Langer R, Domb AJ (2005) Lactic and ricinoleic acid based copolyesters stereocomplexation. Macromolecules 38(13):5634–5639. https://doi.org/10.1021/ma050157h

Slivniak R, Langer R, Domb AJ (2006) Hydrolitic degradation and drug release of ricinoleic acid-latic acid copolyesters. Pharm Res 23(6):1306–1312. https://doi.org/10.1007/s11095-006-0140-x

Soo-Young N (2017) Application of straight vegetable oil from triglyceride based biomass to IC engines – a review. Renew Sustain Energy Rev 69:80–97. https://doi.org/10.1016/j.rser.2016.11.007

Soares IP, Rezende TF, Silva RC, Castro EVR, Fortes ICP (2008) Multivariate calibration by variable selection for blends of raw soybean oil/biodiesel from different sources using Fourier transform infrared spectroscopy (FTIR) spectra data. Energy Fuel 22:2079–2083. https://doi.org/10.1021/ef700531n

Soltani S, Rashid U, Al-Resayes SI, Nehdi IA (2017) Recent progress in synthesis and surface functionalization of mesoporous acidic heterogeneous catalysts for esterification of free fatty acids feedstocks: a review. Energ Convers Manag 141:183–205. https://doi.org/10.1016/j.enconman.2016.07.042

Souza LTA, Mendes AA, Castro HF (2016) Selection of lipases for the synthesis of biodiesel from Jatropha oil and the potential of microwave irradiation to enhance the reaction rate. Biomed Res Int. https://doi.org/10.1155/2016/1404567

Speight JG (2008) Synthetic fuels handbook: properties, process, and performance. McGraw-Hill, New York

Sugami Y, Minami E, Saka S (2017) Hydrocarbon production from coconut oil by hydrolysis coupled with hydrogenation and subsequent decarboxylation. Fuel 197:272–276. https://doi.org/10.1016/j.fuel.2017.02.038

The Statistics Portal (2017) Leading biodiesel producers worldwide in 2016, by country (in billion liters). https://www.statista.com/statistics/271472/biodiesel-production-in-selected-countries. Accessed 08 Jun 2017

Türünç O, Meier MAR (2010) Fatty acid derived monomers and related polymers via thiol-ene (click) additions. Macromol Rapid Commun 31(20):1822–1826. https://doi.org/10.1002/marc.201000291

Türünç O, Montero de Espinosa L, Meier MAR (2011) Renewable polyethylene mimics derived from castor oil. Macromol Rapid Commun 32(17):1357–1361. https://doi.org/10.1002/marc.201100280

Tüzün A, Lligadas G, Ronda JC, Galià M, Cádiz V (2016) Castor oil-derived benzoxazines: synthesis, self-metathesis and properties of the resulting thermosets. Eur Polym J 75:56–66. https://doi.org/10.1016/j.eurpolymj.2015.12.004

Tzen J, Cao Y, Laurent P, Ratnayake C, Huang AHC (1993) Lipids, proteins, and structure of seed oil bodies from diverse species. Plant Physiol 101:267–276. PMCID: PMC158673

USDA (2017) Vegetable oils. https://apps.fas.usda.gov/psdonline/app/index.html#/app/statByCommodity. Accessed 20 May 2017

Van De Mark MR, Sandefur K (2005) Vegetable oils in paint and coatings. In: Erhan SZ (ed) Industrial uses of vegetable oil. AOC Press, Urbana, pp 143–162. https://doi.org/10.1201/9781439822388.ch8

Van Gerpen J (2014) Animal fats for biodiesel production. http://www.articles.extension.org/pages/30256/animal-fats-for-biodiesel-production. Accessed 28 May 2017

Vivek N, Sindhu R, Madhavan A, Anju AJ, Castro E, Faraco V, Pandey A, Binod P (2017) Recent advances in the production of value added chemicals and lipids utilizing biodiesel industry generated crude glycerol as a substrate—metabolic aspects, challenges and possibilities: an overview. Bioresour Technol 239:507–517. https://doi.org/10.1016/j.biortech.2017.05.056

Vrtiska D, Simacek P (2016) Prediction of HVO content in HVO/diesel blends using FTIR and chemometric methods. Fuel 174:225–234. https://doi.org/10.1016/j.fuel.2016.02.010

Wool R, Sun XS (2005) Bio-based polymers and composites. Academic Press, London, p 640. ISBN: 978-0-12-763952-9

Xia Y, Larock RC (2011) Castor-oil-based waterborne polyurethane dispersions cured with an Aziridine-based Crosslinker. Macromol Mater Eng 296(8):703–709. https://doi.org/10.1002/mame.201000431

Yılmaz MT, Karakaya M, Aktas N (2010) Composition and thermal properties of cattle fats. Eur J Lip Sci Technol 112:410–416. https://doi.org/10.1002/ejlt.20099133

Zhang L, Jia B, Tan X et al (2014) Fatty acid profile and unigene-derived simple sequence repeat markers in Tung tree (*Vernicia fordii*). PLoS One 9. https://doi.org/10.1371/journal.pone.0105298

Zhao L, Chen Y, Yan Z, Kong X, Hua Y (2016) Physicochemical and rheological properties and oxidative stability of oil bodies recovered from soybean aqueous extract at different pHs. Foods Hydrocol 61:685–694. https://doi.org/10.1016/j.foodhyd.2016.06.032

Zhao X, Wei L, Cheng S, Julson J (2017) Review of heterogeneous catalysts for catalytically upgrading vegetable oils into hydrocarbon biofuels. Catalysts 7:83–107. https://doi.org/10.3390/catal7030083

Chapter 4
Starch Biomass for Biofuels, Biomaterials, and Chemicals

Susana Marques, Antonio D. Moreno, Mercedes Ballesteros, and Francisco Gírio

Abstract The success of modern biorefineries, including those using starch-based feedstocks, should be based on versatile biomass supply chains and on the production of a wide spectrum of competitive bio-based products. This chapter summarizes the current knowledge of bio-based products obtained mainly from biochemical platforms from starch- and sugar-based feedstocks. After an initial review of starch production sources and starch properties as well as starch-based end applications, this chapter reviews the state of the art of starch hydrolytic enzymes, focusing on a bio-based platform for the main value-added (bio)chemicals, biofuels, and biomaterials that can be obtained from sugar-based feedstocks.

At the present time, food and biofuels applications still dominate most of the uses of starch-based raw materials. Although bio-based chemicals and biomaterials still do not account for a significant share of current biomass use, new bioeconomy sectors are emerging such as biomaterials and green chemistry, and several markets (e.g., bioplastics, biolubricants, biosolvents, and biosurfactants) are expected to grow in the near future. Several examples of biological production routes are described in this chapter, namely, for ethanol, lactic acid, and polylactic acid (PLA), polyhydroxyalkanoates (PHAs), succinic acid, 1,4-butanediol (BDO), farnesene, isobutene, acrylic acid, adipic acid, ethylene, and polyethylene. One example of using a chemical catalytic route to obtain furan-2,5-dicarboxylic acid (FDCA) is also reported.

Keywords Starch • Sugar platform • Bio-based products • Biofuels • Biomaterials • Chemicals • Bioeconomy

---

S. Marques • F. Gírio (✉)
Bioenergy Unit, LNEG, Estrada do Paço do Lumiar 22, 1649-038 Lisbon, Portugal
e-mail: susana.marques@lneg.pt; francisco.girio@lneg.pt

A.D. Moreno • M. Ballesteros
Department of Energy, Biofuels Unit, CIEMAT, Avda. Complutense 40, 28040 Madrid, Spain
e-mail: david.moreno@ciemat.es; m.ballesteros@ciemat.es

S. Vaz Jr. (ed.), *Biomass and Green Chemistry*,
https://doi.org/10.1007/978-3-319-66736-2_4

4.1 Introduction

The successful development and implementation of a bio-based economy are important steps in reaching a sustainable society in the long term. A bioeconomy and bio-based products offer several benefits, particularly focusing on carbon sequestration and storage and the need to replace fossil carbon-based chemicals. Analogous to today's petroleum refineries, biorefineries are conversion facilities where biomass feedstocks can be transformed into several fuels, power, heat, and value-added chemicals (Olsson and Saddler 2013).

Biomass, in the context of bioeconomy, constitutes the organic matter, especially plant matter, that can be converted into food, feeds, chemicals, materials, fuel, and energy, including sugar- and starch-based feedstocks and lignocellulose. In the particular case of starch-based materials, grains (such as corn or wheat) and tubers (such as potatoes and cassava) are the most representative feedstocks. Other raw materials such as sago palm, marine algae, and microalgae can also be considered as alternative choices for starch production (Höfer 2015).

Starch is a soft, white, tasteless, and odorless polymer that is not soluble in cold water, alcohol, or other solvents. It comprises glucose monomers linked through α-1,4-glycosidic bonds and branched by α-1,6-glycosidic linkages. The simplest form of starch is the linear polymer amylase; amylopectin is the branched form (Fig. 4.1). Within the biorefinery context, starch biomass can potentially be converted into a wide range of chemical products via chemical/biological processing, with applications not only in the food and pharmaceutical sectors but also in the paper and adhesive industries (Gozzo and Glittenberg 2009). In its native form, starch is mainly used as a raw material for further processing, because of its poor

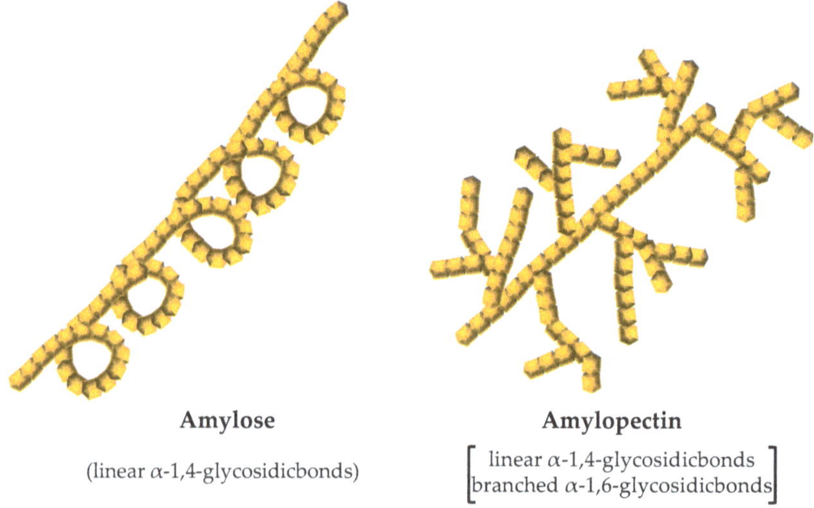

Fig. 4.1 Linear (amylase) and branched (amylopectin) structures of starch

stability, the high viscosity of aqueous solutions, or the formation of precipitates or microgels, or as a cost-effective binder or thickener for industrial applications. When processed, starch-based products include gelatinized starches, thermoplastic biopolymers, starch esters and ethers, maltodextrins, glucose syrups, cyclodextrins, sorbitol, and fermentation-derived products such as ethanol (Gozzo and Glittenberg 2009; Höfer 2015). This chapter reviews the most important advances for the conversion of starch-based feedstocks, from the use of novel raw materials to the investigation of new processing technologies.

4.2 Conventional and Novel Feedstocks for a Starch-Based Biorefinery

The history of industrial conversion of starch-based resources covers more than 200 years, since Kirchhoff first reported the possibility of obtaining glucose and dextrose from potato starch after a cooking process in acidic conditions. The successful development of sugar production from sugar beet, however, initially obstructed further development of the starch industry. This industry is currently growing worldwide, and about 180 million tons of starch and starch derivatives are expected to be produced by 2022 (Global Industry Analysts 2016). In Europe alone, the starch industry produced 10.7 million tons of starch, and the population and other industries consumed 9.3 million tons of starch and starch derivatives, in 2015 (Starch Europe 2017), of which 62% was in food, 1% in feed, and 37% in nonfood applications, mainly paper making (Scarlat et al. 2015).

Starch production is mainly based on cereal grains and tubers. From a quantitative point of view, wheat (*Triticum* spp.), corn (*Zea mays*), rice (*Oryza sativa*, Asian rice; *Oryza glaberrima*, African rice), potato (*Solanum tuberosum*), and cassava (*Manihot esculenta*) are the leading crops for starch production. Wheat is the most important food grain by far, and its cultivation involves more land area than any other commercial crop (Curtis 2002). Wheat grains are good sources for starch production because of the high starch content, which represents about 58–70% of the total dry grain weight (Höfer 2015). Proteins are also important components of wheat grains, because the higher the protein content, the lower the starch content. The starch-to-protein ratio is highly dependent on wheat variety, and those varieties selected for having higher starch content are therefore desirable as raw materials in a biorefinery (Saunders et al. 2011). Wheat cultivation requires warm temperatures (about 25 °C, with maximum growth temperatures of 30–32 °C) and much sunshine (especially during the stage when the grains are filling). From sowing to harvesting, wheat requires about 3.5–4.5 months depending on climate, seed type, and soil conditions. With the aim of promoting wheat as a viable crop, different techniques, including soil preparation through the use of crop rotation, the addition of fertilizers, and extensive mechanization of the harvesting process, have improved plant growth and facilitated the reaping, threshing, and winnowing steps during grain separation. The growth of wheat, as any other crop, is highly dependent on climate conditions. In this context, the very high yielding triticale (*Triticum aestivum* L.), a cross-breed between wheat and rye with a starch content similar to that of wheat, or hulless barley (*Hordeum vulgare* L.) are appropriate crops for starch production in those places with less favorable climate conditions, such as high latitudes, high altitudes, or saline environments (Oettler 2005; Höfer 2015).

Rice is another important source of starch worldwide, containing up to 75% starch depending on the variety. Starch can be isolated from rice with good recovery yields by using alkaline solutions (the so-called wet process) or proteolytic enzymes (Puchongkavarin et al. 2005). Rice starch particles measure only about 7–9 μm and are included among the smallest vegetable powders. This property increases the surface area of rice starches, which results in a soft-touch effect in rice-based products that makes them ideal for use in decorative cosmetics and in skin and hair care products (Mitchell 2009).

Besides wheat and rice, maize (corn) is one of the most important cereal grains in the world. In each individual corn kernel – the fruits of maize – starch is the major chemical component, about 72% of the kernel weight. Also, 1–3% of glucose, fructose, or sucrose can be found in the corn kernel. Corn starch usually comprises up to 25–30% amylose and 70–75% amylopectin in weight. However, corn with as much as 65–80% starch can also be found in commercial varieties (Schwartz and Whistler 2009). As for wheat, a warm climate is required for maize cultivation, which also can grow fairly well in high latitudes. In countries with suitable climatic conditions, maize can now be cultivated with very high yield per unit area as the result of current resources and management techniques. Furthermore, the maturing period of this crop is relatively short. As its main disadvantage, maize requires much water, limiting its cultivation in places with low available water capacity.

Tubers such as cassava and potatoes are also important raw materials for the starch industry. These starchy tuberous roots are native to South America. Cassava, also known as manioc, mandioca, yuca, mogo, tapioca root, or kappa, is cultivated extensively in tropical and subtropical regions. Because the starch granules are locked inside the cells, biochemical or mechanical disruption of the cassava is first needed for collecting the starch. Higher yields and better starch quality can be obtained by the mechanical process. In this method, root cassava is sliced and then rasped, grated, or crushed to get a fine pulp with high starch content (Breuninger et al. 2009). In general, cassava starch is primarily used for bioethanol production or is cooked to form a clear gel with a slightly stringy texture that makes it a suitable food thickener (Shanavas et al. 2011). Potato is another relevant tuber for starch production. The chemical composition of potatoes is quite variable and is greatly influenced by variety, environment, and farming practices. Although 80% of its weight composition is water, starch represents about 65–80% of its dry weight (Höfer 2015). The potato plant is an herbaceous perennial crop that can be adapted to different climates, which has increased its importance from an economic perspective, particularly expanding the suitability of this crop to a larger area, being now cultivated all around the globe. Before mechanization became available in the late nineteenth century, tuber harvesting (both potato and cassava) was tremendously laborious and the work of many helpers was needed. Harvesting machines, however, have boosted this process, shortening it from days to just a few hours.

In 2012, the European starch industry used 23 million tons of agricultural raw materials, which included 7.7 million tons of maize, 7.8 million tons of wheat, and 7.5 million tons of potatoes (Scarlat et al. 2015).

As alternatives to cereal grains and tubers, pulses or grain legumes (belonging to the Leguminosae family) are also significant raw products for the starch industry. Pulses include edible grains and seeds such as beans, peas, or lentils. The starch content in pulse seeds is lower than that for cereals, representing about 22–45% of the dry matter (Höfer 2015). Pulse starches, however, are characterized by their high amylose content (35–39% of weight), conferring to them specific properties for functional texturization and film formation. One of the major benefits of growing pulses is their ability to fix nitrogen, which can be exploited in crop rotation.

In addition to edible crops, sago palm (*Metroxylon sagu*) and different species of algae represent alternative starchy feedstocks. The sago palm is a species of palm native to tropical Southeastern Asia. When mature enough (7–15 years), just one sago palm can yield up to 375 kg starch (Karim et al. 2008). With such starch content, plantations of sago palms might potentially supply about 25 tons of starch per hectare per year. This palm is an extremely hardy plant that can grow in wetlands with acidic or saline soils. However, it requires a moist climate with uniformly high sunshine, heat, and humidity. Another potential starch source that is increasingly gaining attention is algae. Of the algae pool, red algae, green algae, and glaucophytes are known to be capable of synthesizing starch. Starch content in algae varies largely among species, unicellular microalgae being the most interesting option. Among the best starch producers, the green alga *Chlorella* can accumulate starch up to 60% dry weight under specific nutrient-limiting conditions (Brányiková et al. 2011). The main advantage of algae in comparison to terrestrial plants is their more efficient photosynthetic machinery. Moreover, many species of algae can grow in saltwater and/or wastewater, allowing the saving of large amounts of freshwater. Eventually, by using genetic engineering techniques, algae may be manipulated for further development to target starch accumulation (Radakovits et al. 2010).

4.3 Thermoplastic Starch: Challenges and Properties

As already discussed, current research efforts are being directed to replace petroleum-based products to meet sustainability criteria in the future bioeconomy. Among petroleum derivatives, plastics (polypropylene, polyethylene, polystyrene) are polymers highly recalcitrant to microbial degradation because of their high molecular weight, complex three-dimensional structure, and hydrophobic nature (Kale et al. 2015). The global production of plastic is about 150 million tons year^{-1}, and the market is continually growing. Among this, bioplastics consumption and production is foreseen to significantly increase in the near future. Indeed, 5–10% of the plastics currently available on the market could be bio-based plastics and this share can increase up to 70–85% (Scarlat et al. 2015). In this context, starch has been extensively studied as a raw material for the production of alternative biodegradable plastics polymers, as drop-in bioproducts that directly correspond to the petrochemical counterpart. Nevertheless, in spite of the considerable interest from both the academic and industrial sectors, production of starch-based plastics still poses some technical limitations and high costs.

Starch-based thermoplastics have potential applications in vegetable waste composting, in packaging, and in the pharmaceutical sector (e.g., as controlled-release drug carriers) (Leja and Lewandowicz 2010). Thermoplastic starch (TPS) is obtained by subjecting native starch to high temperature and shear stress (extrusion, film blowing, injection molding, etc.) in the presence of a plasticizer (Khan et al. 2016). Several physical and chemical reactions take place, including water diffusion, expansion of granules, gelatinization, melting, and crystallization. The plasticizer is the material that confers flexibility to the plastic polymer and increases its applicability potential. During the process, plasticizers are capable of breaking the intramolecular hydrogen bonds of the starch polymer and thus creating new starch–plasticizer interactions. Water and glycerol are the most commonly used plasticizers, but sorbitol, glycols, maltodextrin, and urea have also been considered (Mohammadi et al. 2013).

TPS can also be blended with other polymers to confer a certain specific property to the final products that allows them to be used in a larger number of applications and to meet market needs (Mohammadi et al. 2013; Khan et al. 2016). For instance, blends with polycapromolactone, polyvinyl alcohol, or other relatively hydrophobic polymers reduce the hydrophilic character of starch-based thermoplastics. In addition, blending biodegradable and nonbiodegradable polymers can effectively reduce the amount of plastic waste by the partial degradation of the biodegradable component. Blends with other plastic materials have therefore more useful application than TPS itself. TPS can be blended with various polymers with different properties and potentials. St-Pierre et al. (1997) prepared blends with low-density polyethylene and linear low-density polyethylene using glycerol as the plasticizer. Blends showed greater elongation properties at the breaking point, even without adding the interfacial modifier. In addition, blends with polyethylene have shown to be relatively easy to process and are capable of decreasing the permeability to water vapor and increasing the toughness and flexibility of the resulting plastic (Khan et al. 2016). Kaseem et al. (2012) blended TPS with polypropylene to study the mechanical and rheological properties of the resulting plastics. The rheological properties showed a reduction in thermoplastic viscosity by increasing the glycerol content, whereas the mechanical properties indicated a lower strain capacity at break in comparison with polypropylene polymer. These properties increase the crystallinity of the resulting plastic, contributing to enhance the tensile strength, hardness, and stiffness. Similarly, Martin and Avérous (2001) studied the mechanical and rheological properties of TPS blends with polylactic acid. It was found that the least plasticized TPS was brittle and rigid, while the most plasticized TPS was ductile and flexible. Furthermore, the authors noticed a lack of affinity between the TPS and polylactic acid, which limited the blending capacity. Nevertheless, blends with polylactic acid represent a promising option—especially for packaging applications—because of low price and availability as compared to other biopolymers. Mihai et al. (2007) reported that blending TPS with polystyrene and using glycerol as plasticizer increased the viscosity ratio between the TPS and polystyrene phase while increasing the glycerol content. Furthermore, the glycerol content in the TPS phase and the TPS content in the overall blend had a firm effect on the blend viscosity and, hence, on the ability to foam the material.

Of the major limitations for the commercialization of starch-based materials, the high costs of the technology and the final products have restricted their applications to those sectors where biodegradability is mandatory. Nevertheless, some products have already reached the market, such as Mater-Bi (Novamont, Italy), a plastic material based on TPS and other polymers that is suitable for composting bags, mulch film, or disposable cutlery.

4.4 Enzymatic Hydrolysis of Starch-Based Feedstocks

In addition to TPS, starch represents an excellent sugar source for the fermentation/chemical industry. Starch can be efficiently hydrolyzed into sugars by acid treatment or by the use of enzymes. Enzymatic hydrolysis is, however, advantageous because of (1) the specificity of enzymes, which allows the production of sugar syrups with well-defined physical and chemical properties, (2) the process takes place at milder reaction conditions and thus results in few side reactions and in formation of no color, and (3) it avoids salts accumulation (Guzmán-Maldonado and Paredes-López 1995; Nigam and Singh 1995). Starch-degrading enzymes, promoting liquefaction and saccharification, include α-amylases (EC 3.2.1.1, also called endo-amylase), β-amylases (EC 3.2.1.2), amylo-glucosidases (EC 3.2.1.3), α-glucosidases (EC 3.2.1.20), pullulanases (EC 3.2.1.41), maltotetraohydrolases (EC 3.2.1.60), and isoamylases (EC 3.2.1.68). α-Amylases catalyze the hydrolysis of internal α-1,4-glycosidic linkages into low molecular weight products, such as glucose, maltose, and short oligomers up to maltohexose units (de Souza and de Oliveira Magalhães 2010). The amylases can be obtained from several sources, including plants, animals, and microorganisms. Among them, a large number of microbial α-amylases, mainly from the fungus *Aspergillus* spp. and the bacterium *Bacillus* spp., are now been commercialized. β-Amylases are exo-hydrolase enzymes that act from the nonreducing end of a polysaccharide chain by hydrolysis of α-1,4-glycosidic linkages to yield successive maltose units (Sundarram and Murthy 2014). Primary sources of β-amylases are sweet potatoes and plant seeds such as wheat, barley, or soybeans. Nevertheless, they can be also obtained from microorganisms such as *Bacillus* spp. or *Pseudomonas* spp. Amyloglucosidases attack α-1,4-glycosidic from the nonreducing end, releasing β-D-glucose (Kumar and Satyanarayana 2009). The huge literature on microbial amyloglucosidases indicates the preference for its production in eukaryotic hosts, being more common the use of fungi strains (particularly *Aspergillus* spp. and *Rhizopus* spp.). α-Glucosidases cleave both α-1,4- and α-1,6-glycosidic bonds on the external glucose residues of amylose or amylopectin from the nonreducing end of the starch molecule (Nigam and Singh 1995; Hii et al. 2012). The main sources for glucosidase production are microorganisms such as *Aspergillus* spp., *Mucor* spp., *Bacillus* spp., or *Pseudomonas* spp. Pullulanases (also known as α-dextrin-6-glucanohydrolases or amylopectin-6-glucanohydrolases) catalyze the hydrolysis of α-1,6 linkages in pullulan, a linear α-glucan polymer consisting essentially of maltotriosyl units connected by α-1,6-glycosidic bonds (Hii et al. 2012). These enzymes are derived from various microorganisms such as *Bacillus acidopullulyticus*, *Klebsiella planticola*, *Bacillus* spp., and *Geobacillus stearothermophilus*. Maltotetraohydrolases hydrolyze α-1,4-glucosidic linkages to remove successive maltotetraose residues from the nonreducing chain ends (Mezaki et al. 2001). Finally, isoamylases, similar to pullulanases, are capable of hydrolyzing α-1,6-glycosidic bonds in amylopectins and are the only known enzymes that completely debranch glycogen completely (Hii et al. 2012). In addition to all mentioned enzymes, starch-active lytic polysaccharide monooxygenases (LPMOs) have recently been described to contribute to starch hydrolysis (Vu et al. 2014; Lo Leggio et al. 2015). LPMOs are recently discovered enzymes that oxidatively deconstruct polysaccharides (Lo Leggio et al. 2015). These enzymes utilize molecular oxygen and an electron donor to oxidize polysaccharides at C1 or C4 positions in the carbohydrate chain, promoting the corresponding breakage and thus rendering the substrate more susceptible to hydrolysis by glycoside hydrolases. All this group of enzymes is needed for the complete hydrolysis of the starch polymer. Moreover, these specific activities have a synergistic action during the hydrolysis process, boosting the release of single sugars.

Enzymatic properties such as optimal temperature, optimal pH value, thermostability, and stability are very much dependent on the enzyme source and are therefore important factors to consider for the proper selection of a certain enzyme. Thus, for starch hydrolysis, enzymes must be active and stable at slightly acidic condition (pH value of 5–7) and at relatively high temperatures (up to 100 °C), given the industrial processes conditions.

4.5 Value-Added Biofuels, Biochemicals, and Biomaterials from a Sugar Platform Perspective

The sugar solutions obtained by enzymatic hydrolysis of starch-rich crops can subsequently be converted into biofuels, biochemicals, or biomaterials through a wide range of biological and chemical technologies.

The simplest biorefineries, typically called "first generation," are based on the use of sugar-rich or starch-rich crops as raw materials. Some of these biorefineries are already maturely established, and the most significant commercial facilities belonging to this category are biofuel-driven biorefineries, producing first-generation (1G) bioethanol, the most common renewable biofuel. Indeed, despite the potential competition with food that favors the development of lignocellulosic-based plants (Searchinger and Heimlich 2015), date sugar and starch crops are still the most often used biomass feedstock, and the main product of sugar- or starch-based biorefineries is ethanol (Gnansounou and Pandey 2017). According to Scarlat et al. (2015), the European Union (EU) is one of the world's largest producers of cereals, with an estimated production of 285 million tons in 2012. The largest share of European cereals is used for feed (more than 60%) and food (23%). However, a significant share of cereals is also used for processing (3.5%) and another for biofuel (ethanol) production (3%).

Novel chemicals and materials produced from starch-based raw materials are also currently available, but the technologies for conversion into some bioproducts are still in R&D, pilot, or demo stage. However, in addition to biofuels, there are some important examples of product-driven biorefineries producing bio-based products that are already commercialized or near market (Gírio et al. 2017). Although bio-based chemicals and materials do not yet account for a significant share of biomass use (Scarlat et al. 2015), new sectors are emerging, such as biomaterials and green chemistry, and this market (including bioplastics, biolubricants, biosolvents, and biosurfactants) is expected to grow in the near future. This growth rate should, however, be lower than in biofuels and biopower because it is not driven by any mandates or incentive schemes (Sir David King et al. 2010). According to Scarlat et al. (2015), it is foreseen that more than 20% of all chemicals coming from the traditional chemistry sector could be produced from biological sources in 2020. Indeed, there are already several bio-based products on the market, and the European chemical industry used approximately 8.6 million tons of renewable raw materials (including starch) in 2011, among a total of 90 million tons of feedstock used (Scarlat et al. 2015). As an example of success, steady increase is reported for lactic acid, with a 10% annual growth, from its use as precursor of PLA (polylactic acid), which is mainly used for production of sustainable biopolymers. European demand for PLA in 2015 was 25,000 tons per year and could reach 650,000 tons per year in 2025 (Mikkola et al. 2016).

In contrast to drop-in bioproducts, which are bio-based versions (chemically equivalent) of existing petrochemical products and thus can easily replace fossil-based products, some bio-based products have novel functionality and slower access to markets. Also, promising novel chemical intermediates are difficult to integrate into current production networks when no established large-scale chemical processes exist for their conversion, such as the case of levulinic acid. In addition, novel products based on new intermediates may offer unique properties unattainable with fossil-based alternatives (e.g., biodegradability), but they also bear higher risks and usually require a complex process until commercialization. This sequence typically happens with bio-based polymers as substitutes of existing polymers. For instance, PLA and PHAs (polyhydroxyalkanoates) have been known for a long time and only recently have been commercialized, but currently several companies are running large-scale production of PLA and PHA with success (King et al. 2010).

The development of a successful biorefinery should be guided by two concepts: (1) to take maximum advantage of intermediate and by-products to manufacture additional chemicals and materials; and (2) to balance high-value/low-volume bio-based chemicals and materials with high-volume/low-value biofuels (King et al. 2010). Indeed, specialty chemicals from biomass, fitted to specific "niche" markets (e.g., for the pharmaceutical, cosmetics, and food sectors), are sold at relatively high prices because of their limited production (<1000 tons per year), "green credentials," and unique quality and properties (Mikkola et al. 2016). In contrast, the production of bulk chemicals (>1000 tons per year) from biomass remains rather limited, with the majority of organic chemicals and polymers still being derived from fossil-based feedstocks (Mikkola et al. 2016).

Among bioproducts, primary building blocks exhibiting multiple functionalities, suitable for further conversion, are really interesting. The US Department of Energy (DOE) carried out a study in 2004, leading to the identification of a group of 12 (albeit it has been known as the DOE "Top 10" report) target sugar-derived building blocks coming to the market (Werpy and Petersen 2004). This list of candidates was revised by Bozell and Petersen (2010) based on the balance among nine criteria boosting their opportunities (listed by approximate order of importance): (1) extensive recent literature demonstrating research activity; (2) multiple product applicability; (3) direct substitute for existing petrochemicals; (4) high-volume product; (5) platform potential; (6) industrial scale-up; (7) existing commercial product; (8) primary building block; (9) commercial bio-based product. Cost evaluation was not included in this evaluation because cost structures will change with the technology development that is rapidly occurring, driven by the intensive ongoing research activities.

More recently, in 2015, *Biofuels Digest* (Lane 2015) has updated this assessment. Taking into account this recent study, based on the previous US DOE's "Top 10" biochemicals (Werpy and Petersen 2004) and IEA Bioenergy Task 42 (2011) reports, and given the current maturity level of industrial activity and market size, this review focused on ten products that can be obtained using sugar derived from starch that were selected as being of particular interest in a previous report (E4tech, RE-CORD, and WUR 2015). Some of the selected bioproducts are primary products (obtained directly from sugars): succinic acid [also produced via glycerol or FDCA (furan-2,5-dicarboxylic acid), and also intermediate for BDO (1,4-butanediol)], BDO (also produced via succinic acid), farnesene, isobutene, polyhydroxyalkanoates (PHAs), and FDCA (also intermediate for succinic acid). In addition, secondary chemicals are produced from sugars via an intermediate: acrylic acid [via 3-HPA (3-hydroxypropionic acid)], adipic acid (via glucaric acid), polyethylene (PE, via ethanol), and polylactic acid (PLA, via lactic acid).

Some of these established bio-based products already dominate global production, such as is the case of ethanol and lactic acid (E4tech, RE-CORD, and WUR 2015), with the largest markets, and others do not have an identical fossil-based substitute (e.g., FDCA, farnesene). Indeed, bio-based FDCA and farnesene have the highest current prices, but these are expected to drop once the relevant conversion technologies have been successfully commercialized (E4tech, RE-CORD, and WUR 2015). The smallest bio-based markets are, as is to be expected, those of the earliest stage products, such as acrylic acid and its precursor 3-hydroxypropionic acid (3-HPA), and adipic acid. Bio-based succinic acid exhibits the fastest growing market at present, because of the level and breadth of industry activity in the product (E4tech, RE-CORD, and WUR 2015).

Bio-based chemicals and materials can be produced via the sugar platform by different potential pathways that can be grouped into biological or chemical transformation pathways, and each of the bioproducts previously assigned as attractive is highlighted under the most commonly applied route for its production.

4.5.1 The Biological Route

Although there are many possible methods for the transformation of sugars, many products that are familiar today are fermentation based (Zwart 2006). Indeed, sugars can be converted through fermentation, by bacteria, fungi, or yeast (genetically modified or not) into alcohols, organic acids, alkenes, lipids, and other chemicals, under diverse process conditions (e.g., low/high pH value, anaerobic/aerobic, nutrient rich/deprived). The product of interest can also be produced intracellularly (requiring lysis/death of the cell to extract the product, usually via solvents), or extracellularly (simply requiring its separation/extraction from the fermentation broth). Given the wide range of organisms now available, many of them having been recently discovered and exploited, the fermentation of sugars has great potential for the development of new bioproducts (Gírio et al. 2017). Indeed, the fermentation of pure C6 sugars liberated from starch is straightforward, yielding a very large number of products. This chapter focuses only on the most significant biofuels and bioproducts, as previously listed. In addition to the target bioproducts, the fermentation-based processing of many starch crops also delivers valuable animal feed rich in protein and energy as a co-product, for example, distiller's dried grains with solubles (DDGS) (Gnansounou and Pandey 2017).

4.5.1.1 Ethanol

Ethanol is the most common representative of the biochemical transformation of biomass into fuel, and it exhibits good performance against the nine criteria proposed by Bozell and Petersen (2010). Ethanol was omitted from the DOE original list because it might be categorized as a supercommodity by its expected high production volume.

Indeed, bioethanol is currently used as a biofuel when mixed with gasoline, but it is also a useful intermediate platform chemical. Ethanol is of interest as a precursor to the corresponding olefin, ethylene, via dehydration, and thus links the biorefinery with petrochemical industry infrastructures (Mikkola et al. 2016). Ethanol dehydration was the source of most ethylene in the early stage of the twentieth century, but this route was later discarded in favor of steam cracking processes using crude oil with the expansion of the petrochemical industry (Bozell and Petersen

2010). More recently, increasing crude oil prices have renewed interest in ethanol dehydration, with large companies, such as Dow, Braskem, and Solvay, building ethanol-to-ethylene plants. Dow and Braskem manufacture "green" polyethylene whereas Solvay produces polyvinyl chloride (Bozell and Petersen 2010). Ethanol can also be oxidized, over gold nanocatalysts or Mo-V-Nb mixed oxides, yielding commodity chemicals such as acetic acid and ethyl acetate (Bozell and Petersen 2010).

Nowadays, commercial ethanol production is predominantly based on edible sugar and starchy biomass: mainly sugarcane in Brazil, corn grains in the USA, and sugar beet (58%) and wheat (19%) in Europe (Mikkola et al. 2016; Scarlat et al. 2015). Although sucrose conversion into ethanol provides higher yields compared to starch, the share of sugar crops and molasses for world ethanol production is limited by the instability of sucrose price, given the competition with refined sugar. Indeed, in the period 2012–2014, starch crops provided 57% of the ethanol produced worldwide, with the major contribution of corn, especially in the USA, the largest producer country (Gnansounou and Pandey 2017). The use of corn undoubtedly dominates in the USA, whereas wheat and rice are used in Europe and China, respectively, together with a minor share of barley and other coarse grains, for ethanol production from starch (E4tech, RE-CORD, and WUR 2015; ETIP Bioenergy n.d.). Indeed, the USA pioneered the corn industry, with implementation in Jersey City (NJ) of the first dedicated cornstarch plant in the world by Wm. Colgate & Company in 1844. Much later, in the 1970s, the multi-product corn wet mills appeared in their current form, prompted by the development of technology for the production of high fructose corn syrup for use in the soft drinks industry (Mikkola et al. 2016). Nowadays, there are 216 plants producing 59.5 billion liters of ethanol per year from sugar/starch in the USA, and 74 of the largest commercial 1G bioethanol facilities in operation in the USA, belonging to eight companies, produce 50% of this capacity using corn as raw material. Besides corn, sorghum is also used as a relevant sugar feedstock in the USA, and less than 0.2% of 1G ethanol produced in the USA is obtained from wastes (e.g., beverage waste or wheat screenings) (Gírio et al. 2017). The importance assigned to biomass-derived energy production in China has been increasing in the past 10 years, and China has licensed five fuel ethanol plants (mostly state refineries) for operation, all of which are based on starch crops (King et al. 2010).

Indeed, China has traditionally had an important role in biomass biorefining, dating to the early implementation of hydrolysis and acetone-butanol-ethanol (ABE) batch fermentation plants, mostly aimed at producing acetone and ethanol solvents from corn, cassava, potato, and sweet potato. The first industrial facility of continuous ABE fermentation in the world was the Russian Dokshukino plant, which started operation in 1960 using starch for fermentation by Clostridia bacteria, and several ABE plants were built in the Union of Soviet Socialist Republics in the period between 1960 and 1980. However, almost all these plants were closed in response to the strong competition with cheap solvents from the petrochemical industry (Bozell and Petersen 2010; Mikkola et al. 2016).

Although bioethanol is the dominant product, there are a few commercial biorefinery facilities producing other sugar platform products via fermentation of starch raw materials, such as the previously identified lactic acid or succinic acid (E4tech, RE-CORD, and WUR 2015). Other primary bio-based products from sugars, such as 1,4-butanediol (BDO), farnesene, and polyhydroxyalkanoates (PHAs), are currently produced at R&D, pilot, or demonstration scale.

4.5.1.2 Lactic Acid and Polylactic Acid (PLA)

Organic acids are easily produced from biorefinery sugar streams and are thus attractive platform chemicals. Lactic acid is a bio-based chemical produced by glucose fermentation using *Lactobacillus rhamnosus*, or other lactic acid bacteria, and biomass feedstocks (Marques et al. 2017a, b; Gírio et al. 2017). Indeed, lactic acid can be manufactured either by chemical synthesis (by reaction of acetaldehyde with acid cyanide, yielding lactonitrile, which is further hydrolyzed to lactic acid) or by fermentative processes, but the latter have been preferentially used in industrial production, accounting for approximately 90% of the total worldwide production (Hofvendahl and Hahn-Hägerdal 2000). The fermentative route allows the selective production of the desired L-lactic acid stereoisomer, with additional advantages in terms of energy efficiency and yield. Current commercial fermentation gives close to 90% yield of calcium lactate based on glucose feed, which is neutralized to give pure lactic acid. This neutralization produces approximately 1 ton of $CaSO_4$ for every ton of lactic acid, posing a waste disposal problem in commercial operation (Bozell and Petersen 2010). Alternative separation and purification technologies, based on desalting and water-splitting electrodialysis, have been examined to eliminate the neutralization step in the conventional chemical downstream processing scheme that typically consumes 50% of the cost of lactic acid manufacture (Pal et al. 2009).

Worldwide demand for lactic acid is expanding, driven by the more conventional use in the paint, cosmetic, pharmaceutical, and food preservatives markets, and mainly for polylactic acid (PLA) production. PLA, or polylactide, is a biodegradable and recyclable thermoplastic polyester resin (Vaidya et al. 2005), suitable for packaging materials, insulation foam, automotive parts, and fibers (textile and nonwoven) (E4tech, RE-CORD, and WUR 2015), and more recently, because of its biocompatibility, for biomedical applications (Vijayakumar et al. 2008). Packaging is likely to remain the key market for PLA, with the expected increase in demand for environmentally friendly starch-based plastics for this application, as a substitute for PE (polyethylene) and PS (polystyrene) products. Although lactic acid can undergo direct polymerization, the process is more effective if lactic acid is first converted to a low molecular weight pre-polymer and then depolymerized to the lactide. A wide range of catalysts is known to promote lactide polymerization. The resulting polymer exhibits performance properties similar to, or exceeding, those of PS, a storage resistance to fatty foods and dairy products equivalent to PET (polyethylene terephthalate), excellent barrier properties for flavors and aromas, and good heat stability (Gallezot 2007).

High optical purity is a prerequisite for synthesis of the homopolymers poly-L-LA and poly-D-LA, which form regular structures in a crystalline phase in contrast to the amorphous material resulting from copolymerization of D(−)- and L(+)-isomers (Wang et al. 2015). The ratio of poly-L-LA and poly-D-LA modulates the properties and the degradability of PLA; however, D-lactic acid is not suitable for use in the food, drink, and pharmaceutical industries because it can cause metabolic problems and is toxic to the human body. Thereby, the major challenge lies in producing optically pure lactic acid achieving high concentrations, yields, and productivity using cheap renewable resources (Abdel-Rahman and Sonomoto 2016). Indeed, most PLA producers also manufacture lactic acid, and this is mostly produced from cornstarch (in the USA), tapioca roots, chips, or starch (in Asia), or sugarcane and sugar beets (in the rest of the world).

The largest global commercial producer of PLA is the US-based NatureWorks (former Cargill Dow), producing PLA resins under the Ingeo brand, with a commercial production plant in Nebraska (with an annual 150-kton capacity, in 2015), using cornstarch as feedstock, and plans for a new plant in Thailand (E4tech, RE-CORD, and WUR 2015). The largest global lactic acid producer is Corbion Purac (Netherlands). The latter produces lactic acid, lactic acid derivatives, and lactides (including lactide resins for high-performance PLA bioplastics), operating five production plants, in the USA, the Netherlands, Spain, Brazil, and Thailand (the largest plant, with an annual 100-kton capacity) (E4tech, RE-CORD, and WUR 2015).

In Europe, there are other PLA (and lactic acid) producers, including Synbra Technology, which operates a commercial (annual 5-kton capacity) plant in the Netherlands and a pilot production plant in Switzerland (annual 1-kton capacity). Besides Corbion Purac, Galactic and Jungbunzlauer are operating commercial lactic acid production plants in Europe. Other PLA producers include the Chinese Zhejiang Hisun Biomaterial (with an annual 5.5-kton capacity to be expanded to 50 kton, using cassava instead of corn). Other lactic acid producers include Henan Jindan Lactic Acid Technology (with an annual 100-kton capacity, the largest in Asia) (E4tech, RE-CORD, and WUR 2015).

In addition to its use in the synthesis of biodegradable polymers, lactic acid can be regarded as a feedstock for the green chemistry in the future as it is a platform chemical for the production of several downstream bio-based products (Gao et al. 2011) (outlined in Fig. 4.2). Lactic acid readily undergoes esterification to yield lactate esters, of interest as new “green” solvents. Catalytic reduction of lactic acid leads to propylene glycol, which can be further dehydrated to give acrylic acid esters, but in practice this conversion proceeds with low yield. Lactic acid can also be spun using wet, dry, and electrospinning techniques to give biodegradable fibers for apparel, furniture, and biochemical materials, such as dissolving sutures. New nanostructural materials prepared from lactic acid using electrospinning have found use in neural tissue engineering (Bozell and Petersen 2010).

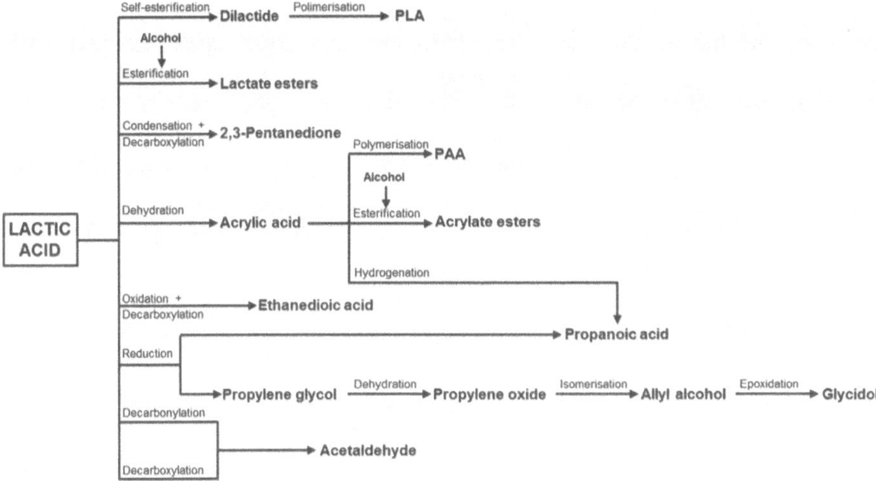

Fig. 4.2 Overview of lactic acid as a platform chemical. *PLA* polylactic acid, *PAA* polyacrylic acid

4.5.1.3 Polyhydroxyalkanoates (PHAs)

Polyhydroxy alkanoates (PHAs) are linear polyesters directly produced by bacterial fermentation of sugars or lipids. PHAs are typically produced by bacteria under physiological stress so as to store carbon and energy. PHAs can be combined with diverse monomers yielding different thermoplastic or elastomeric biodegradable plastics (suitable for home composting) (E4tech, RE-CORD, and WUR 2015). PHAs include polyhydroxybutyrate (PHB) and polyhydroxyvalerate (PHBV). PHB exhibits mechanical, physical, and thermal properties similar to those of polypropylene (PP), polyethylene (PE), low-density polyethylene (LDPE), high-density polyethylene (HDPE), polyvinylchloride (PVC), PS, and PET, depending on its grade. Given its tunable properties, it might have several applications. Besides its use as plastics/foams substitute (e.g., for packaging, injection molding), it can have niche applications, such as for internal sutures, capsules in pharmacology (as a non-toxic, compatible, and naturally absorbed polymer), and also in automotive, design, and high-tech electronics sectors (E4tech, RE-CORD, and WUR 2015).

PHB is intracellularly produced from glucose under aerobic conditions by certain types of bacteria (e.g., *Alcaligens euthrophus* and *Lactobacillus acidophilus*), and thus PHB should be extracted from bacterial cells using different technologies, some of them based on organic solvents, followed by purification. PHAs thus exhibit very high costs and investment, and activity in their production is relatively low as compared to PLA and other bioplastics. Global production capacity in 2014 was estimated as 54 kton. Once feedstock makes up close to 50% of the production cost of PHA, this is expected to decrease by using cheaper feedstocks, such as starch, for example, hydrolyzed cornstarch, which is less than half the price of glucose. In addition, production costs may decrease via integration with sugar and ethanol mills, by assuring on-site energy production by burning the bagasse by-product and co-producing the solvent (iso-pentanol) for cell extraction. In this context, PHB production is beginning to be associated to the Brazilian sugar industry (E4tech, RE-CORD, and WUR 2015).

When the cost barrier is overcome, enabling mass production, the potential market for PHA may be as much as 50% of the current plastics market. However, nowadays the PHA market keeps focused on niche high-value markets, and there are currently few PHA developers based on sugar fermentation in Europe: these include the German-based Biomer, owner of a demonstration-scale plant producing four different grades of PHB; and KNN, which is considering to produce PHA resin via the sugar platform (using pulp and paper starches) in the Netherlands and Sweden. There are a few more PHA developers in China (Tianjin Green Bio, Yikeman, and TianAn Biopolymer Co.) and the Americas (PHB Industrial in Brazil and Metabolix in the USA) (E4tech, RE-CORD, and WUR 2015).

4.5.1.4 Succinic Acid

Succinic acid, a 1,4-diacid, is also a widely investigated chemical building block that is most commonly biochemically produced by low-pH fermentation of biorefinery sugars by adequate yeasts or bacteria, such as *Anaerobiospirillum succiniciproducens* or engineered *Mannheimia succiniciproducens* (Bozell and Petersen 2010). Recombinant *Escherichia coli* strains have been licensed for commercial production of succinic acid, by Reverdia (Roquette/DSM) in Europe (Italy, with 10 kton of annual capacity), and BioAmber (a Canadian company that is also running a plant in France with an annual capacity of 3 kton) (Bozell and Petersen 2010). Other succinic acid producers include Succinity (joint venture between BASF and Corbion Purac) and Myriant (a US-based company) (Lane 2015). Succinity GmbH is operating a plant in Montmelo (Spain) with an annual capacity of 10 kton; Myriant owns several commercial plants in the USA and also a small demonstration facility (1-kton annual capacity) in Leuna (Germany) (E4tech, RE-CORD, and WUR 2015).

As is lactic acid, succinic acid is initially isolated from fermentation broth as a salt, and electrodialysis has also been successfully investigated as an alternative technology for its recovery. Direct hydrogenation of the aqueous fermentation broth has also been studied (Bozell and Petersen 2010). Competing routes to produce succinic acid include its chemical synthesis starting from glycerol (E4tech, RE-CORD, and WUR 2015).

Succinic acid, previously highlighted as exhibiting the fastest growing market, provides high-value niche applications, for example, personal care products and food additives (as an acidity regulator). In addition, it is a powerful platform chemical that is used for the production of bio-based polymers, such as polybutylene succinate (PBS), polyester polyols for polyurethanes, coating and composite resins, phthalate-free plasticizers, and 1,4-butanediol (BDO), for footwear, packaging, and paints, or personal care ingredients. It is a substitute for maleic anhydride (Werpy and Petersen 2004). Succinate esters are also precursors for known petrochemical

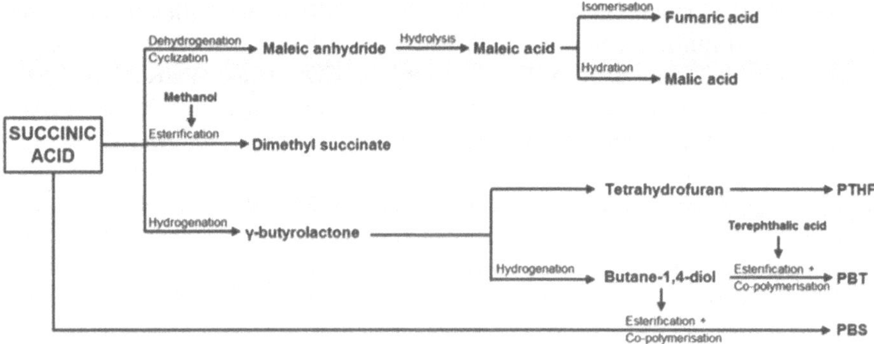

Fig. 4.3 Overview of succinic acid as a platform chemical. *PTHF* polytetrahydrofurane. *PBT* polybutylene terephthalate, *PBS* polybutylene succinate

products such as tetrahydrofuran, γ-butyrolactone, or various pyrrolidinone derivatives (Bozell and Petersen 2010). The use of succinic acid as a platform chemical is summarized in Fig. 4.3 (Bozell and Petersen 2010). Indeed, bio-based succinic acid is a drop-in replacement for petroleum-based succinic acid (Lane 2015) and has the potential to be cheaper than the fossil-based product when produced at large scale (E4tech, RE-CORD, and WUR 2015).

In 2013, global annual production of bio-based succinic acid was 38 kton, corresponding to a market value of 108 million USD, which is comparable to the market of the fossil-based product, but the bio-based product has the potential to significantly increase the share driven by the demand for BDO and PBS. Indeed, by 2020 the bio-succinic acid market is projected to reach 600 kton with annual revenues of 539 million USD (E4tech, RE-CORD, and WUR 2015).

4.5.1.5 1,4-Butanediol (BDO)

BDO is industrially used as a solvent and in the manufacture of some types of plastics, elastic fibers, and polyurethanes, for use in golf balls and skateboard wheels to printing inks and cleaning agents (Lane 2015). The bio-based product can be a direct drop-in replacement for fossil BDO. It can be converted into numerous chemicals, including γ-butyrolactone (GBL), the solvent tetrahydrofuran (THF), and the resin polybutylene terephthalate (PBT) (E4tech, RE-CORD, and WUR 2015). BDO can be produced in the petrochemical industry in various ways from acetylene, maleic anhydride, propylene oxide, and butadiene. Bio-based BDO can be produced via direct fermentation of sugars or via the catalytic hydrogenation of succinic acid, but the first is the dominant industrial route (E4tech, RE-CORD, and WUR 2015).

The market share of bio-based BDO constitutes a very small fraction (less than 0.2% in 2013) of the total BDO market (Lane 2015). The largest application of BDO is in the manufacture of THF, accounting for 30% in 2013, followed by polyurethane at 25%, and PBT, which used about 22% of all BDO worldwide.

The global BDO market is foreseen to reach 2.7 million tons with a market value close to 7000 million USD by 2020, and the bio-based market is expected to increase by 43% from 2014 to 2020 (E4tech, RE-CORD, and WUR 2015).

Genomatica, a California-based company, is the main agent in bio-based BDO production, having patented a GENO BDO process using an engineered microorganism for direct production via fermentation of sugars. Several European companies, such as BASF, Novamont (Mater Biotech), DSM (for PBT production), and Biochemtex (but using cellulosic biomass converted by Proesa technology), are employing the Genomatica technology. Johnson-Matthey-Davy Technologies is producing bio-BDO and THF, in UK, via Myriant's succinic acid. In the USA, Tate & Lyle has also signed a joint development agreement with Genomatica to produce bio-based BDO from dextrose sugars at demonstration scale, whereas BioAmber produces bio-BDO (and bio-THF) from its bio-succinic acid (via chemical processing). The Japanese company Toray produces bio-based BDO (converted into PBT), also based on Genomatica technology (E4tech, RE-CORD, and WUR 2015).

The Genomatica technology enables producing BDO directly from an abundant feedstock, and this company believes that the sugar market is sufficiently robust and can grow to include chemicals production, although currently 50% of global sugar is used for ethanol production (E4tech, RE-CORD, and WUR 2015).

4.5.1.6 Farnesene

Farnesene is a branched-chain alkene with 15 carbon atoms that is found in plants (e.g., in the skin of green apples and other fruits) and some insects. There is no fossil-based substitute, but it is an attractive building block to obtain bio-based products with several niche market applications—solvents, moisturizer emollients, adhesives, fragrances, surfactants, stabilizers, resins, foams, coatings, sealants, emulsifiers, and vitamin precursors. Farnesene has also applicability as a fuel and lubricant feedstock, replacing jet fuel (properties consistent with C15 paraffins but still with higher cost) and diesel and industrial oils (E4tech, RE-CORD, and WUR 2015).

It can be produced from sugars via aerobic microbial fermentation using genetically modified microorganisms (yeasts). The US-based company Amyris, the only player in the market, has developed a pathway enabling the conversion of C6 sugars into farnesene, and its production is already running at demonstration plants in the USA, Spain, and Brazil. The industrial facilities in Brazil are located adjacent to existing sugar and ethanol mills. Farnesene produced by Amyris targets fuel applications, with its use in blends (in diesel and jet fuel) already demonstrated (as diesel for buses in Brazil and as jet fuel in multiple flights) (E4tech, RE-CORD, and WUR 2015).

4.5.1.7 Isobutene

Isobutene (isobutylene or 2-methylpropene) is a four-carbon branched alkene that is a key precursor for chemicals such as fuel and lubricant additives, polymers, and pharmaceuticals. However, because of its toxicity, stringent measures are required to prevent leakage into the environment (E4tech, RE-CORD, and WUR 2015). The addition of isobutene to methanol yields MTBE (methyl tert-butyl ether) and to ethanol gives ETBE (ethyl tert-butyl ether), the most important fuel additives in the market. It is also used to produce isooctane, an additive applied in aviation fuel. In addition, it is used in polymerization reactions, for example, with isoprene to yield butyl rubber for the production of tires, gas masks, baseballs, and chewing gum. It can also yield poly(methyl methacrylate) plastics and tert-butanol and tert-amines for use in various chemical processes and products. A future application of isobutene, with increased demand, could be the production of antioxidants to be used in the food industry (E4tech, RE-CORD, and WUR 2015).

Nowadays, isobutene is commercially produced by petrochemical cracking of crude oil, but it can also be produced by dehydration of isobutanol obtained via biomass digestion. As a biological alternative, the French company Global Bioenergies, owner of three pilot plants in Europe, is developing a process based on direct fermentation of glucose into isobutene using engineered bacterial strains. This completely biological route is advantageous in terms of energy demand for recovery of the gaseous isobutene (instead of isobutanol), which also exhibits lower aqueous solubility than isobutanol and thus decreased product toxicity to the microorganisms. Lanxess is developing in Germany the dehydration process for conversion of bio-based isobutanol into isobutene, targeting production of bio-based butyl rubber for the tire industry. Most isobutanol used by Lanxess is supplied by the US-based biochemicals and biofuels company Gevo, produced from corn-based fermentable sugars. Gevo and Butamax lead the routes via isobutanol (E4tech, RE-CORD, and WUR 2015).

The global annual production of isobutene is around 15 million tons, corresponding to a market of 25–30 million USD, but bio-based isobutene currently accounts only for 10 tons per year produced by a small number of players. Rapid growth of isobutene markets is expected, mostly driven by the aerospace market, together with rubber for the automotive market (E4tech, RE-CORD, and WUR 2015).

4.5.1.8 Acrylic Acid

Acrylic acid is an organic acid with three carbon atoms, systematically named 2-propenoic acid. Acrylic acid and its esters readily combine with themselves (to form, e.g., polyacrylic acid, used mostly in superabsorbent polymers) or other monomers (e.g., acrylamides, acrylonitrile, vinyl, styrene, butadiene) by reacting at their double bond. These homopolymers or copolymers are used in the manufacture of various plastics, coatings, adhesives, fibers and textiles, resins, detergents and cleaners, elastomers (synthetic rubbers), and floor polishes and paints. Acrylic acid is also a drop-in replacement as a chemical intermediate in multiple industrial processes (E4tech, RE-CORD, and WUR 2015; Lane 2015).

Although conventional petrochemical acrylic acid is produced via the oxidation of propylene, bio-based acrylic acid is produced through the dehydration of 3-hydroxypropionic acid (3-HPA), which can be obtained by sugar fermentation (E4tech, RE-CORD, and WUR 2015). As an alternative, acrylic acid can be produced from sugar-derived lactic acid by dehydration. However, none of these processes is yet commercially implemented. The production of acrylic acid in 2013 comprised approximately 5 million tons but the annual production for the bio-based product, still in the pilot phase, accounted only for around 300 tons. The global acrylic acid market is projected to steadily increase, reaching an annual demand of around 7.4 million tons by 2020 (E4tech, RE-CORD, and WUR 2015), partially because of a foreseen increase (4–5%) in the demand for superabsorbent polymers.

Two key strategic partnerships, BASF-Cargill-Novozymes (in Europe) and OPXBio-DOW (in the USA), have been responsible for the development (at pilot scale) of the 3-HPA route for production of bio-based acrylic acid (E4tech, RE-CORD, and WUR 2015). Given the various possible routes to produce bio-based acrylic acid, there is fragmentation among technology developers, with Metabolix together with Cargill-BASF-Novozymes and OPXBio/Dow focusing on the 3-HPA process, and Myriant and SGA Polymers investing on the lactic acid route, also via the sugar platform. In addition, Genomatica have filled a patent from a process to produce acrylic acid via fumaric acid (Genomatica 2009). However, major industrial players have been collaborating toward fast-track commercial development via different processes and feedstocks. The shift from petro-based acrylic acid toward the bio-based equivalent is pulled not only by the demand for reductions in GHG emissions, the result of environmental concerns, but also by the incentive to reduce reliance on crude oil and the associated price volatility (E4tech, RE-CORD, and WUR 2015).

4.5.1.9 Adipic Acid

Adipic acid is the most widely used dicarboxylic acid from an industrial perspective, as a monomer for production of nylon and polyurethane (E4tech, RE-CORD, and WUR 2015). Indeed, 85–90% of adipic acid is used in the production of nylon-6,6, a high performer engineering resin, or is further processed into fibers (polyurethanes, adipic esters) for application in carpeting, automobile tire cord, and clothing. It is also used to manufacture plasticizers and lubricant components. Food-grade adipic acid is used as a gelling aid, an acidulant, and as a buffering agent (E4tech, RE-CORD, and WUR 2015).

As a drop-in chemical (also produced from petrochemicals), bio-based adipic acid can be produced by fermentation, directly from sugars or via hydrogenation of muconic or glucaric acid. Verdezyne has developed a process based on genetically modified microorganisms for direct conversion of glucose into adipic acid. It can also be chemically produced by a two-step chemo-catalytic route involving anaerobic oxidation of glucose into glucaric acid that is finally converted to adipic acid by hydrogenation (E4tech, RE-CORD, and WUR 2015).

The global market of fossil adipic acid was 2.7 million metric tons in 2013 and it is expected to grow, achieving 7240 million USD by 2020, mainly from the increasing demand for nylon-6,6 within the automobile and electronics industries, and for polyurethanes to be used in footwear (E4tech, RE-CORD, and WUR 2015). Despite the environmental benefits, with the potential to provide cost advantage of the equivalent bio-based product, this is not yet commercially produced. Bio-based adipic acid is still at the R&D stage, led by Verdezyne and Rennovia, owners of pilot plants in the USA, and DSM and Biochemtex in Europe. In the USA, BioAmber, Genomatica, Amyris and Aemetis are also developing sugar-based fermentation production processes (E4tech, RE-CORD, and WUR 2015).

4.5.1.10 Ethylene and Polyethylene (PE)

Fossil PE is the most common plastic produced globally [with an annual production volume of 88 million tons and a market share close to 30% (E4tech, RE-CORD, and WUR 2015)], especially used by the automotive industry and manufacturers of cosmetics, packaging, toys, personal hygiene, and cleaning products. There are diverse types and grades of PE, exhibiting different properties and thus applications. Based mostly on its density and branching, PE can be categorized as high-density PE (HDPE), low-density PE (LDPE), and linear low density PE (LLDPE), which results from copolymerization of ethylene with longer polymers (e.g., butylene, hexene, or octane).

Bio-based PE, which can be made by dehydrating bioethanol to ethylene that is further polymerized, is a drop-in equivalent replacement for the fossil product. It is not biodegradable but it can be easily recycled. Nowadays, bioethylene is produced from sugarcane (in Brazil) and sugar beet (in Europe), but also by using cornstarch and wheat starch (in the USA) as raw materials (E4tech, RE-CORD, and WUR 2015). There is no commercial activity of bio-based PE in Europe, and Braskem in Brazil is the only commercial-scale producer (from sugarcane, since 2010) worldwide (E4tech, RE-CORD, and WUR 2015). The production cost of bioethylene (the bio-based PE building block) is obviously dependent on the bioethanol price, and the final polymerization step has a small contribution to the PE production cost. Indeed, the production of bio-PE is expected to increase, reaching 840 kton by 2020, but there are some uncertainties associated with competition with other attractive bio-based packaging polymers, such as PET, PLA, and PEF, as already referred (E4tech, RE-CORD, and WUR 2015).

4.5.2 *The Chemical Catalytic Route*

4.5.2.1 Furan-2,5-Dicarboxylic Acid (FDCA)

Although neither FDCA nor any of its derivatives has yet become a commercial product because of its high price, it is very attractive from its potential as a bio-based replacement for polymers (polyesters). It can substitute terephthalic acid and also PBT and polyamides, providing a new class of polyethylene furanoate (PEF) polymers, with application in drinks bottles with superior gas barrier properties (Werpy and Petersen 2004). FDCA has also the potential to be used in the production of novel solvents, and it can also be converted into levulinic and succinic acid.

It is chemically produced via oxidative dehydration of C6 sugars, such as glucose. Once most of the routes to FDCA proceed via oxidation of 5-hydroxymethylfurfural (HMF), its commercialization might be boosted by the recent achievements on development of more efficient production of HMF by bacterial fermentation (Bozell and Petersen 2010; Lane 2015). Indeed, a significant growth of the current incipient FDCA market is projected, with a volume of 500 kton, corresponding to 498 million USD, foreseen for 2020 (E4tech, RE-CORD, and WUR 2015).

There are only a few companies handling the production and commercialization of FDCA, and the market is dominated by a single key player, Avantium, settled in the EU (Lane 2015). This company has operated, since 2011, a pilot plant in the Netherlands based on development of a two-step catalytic process to convert sugars into FDCA that is used to produce PEF. Corbion Purac, AVA Biochem, and Novozymes are also active in developing FDCA chemical production from glucose in Europe (Bozell and Petersen 2010).

4.6 Concluding Remarks

A large consortium of research organizations has proposed a "Joint European Biorefinery Vision for 2030" (Luguel 2011) based on several key points. The success of modern biorefineries, including starch based, should be based on versatile biomass supply chains and on the production of a wide spectrum of competitive bio-based products. This flexibility – multiple product and multiple feedstock – not only minimizes risks associated with raw material availability and product demand (market) but also allows all-year operation by using feedstocks that mature at different times (E4tech, RE-CORD, and WUR 2015). This flexibility strategy is evident in the typical sugar- and starch-based biorefinery 'Les Sohettes' complex, located in Pomacle (France). This biorefinery consists of a sugar beet processing unit, a wheat refinery and a sugar plant, an ethanol distillery (Cristanol), a research center (ARD), a demo-plant for second-generation ethanol (Futurol), a straw-based paper production pilot unit (CIMV), and a succinic acid pilot plant (BioAmber). This facility is a good example of integration in the product networks as two major crops are being used as feedstocks: sugar beet and wheat. The combination of both crops allows year-long biorefining operations, because the harvesting period of sugar beet is rather short (typically only a few months in a year), which would render the sole use of sugar beet for production of ethanol and other bioproducts uneconomical (Gnansounou and Pandey 2017).

As another example of multi-product strategy, NatureWorks LLC has been operating a corn biorefinery since 2002 (Nebraska, USA). This integrated biorefinery processes corn to produce corn oil, sugar, ethanol, lactic acid, and PLA (Gnansounou and Pandey 2017).

More broadly, cluster-based biorefineries constituted by different value chains site plants aggregated as a cluster shall be more competitive, such as demonstrated by the successful implementation of the Chemical Cluster (five-site plants) in Stenungsund (Sweden)—Aga, AkzoNobel, Borealis, Ineos and Perstorp—developing a joint strategy for producing sustainable products. New technologies are being explored for integrating the production of biomass-derived fuels and other products, such as 1,3-propanediol, polylactic acid, and isosorbide, in a single facility (Gírio et al. 2017).

References

Abdel-Rahman MA, Sonomoto K (2016) Opportunities to overcome the current limitations and challenges for efficient microbial production of optically pure lactic acid. J Biotechnol 236:176–192. https://doi.org/10.1016/j.jbiotec.2016.08.008

Bozell JJ, Petersen GR (2010) Technology development for the production of biobased products from biorefinery carbohydrates: the US Department of Energy's "top 10" revisited. Green Chem 12:539–554. https://doi.org/10.1039/B922014C

Brányiková I, Maršálková B, Doucha J, Brányik T, Bišová K, Zachleder V, Vítová M (2011) Microalgae: novel highly efficient starch producers. Biotechnol Bioeng 108(4):766–776. https://doi.org/10.1002/bit.23016

Breuninger WF, Piyachomkwan K, Sriroth K (2009) Tapioca/cassava starch: production and use. In: BeMiller J, Whistler R (eds) Starch, 3rd edn. Academic Press, San Diego, pp 541–568. ISBN: 978-0-12-746275-2

Curtis BC (2002) Wheat in the world. In: Curtis BC, Rajaram S, Gómez Macpherson H (eds) Bread wheat – improvement and production. FAO plant production and protection series. Food and Agriculture Organization of the United Nations (FAO), Rome. ISBN: 92-5-104809-6

de Souza PM, de Oliveira Magalhães P (2010) Application of microbial α-amylase in industry: a review. Braz J Microbiol 41(4):850–861. https://doi.org/10.1590/S1517-83822010000400004

E4tech, RE-CORD, and WUR (2015) From the sugar platform to biofuels and biochemicals. Final report for the European Commission, contract No. ENER/C2/423–2012/SI2.673791. http://www.qibebt.cas.cn/xscbw/yjbg/201505/P020150520398538531959.pdf. Accessed 25 Apr 2017

ETIP Bioenergy–European Technology and Innovation Platform. Sugar crops for production of advanced biofuels. http://www.etipbioenergy.eu/?option=com_content&view=article&id=256. Accessed 13 Jun 2017

Gallezot P (2007) Process options for converting renewable feedstocks to bioproducts. Green Chem 9:295–302. https://doi.org/10.1039/B615413A

Gao C, Ma C, Xu P (2011) Biotechnological routes based on lactic acid production from biomass. Biotechnol Adv 29:930–939. https://doi.org/10.1016/j.biotechadv.2011.07.022
Genomatica (2009) Methods for the synthesis of acrylic acid and derivatives from fumaric acid. Patent WO 2009045637 A2
Gírio F, Marques S, Pinto F, Oliveira AC, Costa P, Reis A, Moura P (2017) Biorefineries in the world. In: Rabaçal M, Ferreira AF, Silva CAM, Silva MC (eds) Biorefineries: targeting energy, high value products and waste valorisation. Springer International Publishing, Cham, pp 227–281. ISBN: 978-3-319-48286-6
Global Industry Analysts (2016) Starch – a global strategic business report. MCP 1215. Global Industry Analysts, Inc. http://www.strategyr.com/Starch_Market_Report.asp. Accessed 27 Mar 2017
Gnansounou E, Pandey A (2017) Classification of biorefineries taking into account sustainability potentials and flexibility. In: Gnansounou E, Pandey A (eds) Life-cycle assessment of biorefineries. Elsevier Science & Technology Books, North Holland, pp 1–40. ISBN: 978-0-444-63585-3
Gozzo A, Glittenberg D (2009) Starch: a versatile product from renewable resources for industrial applications. In: Höfer R (ed) Sustainable solutions for modern economies. The Royal Society of Chemistry, London, pp 238–263. ISBN: 978-1-84755-905-0
Guzmán-Maldonado H, Paredes-López O (1995) Amylolytic enzymes and products derived from starch: a review. Crit Rev Food Sci Nutr 35(5):373–403. https://doi.org/10.1080/10408399509527706
Hii SL, Tan JS, Ling TC, Ariff AB (2012) Pullulanase: role in starch hydrolysis and potential industrial applications. Enzyme Res 2012:921362. https://doi.org/10.1155/2012/921362
Höfer R (2015) Sugar- and starch-based biorefineries. In: Pandey A, Höfer R, Taherzadeh MJ, Nampoothiri KM, Larroche C (eds) Industrial biorefineries and white biotechnology. Elsevier, Amsterdam, pp 157–235. ISBN: 978-0-444-63453-5
Hofvendahl K, Hahn-Hägerdal B (2000) Factors affecting the fermentative lactic acid production from renewable resources. Enzyme Microb Technol 26:87–107. https://doi.org/10.1016/S0141-0229(99)00155-6
IEA Bioenergy Task 42 Biorefinery (2011) http://www.iea-bioenergy.task42-biorefineries.com/. Accessed 15 Apr 2017
Kale K, Deshmukh AG, Dudhare MS, Patil VB (2015) Microbial degradation of plastic: a review. J Biochem Technol 6(2):952–961. ISSN: 0974-2328
Karim AA, Tie AP-L, Manan DMA, Zaidul ISM (2008) Starch from the sago (*Metroxylon sagu*) palm tree: properties, prospects, and challenges as a new industrial source for food and other uses. Compr Rev Food Sci Food Saf 7(3):215–228. https://doi.org/10.1111/j.1541-4337.2008.00042.x
Kaseem M, Hamad K, Deri F (2012) Rheological and mechanical properties of polypropylene/thermoplastic starch blend. Polym Bull 68(4):1079–1091. https://doi.org/10.1007/s00289-011-0611-z
Khan B, Bilal KNM, Samin G, Jahan Z (2016) Thermoplastic starch: a possible biodegradable food packaging material: a review. J Food Process Eng. https://doi.org/10.1111/jfpe.12447
King D, Inderwildi OR, Williams A (2010) The future of industrial biorefineries. World Economic Forum Reports. http://www3.weforum.org/docs/WEF_FutureIndustrialBiorefineries_Report_2010.pdf. Accessed 10 May 2017
Kumar P, Satyanarayana T (2009) Microbial glucoamylases: characteristics and applications. Crit Rev Biotechnol 29(3):225–255. https://doi.org/10.1080/07388550903136076
Lane J (2015) The DOE's 12 Top Biobased Molecules – what became of them? Biofuels Digest. http://www.biofuelsdigest.com/bdigest/2015/04/30/the-does-12-top-biobased-molecules-what-became-of-them/. Accessed 15 Apr 2017
Leja K, Lewandowicz G (2010) Polymer biodegradation and biodegradable polymers: a review. Pol J Environ Stud 19(2):255–266
Lo Leggio L, Simmons TJ, Poulsen JC, Frandsen KE, Hemsworth GR, Stringer MA, von Freiesleben P, Tovborg M, Johansen KS, De Maria L, Harris PV, Soong CL, Dupree P, Tryfona

T, Lenfant N, Henrissat B, Davies GJ, Walton PH (2015) Structure and boosting activity of a starch-degrading lytic polysaccharide monooxygenase. Nat Commun 6:5961. https://doi.org/10.1038/ncomms6961

Luguel C (chief ed.) (2011) Joint European biorefinery vision for 2030. Star-COLIBRI project – strategic targets for 2020 – collaboration initiative on biorefineries. http://www.star-colibri.eu/files/files/vision-web.pdf. Accessed 18 May 2017

Marques S, Gírio FM, Santos JAL, Roseiro JC (2017a) Pulsed fed-batch strategy towards intensified process for lactic acid production using recycled paper sludge. Biomass Convers Bioref 7:127–137. https://doi.org/10.1007/s13399-016-0211-0

Marques S, Matos CT, Gírio FM, Roseiro JC, Santos JAL (2017b) Lactic acid production from recycled paper sludge: process intensification by running fed-batch into a membrane-recycle bioreactor. Biochem Eng J 120:63–72. https://doi.org/10.1016/j.bej.2016.12.021

Martin O, Avérous L (2001) Poly(lactic acid): plasticization and properties of biodegradable multiphase systems. Polymers 42(14):6209–6219. https://doi.org/10.1016/S0032-3861(01)00086-6

Mezaki Y, Katsuya Y, Kubota M, Matsuura Y (2001) Crystallization and structural analysis of intact maltotetraose-forming exo-amylase from *Pseudomonas stutzeri*. Biosci Biotechnol Biochem 65(1):222–225. https://doi.org/10.1271/bbb.65.222

Mihai M, Huneault MA, Favis BD (2007) Foaming of polystyrene/thermoplastic starch blends. J Cell Plast 43(3):215–236. https://doi.org/10.1177/0021955X07076532

Mikkola J-P, Sklavounos E, King AWT, Virtanen P (2016) The biorefinery and green chemistry. In: Bogel-Lukasik R (ed) Ionic liquids in the biorefinery concept: challenges and perspectives. RSC green chemistry no. 36. The Royal Society of Chemistry, Cambridge, pp 1–37. https://doi.org/10.1039/9781782622598-00001

Mitchell CR (2009) Rice starches: production and properties. In: BeMiller J, Whistler R (eds) Starch, 3rd edn. Academic Press, San Diego, pp 569–578. ISBN: 978-0-12-746275-2

Mohammadi NA, Moradpour M, Saeidi M, Alias AK (2013) Thermoplastic starches: properties, challenges, and prospects. Starch (Stärke) 65(1-2):61–72. https://doi.org/10.1002/star.201200201

Nigam P, Singh D (1995) Enzyme and microbial systems involved in starch processing. Enzyme Microb Technol 17(9):770–778. https://doi.org/10.1016/0141-0229(94)00003-A

Oettler G (2005) The fortune of a botanical curiosity: triticale: past, present and future. J Agric Sci 143(5):329–346. https://doi.org/10.1017/S0021859605005290

Olsson L, Saddler J (2013) Biorefineries, using lignocellulosic feedstocks, will have a key role in the future bioeconomy. Biofuels Bioprod Bioref 7(5):475–477. https://doi.org/10.1002/bbb.1443

Pal P, Sikder J, Roy S, Giorno L (2009) Process intensification in lactic acid production: a review of membrane-based processes. Chem Eng Process 48:1549–1559. https://doi.org/10.1016/j.cep.2009.09.003

Puchongkavarin H, Varavinit S, Bergthaller W (2005) Comparative study of pilot scale rice starch production by an alkaline and an enzymatic process. Starch (Stärke) 57(3-4):134–144. https://doi.org/10.1002/star.200400279

Radakovits R, Jinkerson RE, Darzins A, Posewitz MC (2010) Genetic engineering of algae for enhanced biofuel production. Eukaryot Cell 9(4):486–501. https://doi.org/10.1128/EC.00364-09

Saunders J, Izydorczyk M, Levin DB (2011) Limitations and challenges for wheat-based bioethanol production. In: dos Santos Bernardes MA (ed) Economic effects of biofuel production. InTech, Rijeka, pp 429–452. ISBN: 978-953-307-178-7

Scarlat N, Dallemand J-F, Monforti-Ferrario F, Nita V (2015) The role of biomass and bioenergy in a future bioeconomy: policies and facts. Environ Dev 15:3–34. https://doi.org/10.1016/j.envdev.2015.03.006

Schwartz D, Whistler RL (2009) History and future of starch. In: Bemiller J, Whistler R (eds) Starch, 3rd edn. Academic Press, San Diego, pp 1–10. ISBN: 978-0-12-746275-2

Searchinger T, Heimlich R (2015) Avoiding bioenergy competition for food crops and land. World Resources Institute Working paper, January 2015
Shanavas S, Padmaja G, Moorthy SN, Sajeev MS, Sheriff JT (2011) Process optimization for bioethanol production from cassava starch using novel eco-friendly enzymes. Biomass Bioenergy 35(2):901–909. https://doi.org/10.1016/j.biombioe.2010.11.004
Starch Europe (2017) European Starch Industry. Starch Europe. http://www.starch.eu/european-starch-industry/. Accessed 27 Mar 2017
St-Pierre N, Favis BD, Ramsay BA, Ramsay JA, Verhoogt H (1997) Processing and characterization of thermoplastic starch/polyethylene blends. Polymers 38(3):647–655. https://doi.org/10.1016/S0032-3861(97)81176-7
Sundarram A, Murthy TPK (2014) α-Amylase production and applications: a review. J Appl Environ Microb 2(4):166–175. 10.12691/jaem-2-4-10
Vaidya AN, Pandey RA, Mudliar S, Kumar MS, Chakrabarti T, Devotta S (2005) Production and recovery of lactic acid for polylactide: an overview. Crit Rev Environ Sci Technol 35:429–467. https://doi.org/10.1080/10643380590966181
Vijayakumar J, Aravindan R, Viruthagiri T (2008) Recent trends in the production, purification and application of lactic acid. Chem Biochem Eng Q 22:245–264. Retrieved from http://hrcak.srce.hr/24811
Vu VV, Beeson WT, Span EA, Farquhar ER, Marletta MA (2014) A family of starch-active polysaccharide monooxygenases. Proc Natl Acad Sci USA 111(38):13822–13827. https://doi.org/10.1073/pnas.1408090111
Wang Y, Tashiro Y, Sonomoto K (2015) Fermentative production of lactic acid from renewable materials: recent achievements, prospects, and limits. J Biosci Bioeng 119:10–18. https://doi.org/10.1016/j.jbiosc.2014.06.003
Werpy T, Petersen G (eds) (2004) Top value added chemicals from biomass. Volume I: Results of screening for potential candidates from sugars and synthesis gas. U.S. Department of Energy Report. http://www.osti.gov/bridge. Accessed 14 May 2017
Zwart RWR (2006) Biorefinery: the worldwide status at the beginning of 2006, including the highlights of the 1st International Biorefinery Workshop and an overview of markets and prices of biofuels and chemicals. Wageninger. https://pt.scribd.com/document/63521889/Biorefinery-The-Worldwide-Status-at-the-Beginning-of-2006. Accessed Jul 2016

Chapter 5
Lignocellulosic Biomass for Energy, Biofuels, Biomaterials, and Chemicals

Abla Alzagameem, Basma El Khaldi-Hansen, Birgit Kamm, and Margit Schulze

Abstract The main objective of this chapter is to explore the lignocellulose feedstock (LCF) biorefinery for industrial usage according to green chemistry principles. In particular, the isolation and valorization of lignin as one of the most interesting intermediates of LCF biorefineries is discussed, including lignin isolation, purification, and structure analysis. Structure elucidation involves various chromatographic, spectroscopic, microscopic, and thermochemical methods. Thus, basic structure–property relationships regarding the influence of biomass source and isolation process on lignin amount, constitution, and 3D structure are highlighted. Furthermore, storage effects on lignin structure and degradation effects are presented. Finally, potential applications are discussed, including novel lignin-based hydrogels, composite compounds (hybrids), and nanomaterials. Focus is drawn to antioxidant and antimicrobial activity of lignin for applications in packaging and biomedicine, that is, biomaterials for drug release and tissue engineering.

Keywords Antimicrobial activity • Antioxidant activity • Biomass • Biomaterial • Biorefinery • Cellulose • Lignin • Lignocellulose feedstock • Pulping • Renewable resource

---

The original version of this chapter was revised.
An erratum to this chapter can be found at https://doi.org/10.1007/978-3-319-66736-2_9

A. Alzagameem
Faculty of Environment and Natural Sciences, Brandenburg University of Technology BTU Cottbus-Senftenberg, Cottbus, Germany

Department of Natural Sciences, Bonn-Rhein-Sieg University of Applied Sciences, Rheinbach, Germany

B. El Khaldi-Hansen • M. Schulze
Department of Natural Sciences, Bonn-Rhein-Sieg University of Applied Sciences, Rheinbach, Germany

B. Kamm (✉)
Faculty of Environment and Natural Sciences, Brandenburg University of Technology BTU Cottbus-Senftenberg, Cottbus, Germany

Kompetenzzentrum Holz GmbH, Linz, Austria
e-mail: b.kamm@kplus-wood.at

S. Vaz Jr. (ed.), *Biomass and Green Chemistry*,
https://doi.org/10.1007/978-3-319-66736-2_5

5.1 Introduction

5.1.1 Status Quo of Biorefinery Systems for Biomass Conversion

The use of lignocellulosic feedstock (LCF) in biorefineries has been recognized as the most successful strategy to produce such valuable products as fuels, power, and chemicals (Ringpfeil 2001). Lignocellulosic raw materials comprise cellulose, hemicellulose, and lignin (Fig. 5.1). The main advantages of this strategy are the low costs of the raw materials (straw, reeds, grass, wood, paper waste, etc.) and the utilization of natural polymers "made by nature" (via photosynthesis) and comprising mainly polysaccharides (Kamm et al. 2015).

Currently, the need remains for the technology to be developed and optimized, especially in the area of separating cellulose, hemicellulose, and lignin. Figure 5.2 shows an overview of potential products from LCF (Kamm et al. 2006).

The first step in using LCF in biorefineries is to release cellulose, hemicellulose, and lignin by pretreatment of the biomass. Glucose and other six-carbon sugars can be produced from cellulose via enzymatic hydrolysis. These products are starter materials for the production of biofuels via fermentation (Dyne et al. 1999).

The production of furfural and hydroxymethyl furfural is important because these chemicals are starting materials for nylon-6,6 and nylon-6. Levulinic acid can be obtained directly from cellulose hydrolysis (as shown in Fig. 5.2). The pulp industries represent the first type of nonfood biorefinery producing value-added products (Kamm et al. 2006).

Currently, the development of LCF biorefineries corresponds to the raw materials available and focuses on a complete separation of the cellulose, hemicellulose, and lignin fractions using combinations of mechanical, chemical, and biotechnical methods. Isolation methods such as Organosolv, acid, or alkaline steam-pressure processes are well-known methods that are applied commercially as the Organocell,

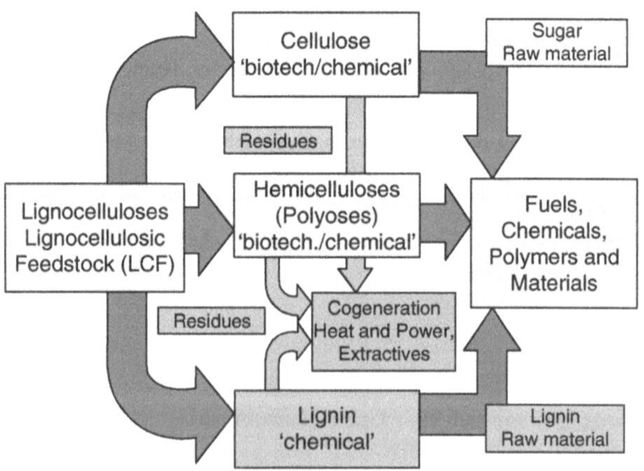

Fig. 5.1 Lignocellulosic feedstock (LCF) biorefinery (Kamm et al. 2015) (Copyright Wiley-VCH Verlag GmbH & Co. KGaA, reproduced with permission)

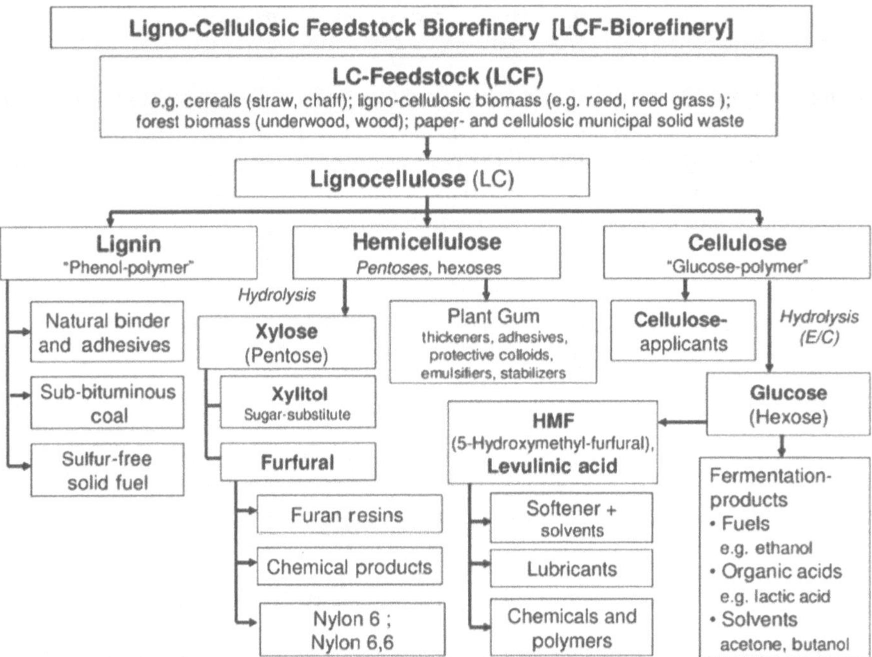

Fig. 5.2 Products of a lignocellulosic feedstock (LCF) biorefinery (Kamm et al. 2015) (Copyright Wiley-VCH Verlag GmbH & Co. KGaA, reproduced with permission)

Alcell, or Soda method (Puls 2009). Thus, various polysaccharides and lignin are available today from lignocellulose feedstocks on an industrial scale, which can be used for further valorization (see Fig. 5.2). In contrast, lignin utilization is mainly limited to energy production. In the past decade, a number of research groups worldwide have started to search for alternatives, focusing on lignin separation, purification, fragmentation, or functionalization to produce lignin-based high-value products. For instance, kraft lignin was used to synthesize novel benign lignin-based polymers for drug-release applications. Richter et al. (2015) reported a class of antimicrobial nanoparticles with biodegradable cores (from Indulin AT lignin) loaded with Ag^+ ions and coated with the cationic polyelectrolyte poly-(dimethyldiallylammonium chloride) (polyDADMAC). The environmentally benign nanoparticles (ebNPs) Ag^+-polyDADMAC exhibit broad-spectrum biocidal action and are capable of neutralizing common gram-negative and gram-positive human pathogens as well as quaternary amine–resistant bacteria, while using at least tenfold less silver than conventional branched polyethylenimine-Ag NPs(bPEI) and $AgNO_3$ aqueous solution. The polyDADMAC coating, which boosts adhesion to microbial cell membranes, may itself have some antibacterial activity. However, both environmentally benign NP controls without polyDADMAC or without Ag^+ ion loading demonstrated much lower antimicrobial efficiencies (Richter et al. 2015).

5.1.2 Consideration of Green Chemistry Principles

Green chemistry refers to the promotion of innovative technologies to minimize the use of hazardous substances. The 12 principles of green chemistry (Table 5.1), developed by Paul Anastas and John Warner (1998), show how this can be achieved.

These principles must be considered as design rules or as a framework to achieve sustainability internationally.

Table 5.1 The 12 principles of green chemistry according to Anastas and Warner (1998)

Principle	Content
1. Prevention	It is better to prevent waste than to treat or clean up waste after it has been created
2. Atom economy	Synthetic methods should be designed to maximize the incorporation of all materials used in the process into the final product
3. Less hazardous chemical synthesis	Wherever practicable, synthetic methods should be designed to use and generate substances that possess little or no toxicity to human health and the environment
4. Design safer chemicals	Chemical products should be designed to affect their desired function while minimizing their toxicity
5. Safer solvents and auxiliaries	The use of auxiliary substances (e.g., solvents, separation agents) should be made unnecessary wherever possible and innocuous when used
6. Design for energy efficiency	Energy requirements of chemical processes should be recognized for their environmental and economic impacts and should be minimized. If possible, synthetic methods should be conducted at ambient temperature and pressure
7. Use of renewable feedstocks	A raw material or feedstock should be renewable rather than depleting whenever technically and economically practicable
8. Reduce derivatives	Unnecessary derivatization (use of blocking groups, protection/deprotection, temporary modification of physical/chemical processes) should be minimized or avoided if possible, because such steps require additional reagents and can generate waste
9. Catalysis	Catalytic reagents (as selective as possible) are superior to stoichiometric reagents
10. Design for degradation	Chemical products should be designed so that at the end of their function they break down into innocuous degradation products and do not persist in the environment
11. Real-time analysis for pollution prevention	Analytical methodologies need to be further developed to allow for real-time, in-process monitoring and control before the formation of hazardous substances
12. Inherently safer chemistry for accident prevention	Substances and the form of a substance used in a chemical process should be chosen to minimize the potential for chemical accidents, including releases, explosions, and fires

5.1.2.1 Lignocellulosic Feedstock for Energy Production

Global Availability and Production Since 1990, renewable energy sources have grown at an average annual rate of 2.2%, which is slightly higher than the growth rate of world total primary energy supply (TPES), 1.9%. According to the International Energy Agency, in 2014, the world TPES was 13,700 million tons oil equivalent (Mtoe), of which 13.8%, or 1894 Mtoe (up 2.6% from 2013), was produced from renewable energy sources. As the consequence of widespread noncommercial use in developing countries, solid biofuels (charcoal) represent the largest renewable energy source (66.2% of the global renewables supply). A second source is hydropower (2.4% of world TPES, 17.7% of renewables), followed by a mixture (geothermal, liquid biofuels, biogases, solar, wind) (International Energy Agency).

Based on data of the National Renewable Energy Action Plans (NREAPs) of the 27 European Union (EU) member states, the estimate of EU biomass availability in 2012 was around 314 Mtoe, expected to increase to 429 Mtoe and then remain at 411 Mtoe in 2020 and 2030, respectively (Balan et al. 2013). Currently, biomass accounts for 7–10% of primary global energy consumption, whereas nuclear and hydroelectric power individually contribute about 7%, and renewable sources (wind, solar) sum to less than 1% of the global energy demand (Hossen et al. 2017). Renewable energy sources, such as lignocellulosic biomass, thermal solar, photovoltaic, hydropower, wind, geothermal, and marine energy sources, are expected to have essential roles in the world's future energy supply. Here, biomass has an advantage over other sources of renewable energy, as it can be stored and is readily available all year round from various sources. In 2008, 10% of the global primary energy was contributed by biomass, which is expected to increase significantly, up to sixfold, until 2050 (Rabaçal et al. 2017).

Regarding number 7 of the Green Chemistry principles, the use of renewable feedstocks is one of the most essential steps for achieving sustainability goals. In principle, biomass as organic matter has always been a source of energy. The structure of the biomass, and thus its chemical and physical properties, determines its value as an energy source. Biomass stores this energy as chemical bonds in cellulose, hemicellulose, and lignin. Ideal energy crops should possess low energy input costs for production, low costs, low yield of impurities, and a low nutrient requirement. Today, the paper industry uses lignin combustion to produce process heat. Lignin is a side product of the pulp production process, available on an industrial scale (Holladay et al. 2007). Gasification processes convert biomass at high temperatures and in the absence of oxygen to synthesize gas ($CO + H_2$) in the pyrolysis method.

5.1.2.2 Lignocellulosic Feedstock for Biofuels Production

Global Availability and Production There is a need for the reformation of global energy policies in support of other clean sources, taking into account ever-increasing energy demands as well as the continuously rising price of fossil fuels. Sustained population growth, as well as economic and social development, will remain a key driver of increasing energy demand. Liquid transportation fuels, such as ethanol and biodiesel, are the most popular biomass-based energy alternatives, albeit these currently comprise only 2% of world biomass energy, but have the potential to contribute substantially.

The share of EU biodiesel in global demand will rise from 42% in 2010 to 74% in 2020, and the bioethanol share will also rise, to 13% in 2030. It must also be observed that meeting 2020 and 2030 EU biomass targets will require a significant import of feedstock from different parts of the world. Implications for direct and indirect land use change are currently under evaluation and discussion in Europe (Elbersen et al. 2012).

In China, cellulosic biomass production, such as agricultural and forestry residues, major portions of municipal solid waste, and dedicated herbaceous and woody crops, is estimated to be 670 million tons per year (Mio t/a) and could provide a significant alternative feedstock for ethanol production roughly equivalent to 4 billion barrels of petroleum (100 million tons). It is estimated that the total annual consumption by the Chinese auto industry is approximately 60 million tons of gasoline. Hence, approximately 5–10 million tons of fuel ethanol must be supplied within the next few years under an E10 (10% ethanol and 90% gasoline blend) standard. The annual consumption of fuel ethanol is approximately 1 million tons, and thus there will be a huge commercial market and very bright prospects for the development of cellulosic ethanol.

The European directive on the promotion of the use of energy from renewable sources (Directive 2009/28/EC) defines binding targets for biofuels and regulates their sustainability. Requirements for the sustainability of biofuels and electricity from liquid biomass have applied since January 2011. The criteria are laid down in the Sustainability Ordinance for Biofuels and Biomass-Electricity. After 2011, biofuels must save at least 35% in greenhouse gas emissions, as well as other sustainability criteria. All member states face a binding target of ensuring that renewable energy sources account for at least 10% of final energy consumption in the transport sector in 2020. In Germany, 57.5 Mio. tons fuel were consumed in 2016 (biodiesel 63.7%; gasoline 30.5%), corresponding to 4.6% of the total energy content. Biodiesel (2.0 Mio. tons) is the most important biofuel in Germany, mainly used in addition to fossil-based fuel. The analogue, bioethanol, is added to gasoline; in all, 1.2 Mio. tons in 2016. Plant oil-based fuels follow in third place (2.000 t in total, corresponding to 0.1% of the total amount) (Fachagentur Nachwachsende Rohstoffe 2017).

Ethanol is the largest volume of biofuel that can be produced via a biochemical conversion process. The conventionally used biomass materials are cornstarch and sugarcane (first generation); after applying more advanced technologies to the biochemical process, more abundant biomass sources such as grasses and wheat straw (second generation) can be used. First-generation bioethanol production requires biomass pretreatment to generate sugar monomers (glucose), which are then converted to ethanol via fermentation processes using microorganisms (yeast, bacteria, etc.). The product, ethanol, undergoes distillation for purification and then can be used directly as a fuel (Naik et al. 2009). Second-generation bioethanol includes lignocellulosic biomass as feedstock. The pretreatment processes for sugar production are more challenging with second-generation fuels, specifically regarding economic viability. The limitation of this process is the extraction of recalcitrant lignin under mild conditions (Galbe and Zacchi 2002). The enzymatic hydrolysis of lignin is also difficult because there are no natural enzymes to degrade it into basic monomers. Therefore, delignification processes are important pretreatments to enhance the enzymatic hydrolysis to produce bioethanol from lignocellulosic biomass. Wi et al. (2015) used a hydrogen peroxide-acetic acid (HPAC) pretreatment to increase the enzymatic digestibility, which showed a high efficiency for removing lignin. There are numerous different pretreatment processes such as ammonia fiber expansion (AFEX), dilute acid, or ionic liquid methods. It must be mentioned that products such as biodiesel, biogas, biomethanol, and biohydrogen are also regarded as biofuels (Agbor et al. 2011; Li et al. 2011; Galbe and Zacchi 2007).

5.1.2.3 Lignocellulosic Feedstock for Biomaterials and Chemicals

Global Availability and Production Lignocellulosic biomass is important not only as a source of energy and biofuel, but also as a source of different valuable chemicals from its chemical composition, mainly containing celluloses, hemicelluloses, and lignin. These components in turn might be used as a raw material for the production of chemicals such as alcohols and organic acids, for which there is a huge market demand. Despite its potential as fuel and raw material, no organized study has been performed on feedstock types, quantities, and characterization of the globally available biomass. Few studies have been reported on biomass, and these focus only on electricity production without considering other applications or the individual contributions of a variety of biomass resources. Even fewer studies have considered the distinct characteristics of different feedstock biomass types. Assessment of biomass availability and quality is an important first step toward utilizing biomass for energy generation and conversion to value-added chemicals. Countrywide assessment of biomass resources has been performed for many countries: the USA, Jordan, Malaysia, Turkey, China, India, and Bangladesh (Hossen et al. 2017; Ozturka et al. 2017).

Lignocellulosic biomass and residues such as wood, grass, and straw are abundant, nonfood raw materials for renewable fuels and products. These substrates are abundant, geographically widely distributed, and do not compete with food, freshwater, and fertile land. However, biorefineries based on lignocellulosic feedstocks do have to consider deviations in seasonal availability. In addition, the chemical industry worldwide defines strong standards and specifications for their products to be met independently on biomass source, climate conditions, and other influencing factors (Singh and Singh 2012). Thus, the use of biomass for the production of chemical products is still more costly compared to fossil-based resources. In particular, petroleum and natural gas refinery technologies have the highest efficiency as the result of more than 50 years of experience. Therefore, strong efforts in fundamental and applied research are required to develop biorefinery systems that are able to compete with conventional technologies.

One of the most important challenges is the handling of multiple biomass feedstock streams. A key step in processing lignocellulosic biomass is the separation of sugars from the lignocellulose. Several pretreatments are applied for this requirement: physical (grinding, milling), chemical (using acidic or basic aqueous media or ionic liquids), physicochemical (steam, hot water, or ammonia fiber expansion), and biological fragmentation (via enzymes or fungi). A number of pilot and demonstration plants have been developed using lignocellulosic feedstock (e.g., wood chips, straw) for the production of bio-based building blocks (ethanol, acetic acid) and green coal in Germany (Leuna), The Netherlands (Bioprocess Pilot Facility), Austria (bioCRACK), Canada (GreenField), the USA (Enchi Corp.), and Australia (Microbiogen) (Panoutsou et al. 2012; Rabaçal et al. 2017).

In addition, the first commercial biorefineries are established mainly to produce ethanol from LCF (e.g., Dupont in Nevada, Liberty Technology in USA, POET-DSM in Iowa, Iogen Corp. in Canada, Raízen/Iogen in Brazil, Cellulac in Ireland) (Downing et al. 2011; Balan et al. 2013; Rabaçal et al. 2017).

As shown in Fig. 5.1, chemicals can be obtained from lignocellulosic biomass. Different methods are used: fermentation of sugars to alcohol, hydrolysis of carbohydrates, oxidation, or pyrolysis. These processing methods can be combined in a running biorefinery system to produce chemicals. As already mentioned, ethanol is one of the most important fermentation products, followed by lactic acid, acetone, and citric acid. Lactic acid, a fermentation product of carbohydrates, is used in the food, cosmetic, and pharmaceutical industries (Dusselier et al. 2013). The worldwide demand for lactic acid is expected to be about 600,000 tons per year by 2020 and will increase because of its use as polylactic acid (PLA) (Zhang and Vadlani 2015). Dehydrated sugars (e.g., furfural) can be obtained from hydrolysis of 5-carbon carbohydrates and hydroxymethyl furfural from 6-carbon carbohydrates. Furfural is an important starter material for substances such as furfural alcohol, furan, or nylon. Levulinic acid is produced by hydrolysis of hydroxymethyl furfural and is used with levulinate esters or 1,4-pentanediol as the building blocks for compounds (van Putten et al. 2013; Dautzenberg et al. 2011). The earliest conversion from biomass to chemicals has been practiced for centuries by pyrolyzing wood to produce charcoal and tar. Today, pyrolysis is used to obtain chemicals such as methanol, acetic acid, and acetone and phenolic compounds.

5.2 Lignin as Agro-industrial Feedstock

Lignin, a principal component of renewable biomass, consists of methoxylated phenylpropane structures, accounting for 15% to 30% of the weight and 40% of the chemical energy of lignocelluloses (Long et al. 2015). Lignin is one of the most abundant naturally occurring polymers on the Earth. However, the majority of lignin that is recovered, typically from kraft pulping processes, is burned as a low-value fuel. In the early 1940s, lignin emerged as a "green" renewable source for vanillin (Zhang et al. 2015; Sana et al. 2017). The structure of this complex amorphous polyphenol molecule depends on the raw material, growing conditions of the plant, and extraction methods used (Gordobil et al. 2015). The main precursors of lignin are three monomer phenylpropanoids: syringyl alcohol (S), guaiacyl alcohol (G), and *p*-coumaryl alcohol (H). These monomers form a randomized structure in a three-dimensional network inside the cell wall. The major inter-unit linkage is an aryl–aryl ether type (García et al. 2009; Lupoi et al. 2015).

Many different functional groups can be found within the structure of lignin, such as carboxylic, aliphatic, and phenolic hydroxyls and methoxyl groups. Figure 5.3 shows the general structure of lignin and its main precursors. Recently, most commercial lignin is obtained as a by-product from the pulp and paper industry. Lignosulfonates have been used as emulsifiers in animal feed and as raw material in the production of vanillin (Dong et al. 2011). Most of this production is used for energy and only a small portion is commercially used as dispersing, binding, and emulsifying agents. Considerable efforts are being made in the search for novel applications of lignin to make it a more profitable product and, by extension, to diversify the biorefinery products (Garcia et al. 2014).

5.2.1 Lignin Isolation Processes

5.2.1.1 Kraft Process

In nature, lignin is associated with carbohydrates such as cellulose and hemicellulose. Lignin is extracted from other lignocellulosic parts via physical or chemical and biochemical treatments. The botanical source, as well as the pulping process (delignification) and extraction procedures, are highly influential for the final structure, purity, and corresponding properties of the lignin produced (Vázquez et al. 1997).

Common pulping processes are based on the cleavage of ester and ether linkages; after this step, the resultant technical lignin differs considerably from the lignin originally found in plants. Figure 5.4 shows the two main categories of extraction processes, sulfur and sulfur-free processes (Laurichesse and Avérous 2014).

Kraft pulping represents the most important process for cellulose production (Rinaldi et al. 2016). The primary waste stream from this technique is black liquor, a strongly alkaline solution that contains various organic and inorganic components (Hamaguchi et al. 2013). Kraft lignin is the predominating organic ingredient in black liquor. The structure of kraft lignin is different from that of native and other technical lignins. As a result of the cleavage of β-aryl bonds during the pulping process, kraft lignin contains an increased amount of phenolic hydroxyl groups. Moreover, some biphenyl and other condensed or repolymerized structures are formed during the kraft pulping process. Therefore, the molecular weight of kraft lignin obtained from black liquor has been reported to be in the range of 200 to

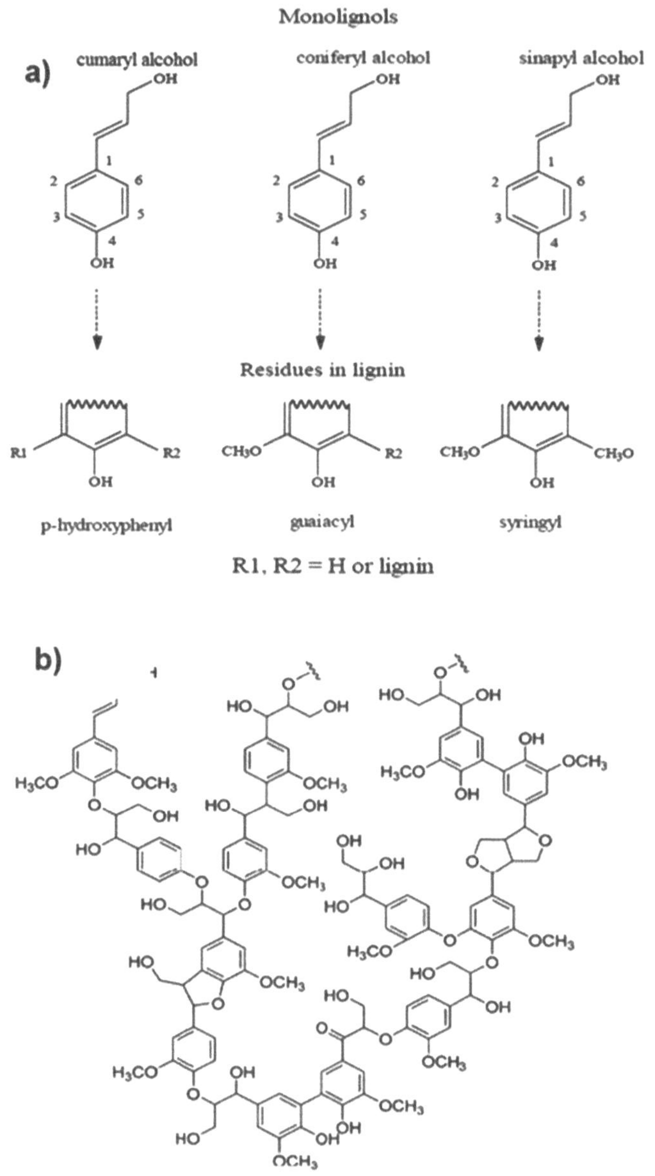

Fig. 5.3 (**a**) Structure of the three lignin precursors (monolignols) and their corresponding fragments in macromolecules. (**b**) Structure of isolated kraft lignin

200,000 g mol^{-1} (Vishtal and Kraslawski 2011). The molecular weight of kraft lignin depends also on the type of wood, the analysis method, and the isolation procedure used (Tolbert et al. 2014). Different methods have been proposed for the fractionation of Kraft lignin. The three main approaches include fractionation based

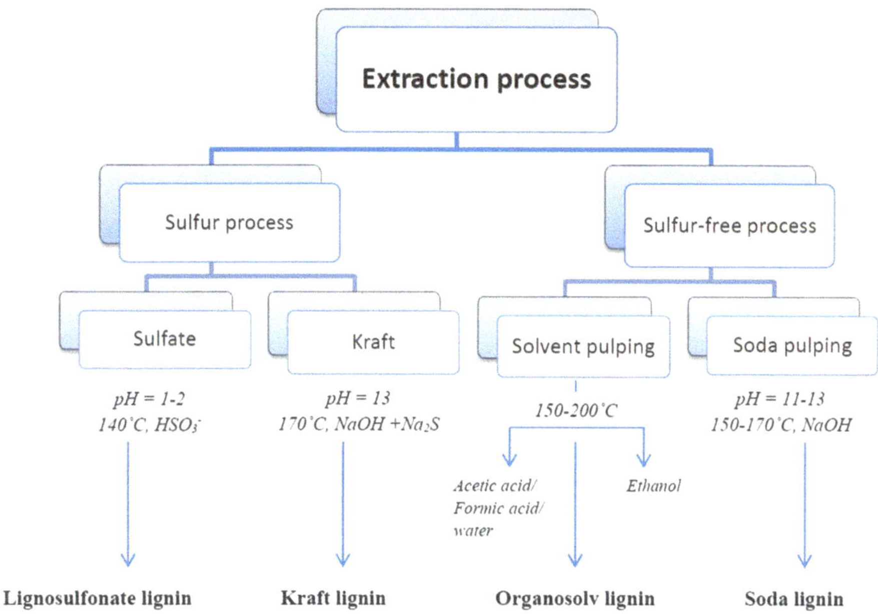

Fig. 5.4 Extraction processes of lignin from lignocellulose-rich feedstocks to produce technical lignins

on solubility in organic solvents, selective acid precipitation at reduced pH values, and membrane ultrafiltration (Alekhina et al. 2015).

Figure 5.5 summarizes our research approach using different renewable resources, including softwoods and hardwoods (e.g., pine and beech) as well as grasses (i.e., *Miscanthus*) to produce novel bio-based products: starting with biomass pulping (i.e., kraft process or lignin isolation via Organosolv treatment), followed by analytical characterization of the lignin structure using a broad variety of spectroscopic and chromatographic methods (Vaz 2014). Applications include those using the isolated lignin or lignin degradation products obtained via fragmentation with ozone or UV irradiation. Today, a huge number of novel lignin-based materials are under investigation for use in construction, packaging, and biomedicine (Constant et al. 2016; Hansen et al. 2016).

The precipitation of kraft lignin via gradual acidification of black liquor was achieved using such acids as HCl or H_2SO_4 at specific pH, temperature, and time of stirring. Results of these precipitation conditions were investigated for yield; the optimum acidification was determined to be that using H_2SO_4 with stirring at room temperature and pH = 2 for 90–180 min to obtain the first fraction of lignin (LE1). Small fragments produced through the pulping process of the biomass, such as small carbohydrates or some lignin fragments, affect the purity of the obtained kraft lignin, limiting the application of kraft lignin as a source of such biomaterials as polyphenols or aromatic compounds. Hence, purification of kraft lignin has a part in enlarging the scope of applications of this rich compound. Selective extraction

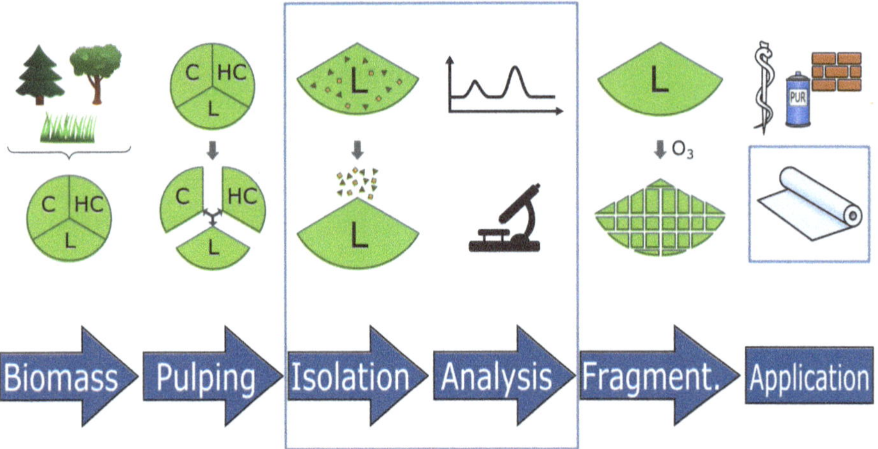

Fig. 5.5 Development of novel lignin-based materials, starting with lignin isolation and structure elucidation, guided fragmentation (via ozone or UV irradiation), and application including lignin-based polyurethanes, resins, and hydrogels for construction, packaging, and biomedicine. *C* cellulose, *HC* hemicellulose, *L* lignin

using different organic solvents, investing the solubility competition of carbohydrates and smaller fragments of lignin, was the key to this purification process. The purification was monitored by thin-layer chromatography (TLC). Double selective extraction using acetone and ethanol, respectively, and diethyl ether as a precipitating solvent, was the best method for purification, giving relatively pure lignin (PL). The fractions produced from the purification were analyzed by spectroscopy, chromatography, and thermal analysis showing the difference between the purification fragments.

5.2.1.2 Organosolv Process

Currently, Organosolv fractionation is a widely studied pretreatment technique for lignocellulosic biomass. Lignocellulosic biomass is known to contain compositional fractions of great value. Fractionation of its components is necessary to optimally exploit the potential of each constituent and has become an important research area. In Organosolv fractionation, biomass is treated with an organic solvent at elevated temperatures to separate out the lignin biopolymer. Lignin, as a polyphenolic network, is separated from the carbohydrate fractions cellulose and hemicellulose. One of its natural functions is to prevent biodegradation of wood (Dashtban et al. 2010). Hence, it tends to inhibit the desired enzymatic attack on saccharification products and inhibit fermentation to ethanol as the targeted end product. The function of the organic solvent in Organosolv fractionation is to solubilize and remove the lignin fraction and thus make the carbohydrates more accessible when subjected to enzymatic hydrolysis (Harmsen et al. 2010). The Organosolv process seems promising as it produces not only a cellulosic pulp displaying good enzymatic digestibility but also sulfur-free lignin fractions potentially usable for material applications (Dababi et al. 2016). Soft wood (mainly spruce and pine) was milled to a 500-µm particle size. Lignin was extracted via the Organosolv process using ethanol according to a modified published procedure (Grisel et al. 2014).

5.2.2 Structure Analysis of Lignin

For a general approach, the quality of the biomass used determines the product quality. Therefore, reliable information is required about the chemical composition of the biomass to establish its best use, which will influence the harvest and the preparation steps. The most widely used analytical techniques for lignocellulosic biomass and products are briefly described next (discussed in detail and reviewed by Vaz 2015; Lupoi et al. 2015; Zhao et al. 2017).

5.2.2.1 Spectroscopy

Various spectroscopic methods are used to identify and quantify lignin including nuclear magnetic resonance (^{1}H, ^{13}C, ^{31}P-NMR), Fourier-transform infrared spectroscopy (FTIR), Raman, and ultraviolet/visible spectrophotometry (UV-Vis). Lignin is a highly branched phenolic polymer providing many active regions for chemical and biological interactions with a wide variety of additional functional groups such as hydroxyl, methoxy, carbonyl, and carboxylic groups. As this can change the chemical and biophysical properties of the lignin, an analysis of various active groups attached to lignin is of importance in a functional classification of lignins. Such analysis can be performed by FTIR analysis (El Mansouri and Salvado 2007). Table 5.2 summarizes the most important IR vibrations and corresponding band assignments for lignin and monolignol units, respectively, according to data analysis for five lignin samples obtained from different isolation processes (Hansen 2015). Assignments are conformed by pyrolysis GC-MS data and NMR analysis (see Figs. 5.8 and 5.9; Table 5.4).

Raman spectroscopy is a vibrational spectroscopic method useful in the investigations of the chemical components of wood and plant materials, and especially lignin, because no sample preparation is needed for it to provide information on the cell wall and its chemical components. It provides also a powerful tool of detecting the lignin chemical structure and its modifications (Larsen and Barsberg 2011). Various subtechniques in the field of Raman spectroscopy have been applied to study lignin and lignin-containing materials. Such techniques have included visible-Raman, micro-Raman, FT–Raman, Raman imaging, UV–resonance Raman, surface-enhanced Raman, and coherent anti-Stokes Raman spectroscopy. In the early days, with excitation in the visible range, it was difficult to obtain Raman spectra of acceptable quality for lignin and lignocellulosics, respectively, because of

Table 5.2 Assignment of the most characteristic Fourier-transform infrared (FTIR) signals determined for lignin

Absorption (cm^{-1})	Assignment
3412–3460	OH stretching
3000–2930	CH stretching in $-CH_2$ and CH_3
2880–2670	$O-CH_3$ stretching
1727–1700	C = O stretching (nonconjugated)
1695–1650	C = O stretching (conjugated)
1605–1593	Aromatic skeleton (S > G)
1515–1505	Aromatic skeleton (S < G)
1460–1455	Asymmetrical deformation, $-CH_2$ and $-CH_3$
1430–1420	Aromatic skeleton
1370–1365	Symmetrical deformation, –CH3
1330–1325	Aromatic skeleton (S)
1272–1266	C=O stretching (G)
1230–1221	C–C, C–O, and C=O stretching (G)
1167–1156	C=O in ester (conjugated) (HGS)
1140	C–H aromatic in-plane deformation (G)
1128–1115	C–H aromatic in-plane deformation (S)
1085–1080	C–O deformation in sec alcohol / aliphatic ether
1038–1028	C–H aromatic in-plane deformation (G > S)
990–965	CH=CH out-of-plane deformation
925–912	C–H aromatic out-of-plane deformation
858–853	C–H out-of-plane deformation position 2, 5, 6 (G)
835–834	C–H out-of-plane deformation position 2,6 (S)
832–817	C–H out-of-plane deformation position 2, 5, 6 (G)

lignin autofluorescence. Although spectra could be obtained using special sampling techniques, the situation was far from satisfactory. In 1986, an FT–Raman spectrometer based on near-IR excitation (1064 nm) was developed and became commercially available. This instrument overcame the obstacle of lignin autofluorescence. Once FT–Raman instruments became commercially available, the use of Raman spectroscopy in the field of wood and other lignocellulosics increased significantly. The only other method that successfully avoided lignin fluorescence was ultraviolet-visible (UV-Vis) resonance Raman spectroscopy. To suppress fluorescence from highly fluorescent samples, the latter method was used in conjunction with an optical Kerr gate (Agarwal et al. 2011).

The ultraviolet (UV) spectrophotometric method is best suited for investigating the topochemistry, concentration, and purity of lignin. As many compounds absorb at the same wavelength, this method is used for quantitative analysis and not for identification of compounds. The absorbance level in UV spectra is directly proportional to the purity level of a compound, in this case lignin. A lower absorbance at higher pH value results from the coprecipitation of non-lignin material such as polysaccharide degradation products, wax, and lipids (Ahuja et al. 2017). LE1 and PL were analyzed via UV-Vis according to González et al. (2006). In particular, the PL sample showed distinct separated peaks at 216 nm, 252 nm, 269 nm, and 351 nm; two main peaks in LE1 appear at 215 nm and 296 nm and two shoulders at 242 nm and 354 nm (Fig. 5.6a). In contrast, the spectra of industrial lignin (KL60) shows two main peaks, one at 216 nm and another at 295 nm, and other shoulders at 245 nm and 342 nm (Fig. 5.6b).

Lignin has a strong UV absorption at 280 nm because of its aromatic nature (Azadi et al. 2013). The hypsochromic effect of NaOH shifts the main peak to 214–222 nm. In Table 5.3, the two main UV-Vis absorption bands of lignin are shown and compared with literature data. Accordingly, the ester or ether bonds between acids, ferulic acids, and lignin were substantially cleaved by the alkali treatment. The intensive absorbance at 279–280 nm relative to 316–320 nm indicates a relatively high content of guaiacyl (G-units), which is similar to that of other monocotyledons and is consistent with a guaiacyl-rich lignin (Vivekanand et al. 2014).

The use of one-dimensional (1D)-NMR for bio-oil analysis is widely reported, for both qualitative and quantitative analysis, but the use of two-dimensional (2D)-NMR is less often described. 2D-NMR and solid-state 2D-NMR have been largely employed for the analysis of lignin and biomass (Michailof et al. 2016). 2D heteronuclear single quantum correlation (HSQC)-NMR techniques could identify the ^{13}C–^{1}H correlations for lignin and provide the important internal structural information about side-chain linkages and aromatic units of lignin. HSQC spectra of aromatic (dC/dH 100–135/5.5–8.5) and aliphatic (dC/dH 50–90/2.5–6.0) regions of lignin sample are presented in Fig. 5.7 according to Chen et al. (2016).

^{31}P-NMR spectroscopy is a facile and direct analytical tool for quantitating the major hydroxyl groups in lignin, and the concentrations of different hydroxyl groups

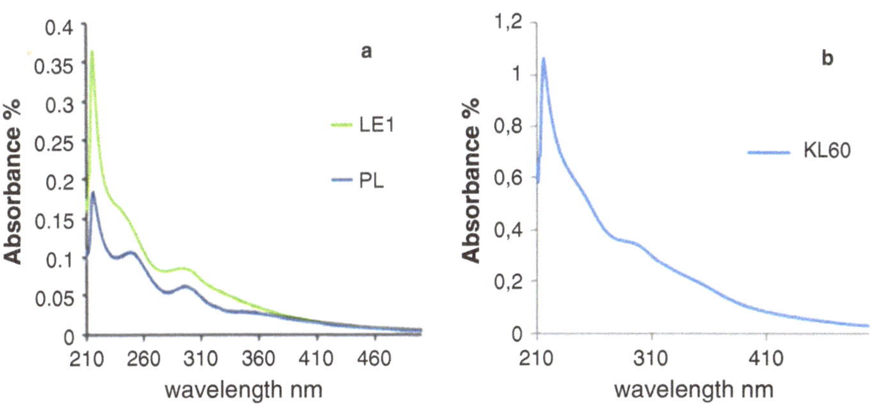

Fig. 5.6 (**a**) Ultraviolet-visible light (UV-Vis) spectrum of two KL extracts (*LE1* first fraction, *PL* purified lignin). (**b**) UV-Vis spectrum of industrial lignin (KL60)

Table 5.3 Ultraviolet-visible light (UV-Vis) absorption data of lignin and their characteristics

λ-experimental (nm)	λ-literature (nm)	Functional group	Intensity	Excitation	Reference
215–222	279–280	Nonconjugated phenolic groups (G-S rich)	High	→ **: anti-bonding molecular orbital	Azadi et al. (2013)
296–303	316–320	Conjugated phenolic groups (*p*-coumaric acid, ferulic acid)	Low	→ *	Vivekanand et al. (2014)

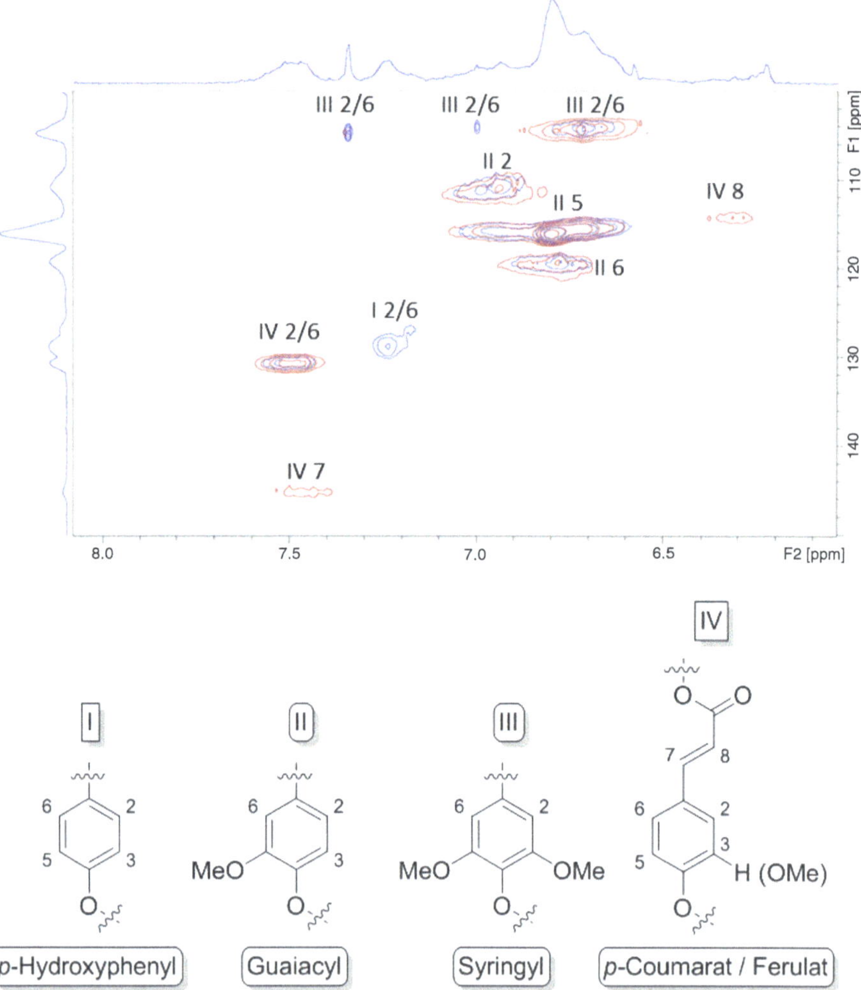

Fig. 5.7 HSQC NMR spectrum, aromatic region (dC/dH 100–150/6.0–8.0) of lignin samples obtained via Organosolv process from *Miscanthus giganteus*. Comparison of leaf lignin (*blue*) and stem lignin (*red*) and corresponding assigned lignin fragments

were calculated on the basis of the internal standard (cyclohexanol) and the integrated peak areas. The results from ^{31}P-NMR can particularly provide an insight into the absolute amounts of aliphatic OH (α-OH and primary OH), phenolic OH (S–OH, G5, 5-OH, G–OH, and *p*-coumarate OH), and carboxylic OH (COOH) in the six lignin fractions. Figure 5.8 provides typical ^{31}P-NMR spectra of lignins obtained from different sources and isolation processes. All lignin samples were phosphorylated with 2-chloro-4,4,5,5-tetramethyl-1,2,3-dioxaphospholane and analyzed via quantitative ^{31}P-NMR spectroscopy (with endo-*N*-hydroxy-5-norbornene-2,3-dicarboximide as internal standard) according to the method described by Granata (Granata and Argyropoulos 1995).

According to the results obtained from ^{31}P NMR spectra, Organosolv-isolated lignin samples exhibit considerable quantities of G and H units and rather low amounts of S units. It shows the presence of aliphatic hydroxyl groups (δ 146–149 ppm, 0.38 mmol g^{-1}), amounts of syringyl units (δ143–144 ppm, 0.12 mmol g^{-1}), guaiacyl units (δ 142–143 ppm, 0.25 mmol g^{-1}), hydroxyphenyl units (δ 139–140 ppm, 0.27 mmol g^{-1}), and low amounts of carboxylic acid (δ 135 ppm, 0.02 mmol g^{-1}). The high amount of guaiacyl hydroxyl units in OBL is a consequence of cleavages of β-O-4 linkages. The resulting ratios of monomeric units (H/G/S) were 32/49/19 for OBL_1 and 43/40/17 for OBL_2. The beechwood lignin data obtained by Faix et al. show monomeric ratios of beech milled wood lignin of 4/56/40 (Faix et al. 1987). El Hage and colleagues investigated Organosolv lignin from *Miscanthus* and found hydroxyl amounts: aliphatic 1.19 mmol g^{-1}, syringyl 0.13 mmol g^{-1}, guaicyl 1.33 mmol g^{-1}, and hydroxyphenyl 0.61 mmol g^{-1} (El Hage et al. 2009). Thus, the beechwood lignins show higher amounts of hydroxyphenyl units arising from the Organosolv isolation method cleaving mainly the β-O-4 bonds. These results were also confirmed by pyrolysis gas chromatography.

5.2.2.2 Chromatography

This set of analytical techniques identifies and quantifies organic compounds using: liquid chromatography coupled with mass spectroscopy (LC-MS) or pyrolysis probe coupled to gas chromatography and mass spectrometry (Py-GC-MS). The molecular weight distributions are determined through size-exclusion chromatography (SEC).

LC-MS is a tool to separate mixtures with multiple components, providing the structural identity of the individual components with high molecular specificity and detection sensitivity. This tandem technique can be used to analyze biochemical, organic, and inorganic compounds commonly found in complex samples of environmental and biological origin such as lignocellulose biomass. LC-MS provides superior capabilities for the detection of higher molecular weight components and

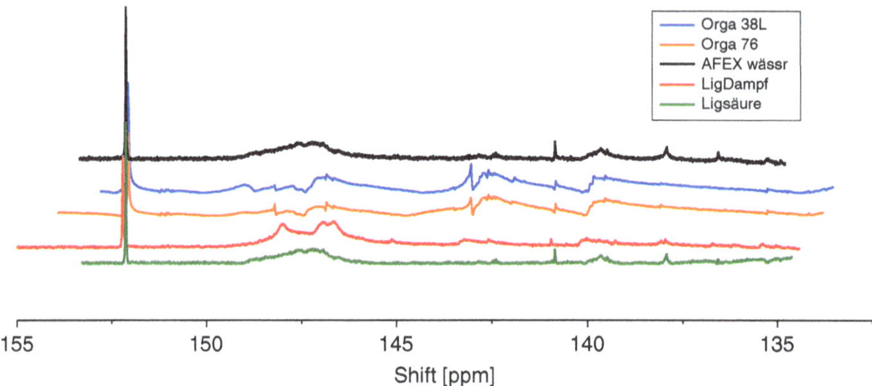

Fig. 5.8 ^{31}P-NMR spectra of five lignins isolated via different processes: AFEX (*black*), Organosolv (*blue* and *orange*), steam explosion (*red*), and hydrolysis (*green*)

reduced fragmentation (Joseph et al. 2016). Widsten et al. investigated the LC-MS of tannin and lignin model compounds, with results showing the type of coupling and specifying the coupling linkages between tannin and lignin (Widsten et al. 2010).

The analytical technique of Py-GC-MS is widely used to research chemical structure and pyrolysis characteristics of biomass and its three main components (i.e., lignin, cellulose, and hemicellulose) to examine the reaction products of biomass thermal degradation (Wu et al. 2012). Py-GC-MS analysis of various hardwood materials gives a good relationship in syringyl/guaiacyl (S/G) and syringaldehyde/vanillin (Sa/Va) ratios between Py-GC-FID and GC analysis of the NBO products (Nakagawa-Izumi et al. 2017). Accordingly, Py-GC-MS has been widely adopted to analyze lignin monomer composition (Chang et al. 2016). Carbohydrates that are not completely removed can be detected by Py-GC-MS. So, pyrolysis coupled to GC-MS gives valuable information about the relative amounts of lignin and carbohydrates. Pyrograms of Organosolv lignin isolated from beechwood are shown in Fig. 5.9. The nature and molecular weight of the released compounds after pyrolysis at a temperature of 550 °C are listed in Table 5.4.

Pyrolysis decomposes the lignin in the absence of oxygen into fragments detected by GC or MS. Basically, two methods are applied in pyrolysis research studies: the so-called continuous mode and the pulse mode. In the continuous mode, the sample is placed into a furnace and the temperature is kept fixed during the pyrolysis procedure. The pulse mode requires an introduction of the sample on a cold pyrolysis probe, which is then rapidly heated to the desired pyrolysis temperature (mainly applied, Curie-point pyrolysis). Curie-point pyrolysis of beech milled wood lignin has been investigated by Genuit and colleagues (Faix et al. 1987). Their results specifically show fragments of syringyl (S), guaicacyl (G), and *p*-hydroxyphenyl units

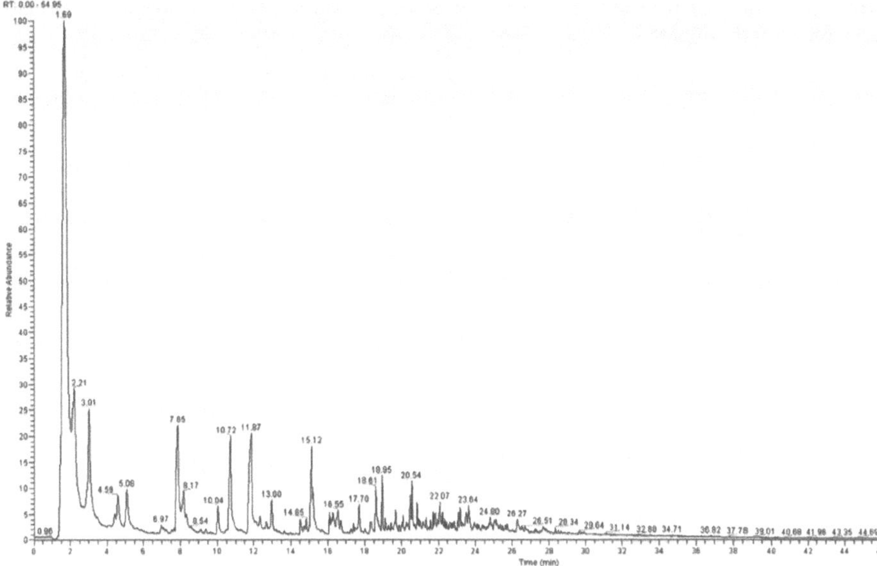

Fig. 5.9 Pyrogram of an Organosolv beechwood lignin (measured at 550 °C)

(H), which is in line with the estimated SGH ratio on the basis of chemical degradation. The pyrogram also reveals fragments derived from carbohydrates. Lignin and polysaccharides are connected through covalent bonds to lignin carbohydrate complexes. Carbohydrates, which are not completely removed, can be detected by Py-GC-MS. So, pyrolysis coupled to GC/MS gives valuable information about the relative amounts of lignin and carbohydrates.

In addition, the molar mass distribution is a key analytical parameter for technical lignins. Various approaches are reported in the literature on how to address the molar mass distributions of lignins. The technique of choice in almost all cases is size-exclusion chromatography (SEC) (Sulaeva et al. 2017). This technique provides weight-average (MW) and number-average (Mn) molecular weights and polydispersity (PD = MW/Mn) of the investigated compounds, basically polymers. The molecular weight varies from 877 to 6117 g mol^{-1}, which corresponds to literature values for comparable samples (Santos et al. 2014). Santos and colleagues studied kraft lignin precipitated from black liquor at pH value of 2 using H_2SO_4. As an analogue to those data, Fig. 5.10 shows the SEC results for two kraft lignins precipitated at a pH value of 2 using H_2SO_4. In Table 5.5, the corresponding data are summarized (MW, Mn, PD).

Table 5.4 Compounds identified by pyrolysis probe coupled to gas chromatography and mass spectrometry (Py-GC-MS)

Compound	Retention time (min)	Molecular weight (g mol^{-1})	Origin
CO_2	1.69	44	
2-Methyl 2-cyclopenten-1-one	3.01	96	C
2,3-Dimethylphenol	4.59	96	C
4-Ethyl-2-methylphenol	5.08	124	L–H
Phenol	7.85	94	L–H
2-Methoxy-4-methylphenol (4-methylguaiacol)	8.17	138	L–G
4-Ethyl-2-methoxyphenol	10.04	122	L–H
2,3-Dimethoxybenzyl alcohol	10.72	138	L–G
2,3-Dimethylhydroquinone	11.87	138	L–G
2-Methylphenol	12.13	108	L–H
4-Ethylphenol	12.49	108	L–H
4-Ethylcatechol	13.00	138	L–H
2,3-Dimethylhydroquinone	14.85	138	L–H
2-Methoxy-4-propenyl-phenol (isoeugenol)	15.12	164	L–G
Phenol, 2-methoxy-4-propyl-(4-propylguaiacol)	16.02	138	L–H
Phenol, 2-methoxy-4-(1-propenyl)-	16.55	138	L–H
3-Methoxy-5-methylphenol	17.03	164	L–G
Phenol, 4-(3-hydroxy-1-propenyl)-2-methoxy-	17.70	164	L–G
Phenol, 2,6-dimethoxy-4-(2-propenyl)-(4-allyl syringol)	18.61	194	L–S
Benzene, 1,2,4-trimethoxy-5-(1-propenyl)-, (*Z*)-	18.95	208	L–S
Phenol, 2,6-dimethoxy-4-(2-propenyl)-	19.74	194	L–S
Ethanone,1-(2,6-dihydroxy-4-meth-oxyphenyl)-	22.07	182	L–G
1,2-Dimethoxy-4-*n*-propyl benzene	23.64	180	L–S
2,6-Dimethoxyphenol (syringol)	24.80	154	L–S
1,2-Dimethoxy-4-*n*-propylbenzene	26.27	180	L–S

C phenylcoumarane substructures, *L–H* lignin *p*-hydroxyphenylic, *L–G* lignin guaiacylic, *L–S* ligin syringylic

5.2.2.3 Thermal Analysis

This analysis method determines the thermal behavior through thermal gravimetric analysis (TGA) and differential scanning calorimetry (DSC). TGA and DSC were used to determine the mass loss of samples from temperature treatment, as indicative of thermal stability and thermal decomposition of a compound. In DSC, the differences in heat flow are recorded as a function of temperature and depend on the behavior of the sample as a result of endothermic or exothermic events during the treatment (Vallejos et al. 2011). Jiang and colleagues studied the fractionation and characterization of kraft lignin by sequential precipitation with various organic

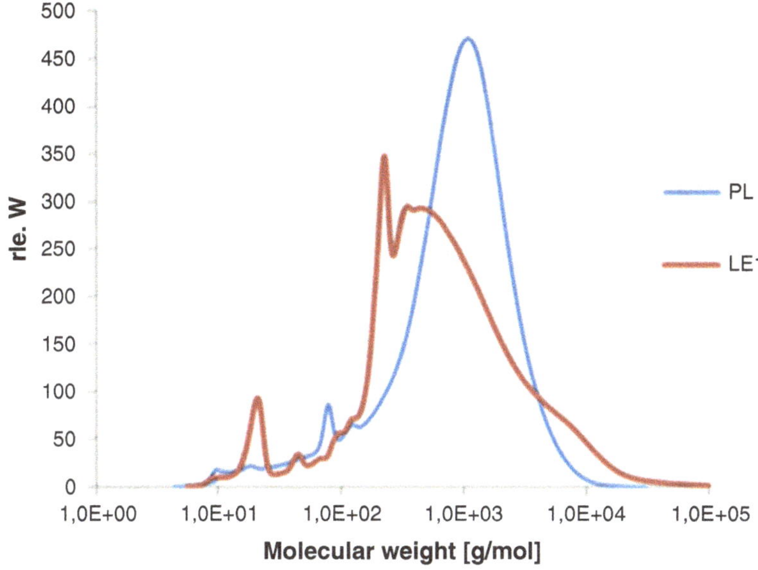

Fig. 5.10 Size-exclusion chromatography (SEC) analysis of two kraft lignins (*LE1* first fraction, *PL* purified lignin) precipitated at pH 2 with H_2SO_4

Table 5.5 Molecular weights (Mn, MW) and polydispersity (PD) of two different kraft lignin extracts (LE1 and PL) obtained via SEC (PS standard)

Extract	Mn (g mol^{-1})	MW (g mol^{-1})	PD
LE1	163	1490	9.1
PL	252	2343	9.3

solvents, finding that the decomposition temperature decreased with decreasing molecular weight (Jiang et al. 2017). Figure 5.11 shows the thermal behavior of five lignin samples obtained from different isolation processes. Obviously, one of the samples still contained solvent residuals.

5.2.2.4 Microscopy and X-ray Techniques

In addition to thermal analyses, microscopic methods are used to study the surface and morphology of biomass components, including via scanning electron microscopy (SEM). SEM is traditionally used to give high-resolution topographical information about the material surfaces. Conventional SEM techniques produce limited material contrast between organic species and require coating of the samples to avoid buildup of negative charge and associated image distortion. Contrast can sometimes be obtained using selective stains. For example, $KMnO_4$ is known to

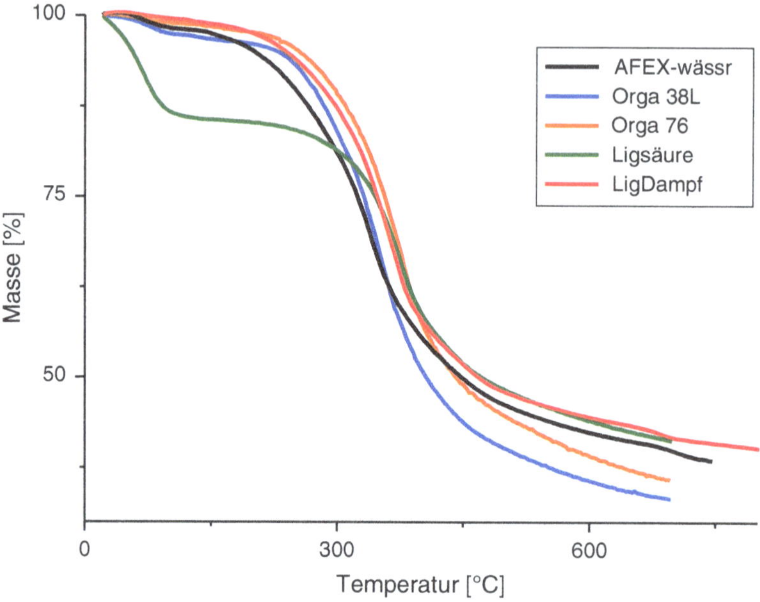

Fig. 5.11 Thermal gravimetric analysis (TGA) curves for lignins isolated via different processes: steam explosion (*red*, "LigDampf"), hydrolysis (*green*, "Ligsäure"), Organosolv (*blue*, *orange*, Orga 38 L, Orga 76), and AFEX (*black*, AFEX-wässr)

Fig. 5.12 Scanning electron microscopy (SEM) image of beechwood lignin isolated via Organosolv process

stain lignin and has been used to determine lignin distribution in plant cell walls (Fromm et al. 2003). Hansen et al. used SEM to study the morphology of beechwood lignins isolated via Organosolv at neutral and acidic conditions (Hansen et al. 2016). As an analogue to these findings, Fig. 5.12 shows one typical surface of a beechwood lignin with rounded semispherical shape and open spaces on the rough surface.

To obtain even more detailed morphological information, various X-ray techniques can be used including X-ray diffraction (XRD), and scattering such as small- and wide-angle X-ray scattering (SAXS, WAXS) (Santos et al. 2014; Giannini et al. 2016). Xu and colleagues studied the structure of biomasses using both WAXS and SAXS, showing that biomass goes through structural change in both crystalline and amorphous domains during acid and alkali pretreatments (Xu et al. 2013). Vivekanand and colleagues studied lignin morphology using XRD on samples isolated at different pH values (Vivekanand et al. 2014). Figure 5.13 shows a typical diffractogram of the amorphous lignin morphology.

5.2.3 *Antioxidant Activity and Total Phenol Content of Lignin*

From their polyphenolic structure, lignins bear a number of interesting functional properties, such as antioxidant activity. Kraft lignin from wood sources in the pulp industry was reported to be as efficient as vitamin E to protect the oxidation of corn oil (Dong et al. 2011). Most antioxidant effects of lignins are considered as derived from the scavenging action of their phenolic structures on oxygen-containing reactive free radicals (Sun et al. 2014). For characterization of the antioxidant activity of

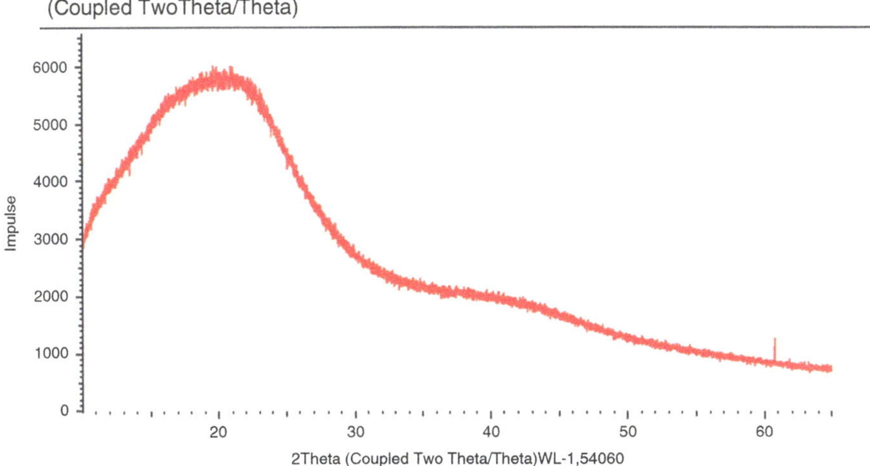

Fig. 5.13 Typical X-ray diffractogram of Kraft lignin

naturally occurring phenolic compounds, a method using 1,1-diphenyl-2-picrylhydrazyl (DPPH) as a reactive free radical is now recognized as allowing search for radical scavenging ability depending on the electronic structure of the flavonoids and catechins. The reactivity of DPPH is far lower than that of oxygen-containing free radicals (OH, RO, ROO, O_2), and in contrast to those the interaction rate is not diffusion controlled. Rather good conformity of the results obtained by the DPPH method and by the tests of the methyl linoleate auto-oxidation and the 2,2′-azino-bis(3-ethylbenzothiazoline-6-sulfonic acid) (ABTS) radical cation scavenging has been reported (Dizhbite et al. 2004). As their free radical scavenging ability is facilitated by their hydroxyl groups, the total phenolic concentration could be used as a basis for rapid screening of antioxidant activity (Baba and Malik 2015). The method of determination of the level of total phenolics is not based on absolute measurements of the amounts of phenolic compounds but rather on their chemical reducing capacity relative to gallic acid. It is very important to point out that there is a positive relationship between antioxidant activity potential and amount of phenolic compounds of the biomass. The content of total phenolic was determined based on the absorbance values of the various extract solutions, reacted with Folin–Ciocalteu reagent, and compared with the standard solutions of gallic equivalents (Amzad and Shah 2015). In our studies we could not only confirm the literature data but also improve the antioxidant activity of Kraft lignin extracts. Thus, the antioxidant activities of the purification fractions, according to the method described by Garcia et al. (2014), are between 60% and 68%, whereas Labidi when using analogue isolation conditions obtained values up to 54.76%. In addition, the kraft lignins (samples LE1 and PL) were compared with grass-based lignins (GL-M) and soft wood lignin, both obtained via Organosolv treatment (OL-SW). The Organosolv lignin DPPH inhibitions were 31% and 42%, respectively. Total phenol contents of LE1 and PL are 29.3% and 30.7%, respectively, whereas the value reported by Labidi et al. was 29.61%. GL-M and OL-SW have TPC values of 34.2% and 34.1%, respectively. Those results show no correlation between the antioxidant activity values and the total phenol content values.

5.2.4 *Storage and Temperature Effects on Lignin Structure*

The macromolecule structure of lignin makes it unstable; it can be fragmented under certain conditions such as storage, temperature, or daylight. Storing lignin purification extracts for 45 days at room temperature caused structural changes that were monitored by thin-layer chromatography (TLC). The solvent system used for the study was 90% ethanol and 10% *n*-hexane. LE1 shows three spots: one dense spot at the baseline is the spot of lignin, the second spot is small and light, located in the middle of the TLC sheet, and the third is a large spot traveling with the mobile phase. After 45 days of storage, TLC was repeated for the fractions, showing that

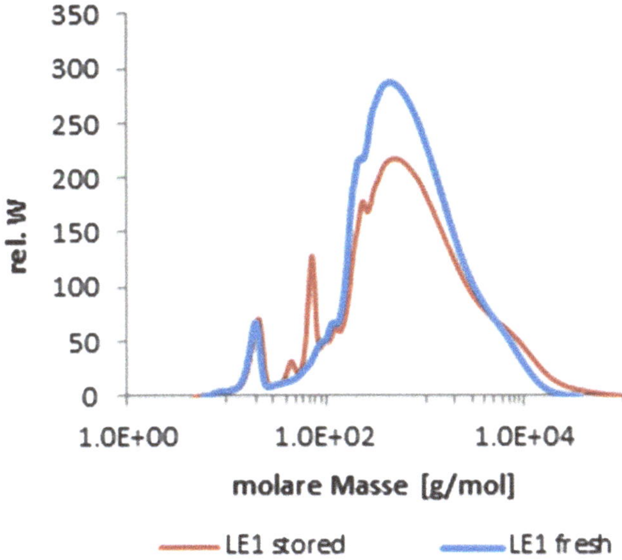

Fig. 5.14 SEC analysis of freshly isolated versus stored lignin (sample LE1)

the spot in the middle for LE1 looked denser and larger than that for fresh LI1, meaning that there are new fragments in lignin caused by storage. SEC (GPC) analysis of both stored and fresh extracts shows this fragmentation clearly (Fig. 5.14) where new peaks at smaller molecular weights (45 g mol^{-1} and 73 g mol^{-1}) appear for the stored samples. The fragmentation was not caused by temperature because the samples were stored at fixed temperature up to 25 °C, nor by any chemical interaction; they also were stored in aluminum foiled vials to prevent the daylight effect. So, most probably it is the storage effect.

The samples were dried at different temperatures: 25, 40, 60, 70, and 90 °C. TLC spotting showed no change in the PL spot at 25 °C and 40 °C, but a new spot for PL was traveling with the solvent system at 60, 70, and 90 °C in addition to the lignin spot. The second spot (in the middle of the TLC plate) of the LE1 fraction became denser and larger than that in the dried fraction at 25 ° and 40 °C. The TGA was measured for the lignin extracts according to Vallejos et al. (2011). The curve of LE1 shows the temperatures where lignin degrades, those temperatures being mainly 60, 380, and 880 °C. The degradation temperatures for PL are 60, 390, and 900 °C. Figure 5.15 shows the TGA curve of LE1.

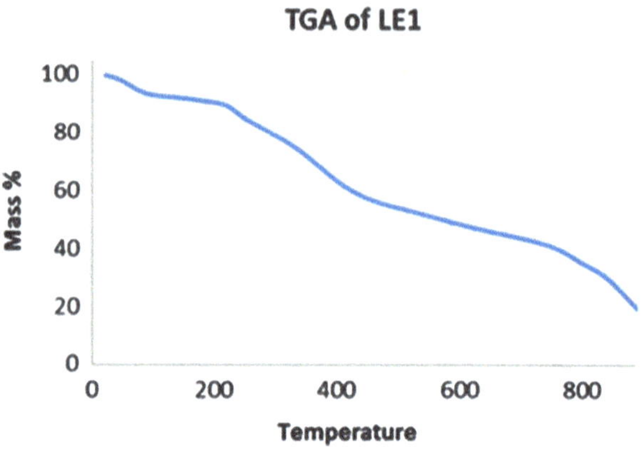

Fig. 5.15 TGA thermal analysis of a kraft lignin (sample LE1)

5.3 Lignin Valorization and Potential Applications

5.3.1 Lignin-Based Polymer Blends, Composites, and Hydrogels

The nontoxic, complex, and natural phenolic nature of lignin has made it increasingly important in many industrial applications. Recent studies have shown that lignins have been used as dispersants, adsorbents in solution, bioplastics, nanocomposites, bio-surfactants, and phenolic resins (Sa'don et al. 2017). Most of the reported lignin-based polymeric materials are highly cross-linked networks resulting from the multifunctionality of the lignin precursors. Recently, thermoplastic materials based on lignin and lignin derivatives have attracted increasing interest for their improved processability and recyclability relative to those of the thermosetting polymer networks. The transformation of lignin into value-added thermoplastic materials is expected to provide cost-effective and biodegradable alternatives for petroleum-based thermoplastics and to address the problem of papermaking waste disposal. However, the rigid nature, relatively low molecular weight, and high polydispersity of technical lignin still hinder the development of high-performance thermoplastic materials. To improve the thermoplasticity of lignin, plasticizers or polymers with low glass transition temperature (Tg) values are usually integrated with lignin by blending or chemical modifications (Wang et al. 2016a, b). Domenek and colleagues studied the influence of lignin based on botanical origin on PLA. Tg was found to be decreased by addition of lignin because of the plasticizing effects of low molecular weight lignin fractions. Mechanical properties such as tensile strength and elongation at break were slightly decreased by addition of lignin based on two effects: "addition of high molecular mass lignin lead to an increase of yield strength and decrease of stress at break, while the plasticization of PLA by low molecular mass lignin lead to a decrease of yield strength and an increase of strain at break" (Domenek et al. 2013). Oxygen permeability was decreased by 20% by the addition of 10% lignin. In starch-based films, lignin was found to enhance plasticity along with better mechanical strength (Wu et al. 2009). In addition, soda and kraft lignin are reported to improve the barrier and physicochemical properties of biopolymer films. Bhat and others studied the effect of lignin from oil palm black liquor waste in comparison to commercial alkali lignin on starch-based films. Addition of lignin in starch-based films resulted in dark brown films with a decreased brightness. An increase in tensile strength and Young's modulus was measured in films by incorporation of lignin isolated from black liquor waste in a concentration up to 3%. Improvement of tensile strength was attributed to "partial miscibility between sago starch and lignin fractions as well as to the presence of hydrophilic groups in lignin" and hydrogen bond, whereas partial miscibility results from low molecular weight lignin fractions (Bhat et al. 2013). Greater amounts of lignin showed a reduction of mechanical properties attributed to weak interfacial adhesion between hydrophilic starch and hydrophobic lignin. However, incorporation of commercial lignin showed a significant decrease of tensile strength and Young's modulus compared to control and isolated lignin films, but an increase of elongation at break. Furthermore, water vapor permeability as well as water solubility were reduced by incorporation of lignin, which had stronger effects than commercial lignin. Wu and colleagues proved synergistic interactions among cellulose, starch, and lignin in mechanical properties of their ternary composites compared to cellulose/starch and cellulose/lignin blended films. At high cellulose content, incorporation of lignin significantly reinforced the mechanical strength of composites. Optimum strength was achieved at higher cellulose and lignin content with a lower level of starch content, whereas optimal flexibility occurred at higher cellulose content with a middle level of lignin and starch content (Wu et al. 2009).

Hydrogels are defined as three-dimensional network structures formed by natural or synthetic polymers able to incorporate large amounts of solvents (water). Biopolymer-based hydrogels have recently received considerable attention for applications in biomedical and biotechnological fields, including enzyme immobilization, tissue engineering, drug delivery, and biosensors, for reasons of their inherent biocompatibility and biodegradability (Park et al. 2015; El Khaldi-Hansen et al. 2017). Ciolacu and colleagues studied the combination of biocompatibility with tissues and blood, nontoxicity, and low price of cellulose with the antioxidant and free radical scavenger properties of lignin. Thus, new cellulose–lignin hydrogels were obtained with the aim for testing in cosmetic and pharmaceutical applications (Ciolacu et al. 2012). Raschip and others developed a hydrogel complex containing xanthan and lignin. They found that the addition of different lignin types can result in different surface adhesion morphology and structure, which lead to different thermal properties of the final products (Raschip et al. 2013).

5.3.2 *Composites Based on Lignin and Cellulose*

Cellulose is the most abundant natural polymer on earth and is a linear polymer of anhydroglucose. Resulting from its chemical structure, cellulose is hydrophilic in nature, and a highly crystalline, fibrous, and insoluble material. Cellulose-based materials offer advantages such as "edibility, biocompatibility, barrier properties, aesthetic appearance, being non-toxic, non-polluting and having low cost" (Tajeddin 2015). Hydroxypropyl methylcellulose (HPMC) is an edible plant derivative shown to form transparent, odorless, tasteless, oil-resistant, and water-soluble films. In addition, they are biodegradable and provide very efficient oxygen, carbon dioxide, and lipid barriers (Imran et al. 2010). However, the high water sensitivity of HPMC produces a loss of barrier properties of the packaging or even solubilization into foods with high water activities. This behavior prevents its industrial application. Additionally, HPMC has poor mechanical resistance, low elasticity, and a rigid and breakable behavior (Sebti et al. 2007), but is still suitable for edible coating purposes. Up to now, HPMC has been used in food industry as an emulsifier, film former, protective colloid, stabilizer, suspending agent, or thickener, and is approved for food uses in Europe (EC 1995).

After cellulose, chitosan is the second most abundant natural polymer on Earth, and known to possess excellent mechanical and thermal properties. It is extracted from crustaceous shells such as crabs, shrimp, and prawns. Because of their high crystallinity, chitosan fibers have been used as a reinforcement in composite materials, although these are not as widespread as synthetic fibers. Chitosan has been found to be nontoxic, biodegradable, biofunctional, and biocompatible. In addition to film-forming properties, it has antibacterial and antifungal activities (Tanjung et al. 2016). Being able to form active, edible, or biodegradable films, chitosan films or coatings offer a moderate barrier against oxygen and good carbon dioxide barrier properties. They tend to exhibit fat and oil resistance and selective permeability to gases, but lack resistance to water transmission; this results from the strong hydrophilic character of polysaccharides, leading to an interaction with water molecules. Furthermore, chitosan coatings limit the contamination on the food surface. Compared to HPMC, chitosan films have better mechanical properties, shown in higher tensile strength, Young's modulus, and elongation at break, but a higher affinity to water, resulting in greater water vapor permeability (Sánchez-González et al. 2011).

Incorporation of lignin to cellulose and chitosan polymers enhances the mechanical properties and water vapor permeability because it has multiple functionalities (Wang et al. 2016a, b). Wu and colleagues proved synergistic interactions among cellulose, starch, and lignin in mechanical properties of their ternary composites compared to cellulose/starch and cellulose/lignin blended films (Wu et al. 2009).

In our studies, we found that at high cellulose content incorporation of lignin significantly reinforces the mechanical strength of corresponding cellulose/lignin composites. Optimum strength was achieved at higher cellulose and lignin content with a lower level of starch content, whereas optimal flexibility was found at higher cellulose content with a middle level of lignin and starch content. Accordingly, we currently investigate composites based on cellulose, lignin, and chitosan: HPMC film-forming solutions are prepared using the procedure described by Sebti et al., dissolving 3 parts of HPMC in 200 parts of 0.01 mol l^{-1} HCl solution, 100 parts of absolute ethanol, and 10% (w/w HPMC) PEG 400 (Sebti et al. 2007): 25 ml of the solution was plated onto a glass slide and dried at room temperature for 36 h.

HPMC-lignin film-forming solutions were prepared by dissolving 5%, 10%, 15%, 20%, 25%, and 30% of the first fraction of the purification of kraft lignin (LE1) in the smallest amount of DMSO, then added to the HPMC-film forming solution mentioned previously, and stirred for 15 min at room temperature (RT): 25 ml polymer solution was plated onto a glass and dried at RT for 36 h. HPMC-Organosolv lignin (OL-SW) was prepared the same way and compared with the Organosolv grass lignin (GL-M). The HPMC film is transparent and water soluble. Water solubility decreases with the addition of lignin. The color of the films was light honey and became denser with the addition of lignin. The brittleness appears at the addition of 25% of LE1 and GL-M and at 20% OL-SW, whereas lignin release starts at 30% LE1 and 20% OL-SW; GL-M films did not exhibit any release. Primary antimicrobial tests on the films against gram-positive bacteria prove the activity of the films against *Staphylococcus aureus*. The films were not active against gram-negative bacteria, for example, *Escherichia coli*, and hence, the addition of chitosan was the key. Chitosan (85% deacetylated) was added to the HPMC-lignin film solutions at 5%. Lignin release appeared at lower percentages starting from 20% in kraft lignins and 15% in Organosolv lignins, whereas at higher percentages the films exhibited chitosan release.

5.3.3 Antimicrobial Activity of Lignin to Be Used in Biomedicine

Technical lignins contain numerous chemical functional groups, such as phenolic hydroxyl, carboxylic, carbonyl, and methoxy groups, analytically studied in detail by numerous researchers and recently reviewed by Lupoi and colleagues (Lupoi et al. 2015). The phenolic hydroxyl and methoxy groups contained in lignin have been reported to be biologically active. Different types of lignins possess antimicrobial, antioxidant, and UV absorption properties. Various investigations have suggested that lignins can be applied to stabilize food and feedstuffs because of their antioxidant, antifungal, and antiparasitic properties (Yang et al. 2016). The literature describing the biological properties of technical lignins has grown rapidly in the past 10 years, as reviewed by Espinoza-Acosta (Espinoza-Acosta et al. 2016).

In 2015, Richter and colleagues reported the synthesis of environmentally benign antimicrobial nanoparticles based on a silver-infused lignin core. The authors emphasize that their results illustrate "how green chemistry principles including atom economy, the use of renewable feedstocks and design for degradation can be applied to design more sustainable nanomaterials with increased activity and decreased environmental footprint" (Richter et al. 2015).

Additionally, other properties such as anticarcinogenic, apoptosis-inducing, antibiotic, and anti-HIV activities have been reported in lignin–carbohydrate complexes (LCCs). Commodity products with antioxidant or antimicrobial properties, such as sunscreen lotions, biocomposites, and clothes that use lignin as a natural ingredient have been prepared, and their characterization has shown promising results (Vinardell and Mitjans 2017).

The biocidal activity of lignin allows reducing the environmental problems related to use of silver nanoparticles and could improve, as an example, the release of active principles in agriculture. Some studies have shown that lignin has an antimicrobial effect, even when it was incorporated into polyethylene film and applied in the finishing processes of textiles. However, none of these investigated the mechanism of the antibacterial activity of lignin when included in polymeric matrices. Lignin nanoparticles that are incorporated into PLA systems containing nanocrystals revealed an innovative capacity to inhibit bacterial growth (Yang et al. 2016).

Besides lignin-based polyurethane foams and resins for construction and insulation, in future the target of lignin valorization will particularly include the development of lignin-based hybrids and nanoparticles for tissue engineering and drug-release applications (Beisl et al. 2017).

5.3.4 Structure–Property Relationship: Effects of Resource, Purification, and Storage on Antioxidant and Antimicrobial Activity

Extraction and purification of kraft lignin from black liquor (BL) lead to differences in properties such as antioxidant activity and antimicrobial activity. Lignin was extracted and purified from three different batches of BL, from the same biomass and the same pulping process. The DPPH assay and the TPC assay were used for the lignin extracts showing that there is a difference in the activities. The results prove that the harvesting time and age of the plants, as a source of the BL, have an influence on the properties of the extracts. Figure 5.15 shows the DPPH inhibitions and the TPC of four extracts of KL from three different batches of BL. LE1 is the product of acidic precipitation by H_2SO_4 at a pH value of 2. The ether soaking product of LE1 is LE1 DEES. LES2a is the product of selective extraction by acetone and diethyl ether and PL is the product of selective extraction by ethanol. In the same Fig. 5.16, the effect of purification on the antioxidant activity and the total phenol content is clear. The available report on lignin model compounds suggests that free phenolic hydroxyl groups are crucial for antioxidant activity, while the aliphatic hydroxyl groups show a negative correlation. The radical scavenging activity of phenolic compounds depends on the rate of abstracting the hydrogen atom from a phenol molecule by a free radical and also on the stability of the radical formed. This abstracting ability was increased if some additional conjugation with substituents took place (Barapatre et al. 2016).

Son and Lewis observed that the inhibition of DPPH radicals scavenging activity occurred when the hydroxyl group was replaced by the methoxy group (di-orthophenolic motif, or a methoxyphenol motif or a dimethoxy phenyl ring). They found the most potent radical scavengers are the antioxidants with di-orthophenolic struc-

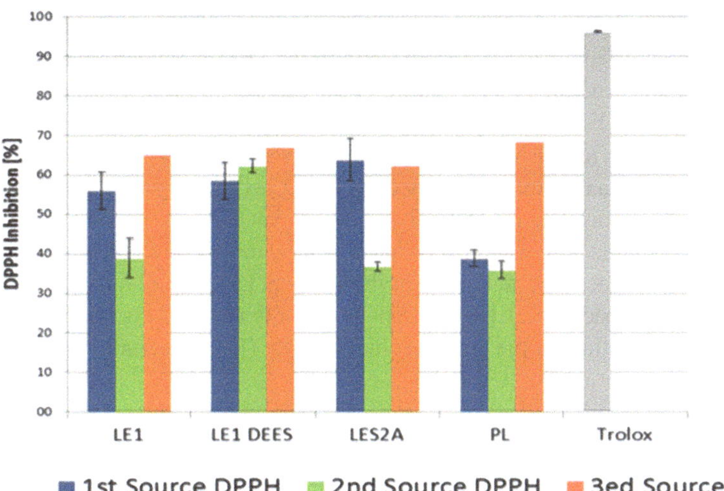

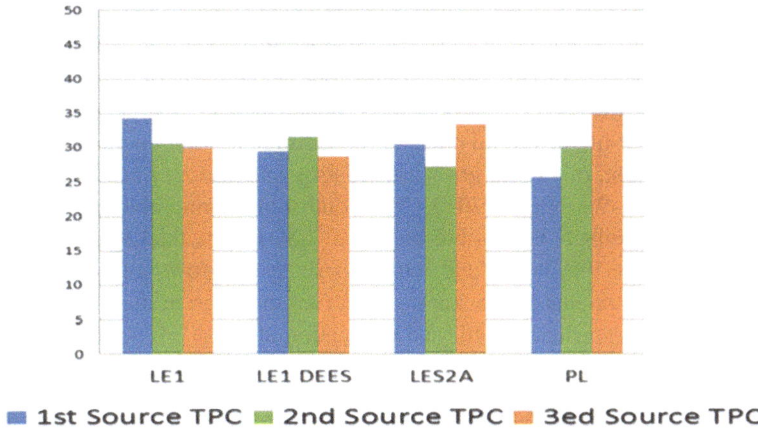

Fig. 5.16 Antioxidant activity of lignins depending on isolation and purification conditions: DPPH inhibitions and total phenol content (TPC) of three sources of KL extracts

ture, then the mono-phenolic compounds, and finally, with very low radical inhibition capacity, the compounds with both hydroxyl groups methylated (Son and Lewis 2002).

The storage of KL extracts has an influence on the SEC analysis as discussed earlier in this chapter, causing fragmentation of lignin and producing smaller compounds. This activity may also have an effect on the antioxidant and antimicrobial activity, depending on the structure of those fragments. Further studies on the storage effect on both antioxidant activity and total phenol content are ongoing.

Dumitriu and Popa (2013) have found that the main determining factor of the antimicrobial effect of lignin correlates with the phenolic components, specifically the side-chain structure and the nature of the functional groups. Typically, the presence of a double bond in α-, β-positions of the side chain and a methyl group in the γ-position grants the phenolic fragments the greatest potency against microorganisms. A primary antimicrobial study done on KL extracts against gram-positive and gram-negative bacteria, respectively, showed that LE1 is the only extract of the purification that is active against *Staphyloccus aureus* and *Listeria monocytogenes* (gram-positive bacteria), and none of the extracts is active against *Escherichia coli* (gram-negative bacteria). LE1 was kept as the only active extract to demonstrate the effect of purification on the antimicrobial activity (Dumitriu and Popa 2013).

5.3.5 Applications of Lignin-Based Polymers in Packaging

Packaging is one of the most demanding industries all over the world. Up to now, packaging materials have been increasingly made of petrochemical-based polymers such as poly(ethylene) (PE), poly(propylene) (PP), or poly(ethylene terephthalate) (PET). As a central focus to reduce waste and to save resources, there is a high interest in packaging materials based on renewable resources. Similarly to the protection of the food or product from the surroundings, one of the further interests is to maintain the quality of the food throughout the product shelf life (Siracusa et al. 2012). Extending the shelf life in several types of food, edible films or coatings can be used to prevent hydration, microbial growth, oxidative rancidity, surface browning, and oil diffusion (Silva-Weiss et al. 2013). The control of the microbiological decay of perishable food products is possible by using "active food packaging" containing antimicrobial agents that can reduce, inhibit, or delay the growth of microorganisms in food. The requirements for packaging materials are appropriate mechanical performance (such as tensile strength and elongation at break), functional properties (such as barrier to oxygen, carbon dioxide, water vapor, anhydride and aromatic compounds), and physical properties (i.e., appearance such as opacity and color) (Siracusa et al. 2012; Silva-Weiss et al. 2013).

Furthermore, active packaging materials need bioactivity such as antioxidant, antimicrobial, and anti-browning activities. Plastic packaging does have some disadvantages, however. It can often be contaminated by foodstuffs or biological substances. Because of such contamination, recycling these materials is not effortless, and the recovery cost is often higher than the cost of producing virgin plastic. As a consequence, several thousand tons of goods made from plastic materials are sent to landfills every day, increasing the amount of municipal waste. Hence, biodegradable polymers based on renewable resources have been studied to overcome the plastic waste problem. In general, biodegradable polymers have attracted attention in recent years as a replacement for conventional petroleum-based plastics. However, commercial utilization of bio-based polymers in packaging is still limited to a rather small percentage as these are more expensive than conventional fossil-based plastics (Sam et al. 2014).

5.4 Conclusions

Lignocellulosic biomass serves as one of the most important and relevant renewable resources. Thus, appropriate LCF biorefinery concepts are under intensive investigation for the production of bioenergy, biofuels, and a number of bio-based chemicals including building blocks, intermediates, and fine chemicals. Special focus is drawn to lignin isolation from LC feedstock because of its importance as a unique source for aromatic building blocks. For large-scale industrial applications, it would be beneficial to use kraft lignin extracted from black liquor, available from the pulp and paper industry. The purification of kraft lignin extracted from black liquor has been proofed through detailed analytical characterization. In addition, antioxidant and antimicrobial activities of lignin extracts are reported. Thus, lignin is a potential additive to be incorporated in commercially interesting biopolymers for packaging applications, such as cellulose or chitosan materials. Future work will focus on the effect of lignin storage on both antimicrobial and antioxidant activities. In addition to packaging, antimicrobial active lignin-based composites are very attractive for biomedical applications such as drug-release materials.

Acknowledgements Financial support (scholarship) was given to Abla Alzagameem by the Avempace-II Erasmus-Mundus Programme and the Graduate Institute of the Bonn-Rhein- Sieg University of Applied Sciences.

References

Agarwal UP, McSweeny JD, Ralph SA (2011) FT-Raman investigation of milled-wood lignins: softwood, hardwood, and chemically modified black spruce lignins. J Wood Chem Technol 31:324–344

Agbor VB, Cicek N, Sparling R, Berlin A, Levin DB (2011) Biomass pretreatment: fundamentals toward application. Biotechnol Adv 29:675–685

Ahuja D, Kaushik A, Chauhan GS (2017) Fractionation and physicochemical characterization of lignin from waste jute bags: effect of process parameters on yield and thermal degradation. Int J Biol Macromol 97:403–410

Alekhina M, Ershova O, Ebert A, Heikkinen S, Sixta H (2015) Softwood kraft lignin for value-added applications: fractionation and structural characterization. Ind Crop Prod 66:220–228

Amzad HM, Shah M (2015) A study on the total phenols content and antioxidant activity of essential oil and different solvent extracts of endemic plant *Merremia borneensis*. Arab J Chem 8:66–71

Anastas PT, Warner JC (1998) Green chemistry: theory and practice. Oxford University Press, New York

Azadi P, Inderwildi OR, Farnood R, King DA (2013) Liquid fuels, hydrogen and chemicals from lignin: a critical review. Renew Sust Energ Rev 21:506–523

Baba SA, Malik SA (2015) Determination of total phenolic and flavonoid content, antimicrobial and antioxidant activity of a root extract of *Arisaema jacquemontii* Blume. J Taibah Univ Sci 9:449–454

Balan V, David Chiaramonti D, Kumar S (2013) Review of US and EU initiatives toward development, demonstration, and commercialization of lignocellulosic biofuels. Biofuels Bioprod Biorefin 7:732–759

Barapatre A, Meena AS, Mekala S, Das A, Jha H (2016) In vitro evaluation of antioxidant and cytotoxic activities of lignin fractions extracted from *Acacia nilotica*. Int J Biol Macromol 86:443–453

Beisl S, Miltner A, Friedl A (2017) Lignin from micro- to nanosize: production methods. Int J Mol Sci 18:1244–1271

Bhat R, Abdullah N, Din RH, Tay GS (2013) Producing novel sago starch based food packaging films by incorporating lignin isolated from oil palm black liquor waste. J Food Eng 119:707–713

Chang G, Huang Y, Xie J, Yang H, Liu H, Yin X, Wu C (2016) The lignin pyrolysis composition and pyrolysis products of palm kernel shell, wheat straw, and pine sawdust. Energy Convers Manag 124:587–597

Chen J, Liu C, Wu SH, Liang J, Lei M (2016) Enhancing the quality of bio-oil from catalytic pyrolysis of kraft black liquor lignin. RSC Adv 6:107970–107976

Ciolacu D, Oprea AM, Anghel N, Cazacu G, Cazacu M (2012) New cellulose-lignin hydrogels and their application in controlled release of polyphenols. Mat Sci Eng C Mater 32:452–463

Constant S, Wienk HLJ, Frissen AE, de Peinder P, Boelens R, van Es DS, Grisel RJH, Weckhuysen BM, Huijgen WJJ, Gosselink RJA, Bruijnincx PCA (2016) New insights into the structure and composition of technical lignins: a comparative characterisation study. Green Chem. 18 (9):2651–2665

Dababi I, Gimello O, Elaloui E, Quignard F, Brosse N (2016) Organosolv lignin-basedwood adhesive influence of the lignin extraction conditions on the adhesive performance. Polymers (Basel) 8:340/1–340/15

Dashtban M, Schraft H, Syed TA, Qin W (2010) Fungal biodegradation and enzymatic modification of lignin. Int J Biochem Mol Biol 1:36–50

Dautzenber G, Gerhardt M, Kamm B (2011) Biobased fuels and fuel additives from lignocellulose feedstock. In: Biorefineries: industrial processes and products, status quo and future directions, vol 1. Wiley, Weinheim, pp 139–164

Directive 2009/28/EC of 28 April. 2009 on the promotion of the use of energy from renewable sources [online] Available at: http://eur-lex.europa.eu/legal-content/EN/ALL/?uri=celex%3A32009L0028. [July 03, 2017]

Dizhbite T, Telysheva G, Jurkjane V, Viesturs U (2004) Characterization of the radical scavenging activity of lignins—natural antioxidants. Bioresour Technol 95:309–317

Domenek S, Louaifi A, Guinault A, Baumberger S (2013) Potential of lignins as antioxidant additive in active biodegradable packaging materials. J Polym Environ 21:692–701

Dong X, Dong M, Lu Y, Turley A, Jin T, Wu C (2011) Antimicrobial and antioxidant activities of lignin from residue of corn stover to ethanol production. Ind Crop Prod 34:1629–1634

Downing M, Eaton LM, Graham RL, Langholtz MH, Perlack RD, Turhollow AF Jr, Stokes B, Brandt CC (2011) US billion-ton update: biomass supply for a bioenergy and bio- products industry. Oak Ridge National Laboratory, Oak Ridge

Dumitriu S, Popa VI (2013) Polymeric biomaterials, vol 1. CRC Press, Boca Raton, pp 551–578

Dusselier M, Van Wouwe P, Dewaele A, Makshina E, Sels BF (2013) Lactic acid as a platform chemical in the iobased economy: the role of chemocatalysis. Energy Environ Sci 6:1415–1442

Dyne DL, et al (1999) Estimating the economic feasibility of converting Lifno-cellulosic feedstocks to ethanol and higher value chemicals under the refinery concept: a phase II study. OR22072–58, University of Missouri

EC (1995) European Parliament and Council Directive No 95/2/EC of 20 February 1995 on food additives other than colours and sweeteners. Official Journal of the European Communities No. L61 18.3.1995. http://ec.europa.eu/food/fs/sfp/addit_flavor/flav11_en.pdf
El Hage R, Brosse N, Chrusciel L, Sanchez C, Sannigrahi P, Ragauskas A (2009) Characterization of milled wood lignin and ethanol organosolv lignin from *Miscanthus*. Polym Degrad Stab 94:1632–1638
El Khaldi-Hansen B, El-Sayed F, Schipper D, Tobiasch E, Witzleben S, Schulze M (2017) Functionalized 3D scaffolds for template-mediated biomineralization in bone regeneration. Front Stem Cell Regen Med Res 4:3–58
El Mansouri NE, Salvado J (2007) Analytical methods for determining functional groups in various technical lignins. Ind Crop Prod 26:116–124
Elbersen B, Staritsky I, Hengeveld G, Schelhaas MJ, Naeff H, Böttcher H (2012) Atlas of EU biomass potential. Deliverable 3.3: spatially detailed and quantified overview of EU biomass potential taking into account the main criteria determining biomass availability from different sources. [online] Available at: http://www.biomassfutures.eu/public_docs/final_deliverables/WP3/D3.3%20%20Atlas%20of%20technical%20and%20economic%20biomass%20potential.pdf. [July 03, 2017]
Espinoza-Acosta JL, Torres-Chavez PI, Ramirez-Wong B, Lopez-Saiz CM, Montano-Leyva B (2016) Antioxidant, antimicrobial, and antimutagenic properties of technical lignins and their applications. Bioresources 11:1–30
Fachagentur Nachwachsende Rohstoffe [online] Available at: https://mediathek.fnr.de/grafiken/daten-und-fakten/stoffliche-einsatzmengen-nachwachsender-rohstoffe-in-deutschland.html. [July 03, 2017]
Faix O, Genuit W, Boon JJ (1987) Characterization of beech milled wood lignin by pyrolysis-gas chromatography photoionization mass spectrometry. Anal Chem 59:508–513
Fromm J, Rockel B, Lautner S, Windeisen E, Wanner G (2003) Lignin distribution in wood cell walls determined by TEM and backscattered SEM techniques. J Struct Biol 143:77–84
Galbe M, Zacci G (2002) A review of production of ethanol from softwood. Appl Microbiol Biotechnol 59:618–628
Galbe M, Zacchi G (2007) Pretreatment of lignocellulosic materials for efficient bioethanol production. Adv Biochem Eng Biotechnol 108:41–65
Garcia A, Toledano A, Serrano L, Egues I, Gonzalez M, Marin F, Labidi J (2009) Characterization of lignins obtained by selective precipitation. Sep Purif Technol 68:193–198
Garcia A, Gonzalez M, Labidi AJ (2014) Evaluation of different lignocellulosic raw materials as potential alternative feedstocks in biorefinery processes. Ind Crops Prod 53:102–110
Giannini C, Ladisa M, Altamura D, Siliqi D, Sibillano T, De Caro L (2016) X-ray diffraction: a powerful technique for the multiple-length-scale structural analysis of nanomaterials. Crystals 6:87/1–87/22
González Arzola K, Polvillo O, Arias ME, Perestelo F, Carnicero A, González-Vila FJ, Falcón MA (2006) Early attack and subsequent changes produced in an industrial lignin by a fungal laccase and a laccase-mediator system: an analytical approach. Appl Microbiol Biotechnol 73:141–150
Gordobila O, Delucisb R, Egüésa I, Labidia J (2015) Kraft lignin as filler in PLA to improve ductility and thermal properties. Ind Crop Prod 72:46–53
Granata A, Argyropoulos DS (1995) 2-Chloro-4,4,5,5-tetramethyl-1,3,2-dioxaphospho-lane, a reagent for the accurate determination of the uncondensed and condensed phenolic moieties in lignins. J Agric Food Chem 43:1538–1544
Grisel RJH, van der Waal JC, de Jong E, Huijgen WJJ (2014) Acid catalysed alcoholysis of wheat straw: towards second generation furan-derivatives. Catal Today 223:3–10
Hamaguchi M, Kautto J, Vakkilainen E (2013) Effects of hemicellulose extraction on the kraft pulp mill operation and energy use: review and case study with lignin removal. Chem Eng Res Des 91:1284–1291
Hansen B (2015) Dissertation, Brandenburgisch-Technische Universität (BTU), Cottbus-Senftenberg

Hansen B, Kusch P, Schulze M, Kamm B (2016) Qualitative and quantitative analysis of lignin produced from beech wood by different conditions of the Organosolv process. J Polym Environ 24:85–97

Harmsen P, Huijgen W, Bermudez L, Bakker R (2010) Literature review of physical and chemical pretreatment processes for lignocellulosic biomass, vol. 9.Tech. Rep. 1184, Biosynergy, Wageningen UR Food & Biobased Research, pp 170–174

Holladay JE, Bozell JJ, White JF, Johnson D (2007) Top value-added chemicals from biomass. Pacific Northwest National Laboratory, Richland

Hossen M, Rahman S, Kabir AS, Hasan MMF, Ahmed S (2017) Systematic assessment of the availability and utilization potential of biomass in Bangladesh. Renew Sust Energ Rev 67:94–105

Imran M, El-Fahmy S, Revol-Junelles AM, Desobry S (2010) Cellulose derivative based active coatings: effects of nisin and plasticizer on physico-chemical and antimicrobial properties of hydroxypropyl methylcellulose films. Carbohydr Polym 81:219–225

International Energy Agency (2014) Renewables information (2016 edition). www.iea.org/statistics/topics/renewables/. IEA bioenergy task 42 report

Jiang X, Savithri D, Du X, Pawar S, Jameel H, Chang H, Zhou X (2017) Fractionation and characterization of Kraft lignin by sequential precipitation with various organic solvents. ACS Sustain Chem Eng 5:835–842

Joseph J, Rasmussen MJ, Fecteau JP, Kim S, Lee H, Tracy KA, Jensen BL, Frederick BG, Stemmler EA (2016) Compositional changes to low water content bio-oils during aging: an NMR, GC/MS, and LC/MS study. Energy Fuel 30:4825–4840

Kamm B, Kamm M, Schmidt M, Starke I, Kleinpeter E (2006) Chemical and biochemical generation of carbohydrates from lignocellulose-feedstock (*Lupinus nootkatensis*): quantification of glucose. Chemosphere 62:97–105

Kamm B, Gruber PR, Kamm M (2015) Biorefineries: industrial processes and products. Wiley, Weinheim

Larsen KL, Barsberg S (2011) Environmental effects on the lignin model monomer, Vanillyl alcohol, studied by Raman spectroscopy. J Phys Chem B 115:11470–11480

Laurichesse S, Avérous L (2014) Chemical modification of lignins: towards biobased polymers. Prog Polym Sci 39:1266–1290

Li C, Cheng G, Balan V, Kent MS, Ong M, Chundawat SP, Sousa LD, Melnichenko YB, Dale BE, Simmons BA, Singh S (2011) Influence of physico-chemical changes on enzymatic digestibility of ionic liquid and AFEX pretreated corn stover. Bioresour Technol 102:6928–696936

Long J, Xu Y, Wang T, Yuan Z, Shu R, Zhang Q, Ma L (2015) Efficient base-catalyzed decomposition and in situ hydrogenolysis process for lignin depolymerization and char elimination. Appl Energy 141:70–79

Lupoi JS, Singh S, Parthasarathi R, Simmons BA, Henry RJ (2015) Recent innovations in analytical methods for the qualitative and quantitative assessment of lignin. Renew Sust Energ Rev 49:871–906

Michailof CM, Kalogiannis KG, Sfetsas T, Patiaka DT, Lappas AA (2016) Advanced analytical techniques for bio-oil characterization. WIREs Energy Environ 5:614–639

Nak SN, Goud VV, Rout PK, Dalai AK (2009) Production of first and second generation biofuels: a comprehensive review. Renew Sust Energ Rev 14:578–597

Nakagawa-Izumi A, H'ng YY, Mulyantara LT, Maryana R, Do Vu T, Ohi H (2017) Characterization of syringyl and guaiacyl lignins in thermomechanical pulp from oil palm empty fruit bunch by pyrolysis-gas chromatography-mass spectrometry using ion intensity calibration. Ind Crop Prod 95:615–620

Ozturk M, Saba N, Altay V, Iqbal R, Hakeem KR, Jawaid M, Ibrahim FH (2017) Biomass and bioenergy: an overview of the development potential in Turkey and Malaysia. Renew Sust Energ Rev 79:1285–1302

Panoutsou C, Uslu A, van Stralen J, Elbersen B, Bottcher H, Fritsche U, Kretschmer B (2012) How much can biomass contribute to meet the demand for 2020 & which market segments are more promising? Proceedings of the 20th EU biomass conference and exhibition, Milan, Italy

Park S, Kim SH, Kim JH, Yu H, Kim HJ, Yang YH, Kim H, Kim YH, Ha SH, Lee SH (2015) Application of cellulose/lignin hydrogel beads as novel supports for immobilizing lipase. J Mol Catal B-Enzym 119:33–39

Puls J (2009) Stoffliche Nutzung von Lignin. Gülzower Fachgespräche Band 31, Fachagentur Nachwachsende Rohstoffe e.V. (FNR), pp 18–41

Rabaçal M, Ferreira AF, Silva CAM, Costa M (eds) (2017) Biorefineries: targeting energy,high value products and waste valorisation. Springer International Publishing, Cham

Raschip IE, Hitruc GE, Vasile C, Popescu MC (2013) Effect of the lignin type on the morphology and thermal properties of the xanthan/lignin hydrogels. Int J Biol Macromol 54:230–237

Richter AP, Brown JS, Bharti B, Wang A, Gangwal S, Houck K, Cohen H, Elaine A, Paunov VN, Stoyanov SD, Velev OD (2015) An environmentally benign antimicrobial nanoparticle based on a silver-infused lignin core. Nat Nanotechnol 10:817–823

Rinaldi R, Jastrzebski R, Clough MT, Ralph J, Kennema M, Bruijnincx PCA, Weckhuysen BM (2016) Paving the way for lignin valorisation: recent advances in bioengineering, biorefining and catalysis. International Edition 55:8164–8215

Ringpfeil M (2001) Biobased industrial products and biorefinery systems. Industrielle Zukunft des 21. Jahrhunderts? Brandenburgische Umwelt Berichte, BUB 10, ISSN 1434-2375

RL W, Wang XL, Li F, Li HZ, Wang YZ (2009) Green composite films prepared from cellulose, starch and lignin in room-temperature ionic liquid. Bioresour Technol 100(9):2569–2574

Sa'don NA, Abdul Rahim A, Hussin MH (2017) The effect of p-nitrophenol toward the structural characteristics and antioxidant activity of oil palm fronds (OPF) lignin polymers. Int J Biol Macromol 98:701–708

Sam ST, Nuradibah MA, Ismail H, Noriman NZ, Ragunathan S (2014) Recent advances in polyolefins/natural polymer blends used for packaging application. Polym Plast Technol 53:631–644

Sana B, Raghavan SS, Ghadessy FJ, Chia KHB, Nagarajan N, Ramalingam B, Seayad J (2017) Development of a genetically programed vanillin-sensing bacterium for high-throughput screening of lignin-degrading enzyme libraries. Biotechnol Biofuels 10:32

Sánchez-González L, Chiralt A, González-Martínez C, Cháfer M (2011) Effect of essential oils on properties of film forming emulsions and films based on hydroxypropyl methylcellulose and chitosan. J Food Eng 105:246–253

Santos P, Erdocia X, Gatto DA, Labidi J (2014) Characterisation of Kraft lignin separated by gradient acidprecipitation. Ind Crop Prod 55:149–154

Sebti I, Chollet E, Degraeve P, Noel C, Peyrol E, Agri J (2007) Water sensitivity, antimicrobial, and physicochemical analyses of edible films based on HPMC and/or chitosan. J Agric Food Chem 55:693–699

Silva-Weiss A, Ihl M, Sobral PJ, Gómez-Guillén MC, Bifani V (2013) Natural additives in bioactive edible films and coatings: functionality and applications in foods. Food Eng Rev 5:200–216

Singh BR, Singh O (2012) Global trends of fossil fuel reserves and climate change in the 21st century (Khan S, ed). Fossil Fuel Environ 2012:168

Siracusa V, Blanco I, Romani S, Tylewicz U, Rocculi P, Rosa MD (2012) Poly(lactic acid)-modified films for food packaging application: physical, mechanical, and barrier behavior. J Appl Polym Sci 125:E390–E401

Son S, Lewis BA (2002) Free radical scavenging and antioxidative activity of caffeic acid amide and ester analogues: structure–activity relationship. J Agric Food Chem 50:468–472

Sulaeva I, Zinovyev G, Plankeele JM, Sumerskii I, Rosenau T, Potthast A (2017) Fast track to molar-mass distributions of technical lignins. ChemSusChem 10:629–635

Sun SN, Cao XF, Xu F, Sun RC, Jones GL (2014) Structural features and antioxidant activities of lignins from steam-exploded bamboo (*Phyllostachys pubescens*). J Agric Food Chem 62:5939–5947

Tajeddin B (2015) Cellulose-based polymers for packaging applications. In: Lignocellulosic polymer composites. Scrivener, New York, pp 477–498

Tanjung FA, Husseinsyah S, Hussin K, Hassan A (2016) Mechanical and thermal properties of organosolv lignin/sodium dodecyl sulphate binary agent-treated polypropylene/chitosan composites. Polym Bull (Berl) 73:1427–1445

Tolbert A, Akinosho H, Khunsupat R, Naskar AK, Ragauskas AJ (2014) Characterization and analysis of the molecular weight of lignin for biorefining studies. Biofuels Bioprod Biorefin 8:836–856

Vallejos ME, Felissia FE, Curvelo AAS, Zambon MD, Ramos L, Area MC (2011) Chemical and physico-chemical characterization of lignins obtained from ethanol-water fractionation of bagasse. Bioresources 6:1158–1171

Van Putten RJ, van der Waal JC, de Jong E, Rasrendra CB, Heeres HJ, de Vries JG (2013) Hydroxymethylfurfural, a versatile platform chemical made from renewable resources. Chem Rev 113:1499–1597

Vaz S Jr (2014) Analytical techniques for the chemical analysis of plant biomass and biomass products. Anal Methods-UK 6:8094–8105

Vaz S Jr (2015) An analytical chemist's view of lignocellulosic biomass. Bioresources 10:3815–3817

Vázquez G, Antorrena G, González J, Freire S (1997) The influence of pulping conditions on the structure of acetosolv eucalyptus lignins. J Wood Chem Technol 17:147–162

Vinardell MP, Mitjans M (2017) Lignins and their derivatives with beneficial effects on human health. Int J Mol Sci 18:1219–1234

Vishtal A, Kraslawski A (2011) Challenges in industrial applications of technical lignins. Bioresources 6:3547–3568

Vivekanand V, Chawade A, Larsson M, Larsson A, Olsson O (2014) Identification and qualitative characterization of high and low ligninlines from an oat tilling population. Ind Crop Prod 59:1–8

Wang C, Kelley SS, Venditti RA (2016a) Lignin-based thermoplastic materials. ChemSusChem 9:770–783

Wang K, Loo LS, Goh KL (2016b) A facile method for processing lignin reinforced chitosan biopolymer microfibres: optimising the fibre mechanical properties through lignin type and concentration. Mater Res Express 3:035301/1–035301/13

Wi SG, Cho EJ, Lee DS, Lee SJ, Lee YJ, Bae HJ (2015) Lignocellulose conversion for biofuel: a new pretreatment greatly improves downstream bocatalytic hydrolysis of various lignocellulosic materials. Biotechnol Biofuels 8:228

Widsten P, Heathcote C, Kandelbauer A, Guebitz G, Nyanhongo GS, Prasetyo EN, Kudanga T (2010) Enzymatic surface functionalisation of lignocellulosic materials with tannins for enhancing antibacterial properties. Process Biochem 45:1072–1081

Wu S, Lv G, Lou R (2012) Applications of chromatography hyphenated techniques in the field of lignin pyrolysis. Appl Gas Chromatogr 2012:41–64

Xu F, Shi YC, Wang D (2013) Towards understanding structural changes of photoperiod-sensitive sorghum biomass during sulfuric acid pretreatment. Bioresour Technol 135:704–709

Yang W, Fortunati E, Dominici F, Giovanale G, Mazzaglia A, Balestra GM, Kenny JM, Puglia D (2016) Effect of cellulose and lignin on disintegration, antimicrobial and antioxidant properties of PLA active films. Int J Biol Macromol 89:360–368

Zhang Y, Vadlani PV (2015) Lactic acid production from biomass-derived sugars via co-fermentation of *Lactobacillus brevis* and *Lactobacillus plantarum*. J Biosci Bioeng 119:694–699

Zhang J, Chen Y, Sewell P, Brook MA (2015) Utilization of softwood lignin as both crosslinker and reinforcing agent in silicone elastomers. Green Chem 17:3176

Zhao W, Xiao LP, Song G, Sun RC, He L, Singh S, Simmons BA, Cheng G (2017) From lignin subunits to aggregates: insights into lignin solubilization. Green Chem 19(14):3272–3281

Chapter 6
Microalgae for Industrial Purposes

Mario Giordano and Qiang Wang

Abstract The use of microalgae for the production of compounds of commercial relevance has received substantial interest in recent years, mostly because these organisms contain a plethora of valuable compounds and their high turnover rate and functional plasticity make them relatively easy to cultivate for the production of biomass and added-value molecules. The metabolic flexibility of algae allows using them for many commercial applications, but it also makes it easy for cultures to diverge from the intended biomass quality. A thorough comprehension of the principles that control growth and carbon allocation is therefore of paramount importance for effective production of algal biomass and derived chemicals. In this review, we intend to provide basic but exhaustive information on how algae grow and on their biotechnological potential. In addition to this primary goal, we also give the reader a succinct panorama of culturing systems and possible applications.

Keywords Algae • Carbon allocation • Genetics • Culture • Cell composition • Stoichiometry • Biofuel • Food • Feed

6.1 Introduction

The use of microalgal biomass for capturing CO_2, for the production of biofuel and pharmaceutical and nutraceutical products, as food and animal feed, and for waste-water treatment has recently become a focal point for research and the object of public interest (e.g., Rasala et al. 2014). Part of this growing popularity is due to the

M. Giordano (✉)
Dipartimento di Scienze della Vita e dell'Ambiente, Università Politecnica delle Marche, Ancona, Italy

Università Politecnica delle Marche and Shantou University Joint Algal Research Center, Shantou, Guangdong, China
e-mail: m.giordano@univpm.it

Q. Wang
Key Laboratory of Algal Biology, Institute of Hydrobiology, the Chinese Academy of Sciences, Wuhan, Hubei, China
e-mail: wangqiang@ihb.ac.cn

S. Vaz Jr. (ed.), *Biomass and Green Chemistry*,
https://doi.org/10.1007/978-3-319-66736-2_6

fact that microalgae are very versatile in their utilization. Furthermore, they represent an excellent alternative to land crops: with their fast turnover rates, microalgae can produce large amounts of biomass in a relatively small volume and within a relatively short time. Furthermore, microalgae cultivation can be carried out on land of low agricultural value, thus avoiding competition with traditional crops. Because of the unique metabolic plasticity of microalgae, it is relatively easy (and certainly easier than for most terrestrial embryophytes) to optimize resource allocation in the cell to favor the production of selected compounds, by controlling culture conditions. Their high photosynthetic yield and efficiency make microalgae potentially capable of capturing substantial amounts of CO_2, thereby contributing to the mitigation of global climate change, especially if microalgae cultures can be fed with flue gases from fossil fuel combustion.

In spite of the foregoing considerations and although the commercial production of microalgae has a rather long history (e.g., Ben-Amotz 2004), progress in the design of production plants and, in general, in the technologies (both biological and engineering) used to commercially exploit algae (Table 6.1) has been rather slow. The causes of this are various, with a great contribution attributable to the fact that players on the market tend not to share information (Grobbelaar 2009). Furthermore, algal cultivation is often conducted by operators who lack a thorough understanding of the physiology of these organisms, thus having limited ability to identify and solve the problems that may arise from their functional complexity. This chapter intends to highlight the crucial aspects of algal physiology and to suggest ways to avoid difficulties and facilitate the attainment of biomass of a defined quality or of specific compounds. To do so, we provide some basic information on how algae fix and allocate carbon and on how culture conditions can influence the organic composition of cells. We also discuss how the chemical composition of culture media affects algal growth and biomass quality. Finally, we provide examples of applications in which physiological knowledge may bring obvious advantages.

6.2 What Are Algae?

In the words of Raven and Giordano (2014), "algae frequently get a bad press. Pond slime is a problem in garden pools, algal blooms can produce toxins that incapacitate or kill animals and humans and even the term seaweed is pejorative, a weed being a plant growing in what humans consider to be the wrong place. Positive aspects of algae are generally less newsworthy." Giving a scientific sound definition of "algae" is not an easy task: algae are organisms that produce O_2 as a waste product of their photosynthesis, but are not "higher plants" (embryophytes) (Raven and Giordano 2014). According to this definition, prokaryotic (cyanobacteria) and eukaryotic photosynthetic organisms are algae. In the group of organisms included in this definition, we find organisms in an approximate size range from 1 μm to 1 mm,

Table 6.1 Main algal species used in biotechnological applications (the information reported in this table was mainly obtained from Enzing et al. 2014 and Borowitzka 2016)

Species	Phylum and class	Product	Production status
Artrospira platensis	Cyanophyta, Cyanophyceae	Biomass as dietary supplement	Mass production
Chaetoceros muellerii	Bacillariophyta, Bacillariophyceae	Biomass as dietary supplement	Small scale
Chlorella vulgaris	Chlorophyta, Chlorophyceae	Canthaxanthin, astaxanthin, β-carotene, biomass as dietary supplement	Mass production
Crypthecodinium cohnii	Miozoa, Dinophyceae	Docosahexanoic acid	Mass production
Dunaliella salina	Chlorophyta, Chlorophyceae	β-Carotene, glycerol	Mass production
Hematococcus pluvialis	Chlorophyta, Chlorophyceae	Astaxanthin, cantaxanthin, lutein	Mass production
Isochrysis spp.	Chlorophta, Chrysophyceae	Biomass as dietary supplement	Mass production
Nannochloropsis spp.	Ochrophyta, Eustigmatophyceae	Eicosapentanoic acid	Small scale
Nitzschia closterium	Bacillariophyta, Bacillariophyceae	Eicosapentanoic acid	Research
Nostoc commune	Cyanophyta, Cyanophyceae	Biomass as dietary supplement	Collected, not cultivated
Nostoc flagelliforme	Cyanophyta, Cyanophyceae	Biomass as dietary supplement	Collected, not cultivated
Nostoc sphaeroids	Cyanophyta, Cyanophyceae	Biomass as dietary supplement	Collected, not cultivated
Odontella	Bacillariophyta, Mediophyceae	Eicosapentanoic acid, docosahexanoic acid	Small scale
Pavlova lutherii	Haptophyta, Pavlovophyceae	Biomass as dietary supplement	Research
Phaeodactylum tricornutum	Bacillariophyta, Bacillariophyta incertae sedis	Eicosapentanoic acid	Small scale
Porphyridium cruentum	Rhodophyta, Protoflorideae	Biomass as dietary supplement, arachidonic acid, triacylglycerols	Small scale
Scenedesmus almeriensis	Chlorophyta, Chlorophyceae	Lutein, β-carotene	Research
Skeletonema spp.	Bacillariophyta, Bacillariophyceae	Biomass as dietary supplement	Small scale
Tetraselmis spp.	Chlorophta, Prasinophyceae	Biomass as dietary supplement	Research

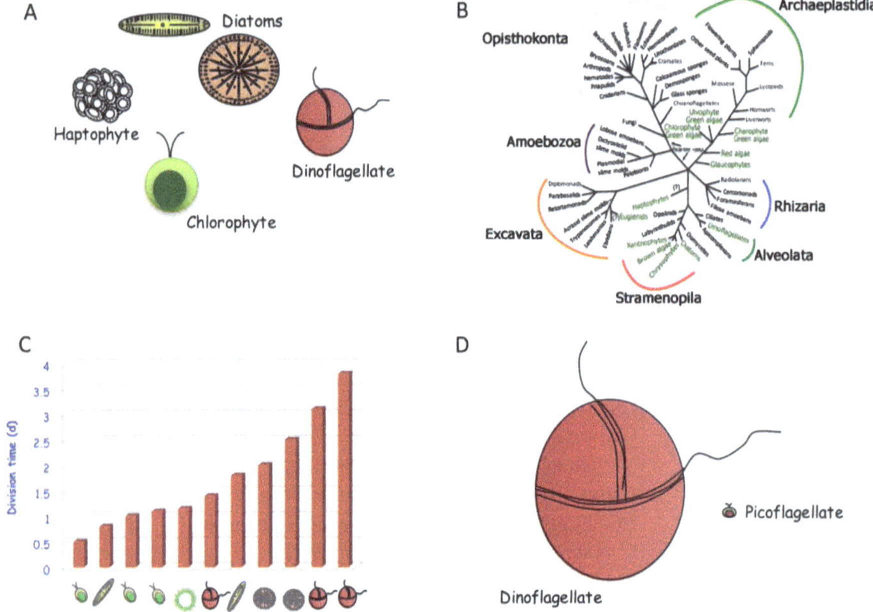

Fig. 6.1 The algae are extremely diverse in morphology (**a**), phylogeny (**b**), growth potential (**c**), and size (**d**). The information in **b** refers to eukaryotic algae and was derived from Knoll (2003) and Brodie et al. (2017). The growth rates in **c** are unpublished data; the species are not identified because of non-disclosure agreements

highly phylogenetically and morphologically diverse (and this diversity has only been minimally explored; de Vargas et al. 2015), with an extremely broad range of growth rates (Fig. 6.1).

6.3 Algae Produce Biomass Through Photolithotrophy, Heterotrophy, or Mixotrophy

Although the ability to carry out photosynthesis is certainly a key characteristic of algae, they are also capable of various degrees and various modes of heterotrophy. For some application, the ability of algae to combine autotrophy and heterotrophy can be advantageous, offering the ability to stimulate growth by supplying organic nutrients when light is not present or insufficient and when CO_2 is subsaturating. The ability to grow on organic substrates can also be exploited in phytodepuration processes that require the treatment of dissolved or particulate organic matter, in addition to that of inorganic nutrients. A brief description of algal trophisms may help the comprehension of the ensuing topics.

6.3.1 Photolitothrophy

Photosynthesis is a mode of nutrition that uses inorganic carbon in a light-dependent process to generate organic compounds (photolithotrophy). In photosynthesis the generation of energy in the form of reductants and ATP is conducted on the photosynthetic membranes, which in eukaryotes are located in the chloroplast (Chen et al. 2015a). The photosynthetic generation of reductants is the consequence of the excitation of specialized chlorophyll *a* molecules, which initiate an electron transfer along a redox gradient. This redox energy is employed chemo-osmotically (Mitchell 1961) to produce ATP, whereas the electrons are finally allocated onto soluble molecules (e.g., ferredoxins and NADPH) that can then be employed in metabolism (Schmollinger and Merchant 2014). The production of ATP and reductants is thus a function of the amount of photons captured by the so-called antenna systems, partially modulable ensembles of pigment–protein complexes. The energy made available by the processes occurring on photosynthetic membranes is then employed for the assimilation of primarily inorganic carbon. Appreciable proportions of reducing power and ATP are also used to acquire and assimilate other nutrients (N, P, S, etc.) and for various metabolic processes. CO_2 assimilation is carried out through the carboxylation of pentose ribulose bisphosphate (RuBP) with the catalysis of the enzyme ribulose bisphosphate carboxylase/oxygenase (Rubisco) (Marcus et al. 2011). The enzyme rubisco can also catalyze the oxygenation or RuBP, and O_2 and CO_2 compete with each other at the active sites of rubisco (Bowes et al. 1971). The presence of O_2 in the gas phase decreases the efficiency and yield of photosynthesis and leads to a conspicuous increase in the energetic cost of CO_2 fixation. Although in the course of evolution the ability of rubisco to favor carboxylation versus oxygenation has increased, no rubisco present today on the planet is capable of conducting carboxylation in the absence of oxygenation (Giordano et al. 2005). Most algae (Raven et al. 2005, 2008, 2012) overcome the difficulties associated to the double activity of rubisco (carboxylation and oxygenation) and the low CO_2:O_2 ratio in the extant atmosphere by expressing CO_2-concentrating mechanisms (CCMs) that pump CO_2 toward rubisco in an energy-dependent matter (Giordano et al. 2005; Raven et al. 2014). The CCMs are of various sorts and their modulation is strongly responsive to the CO_2:O_2 ratio in the environment. The activity of CCMs may also depend on factors that are distinct from the availability of CO_2 and O_2: the presence of CCMs allows algae to decrease the amount of rubisco in the cell and this leads to substantial savings in N, S, and also to an enhanced Fe and light use efficiency (Beardall and Giordano 2002; Raven et al. 2014); the availability of these resources can therefore have a role in the modulation of CCMs (Beardall and Giordano 2002; Raven et al. 2008, 2012, 2014). Elevated CO_2 leads to the downregulation of CCM (Giordano et al. 2005). CCM downregulation allows cells to save the energy they would otherwise have invested into pumping inorganic carbon toward rubisco. However, this does not necessarily create a higher growth rate, because the energy saved is not always invested in growth, or growth limitation (before the CO_2 increase) may not reside in CO_2 (Giordano and Ratti 2013).

6.3.2 Mixotrophy

Mixotrophy can be defined as the concomitant occurrence of photolithotrophy and chemo-organotrophy (use of exogenous organic C as the source of both energy and C for metabolism; also see Raven and Beardall 2016). The uptake of organic C can be carried out on a molecule-by-molecule basis (osmotrophy) or through the acquisition of particles (phagotrophy) (Flynn et al. 2010; Raven et al. 2013). Algae are, for the most part, primarily photolithotrophs and recur to mixotrophy only when photolithotrophy is hindered by the scarcity of inorganic nutrients or light. There are, however, protists that, although they fall within the definition of "algae" sensu Raven and Giordano (2014), are primarily phagotrophic, but can photosynthesize under prey limitation. Finally, some organisms are obviously derived from photosynthetic organisms but are obligate chemo-organitrophs (Mitra et al. 2016). The ancestral condition in algae is probably that of obligate photolithotrophs (Beardall and Raven 2016), which is confirmed by the fact that basal cyanobacteria such as *Glaeobacter violaceous* (Blank and Sanchez-Baracaldo 2010) are incapable of chemo-organotrophy (Beardall and Raven 2016, and references therein). On the other hand, Raven et al. (2009) wrote that "without phagotrophy at the cell level there would be no photosynthesis in eukaryotes"; in other words, as the acquisition of the chloroplast is the consequence of a phagotrophic event, phagotrophy/mixotrophy appears to be an inherent property of eukaryotic photosynthetic organisms. It must be considered that primary endosymbiotic events, that is, those in which a heterotrophic prokaryote engulfed a photosynthetic prokaryote, have most likely been relatively rare, whereas subsequent secondary and tertiary endosymbiotic events, in which an eukaryote engulfed a photosynthetic eukaryote, may have occurred more readily (McFadden 2001; Gentil et al. 2017; Lane 2017). Among extant algae, clades deriving from primary endosymbiosis are rarely mixotrophic, whereas mixotrophy is much more frequent in algae originated from secondary and tertiary endosymbiotic events (Beardall and Raven 2016). According to Raven (1995, 1997), the cost of the photosynthetic apparatus and the uptake systems for nutrients different from C sums up to about 50% of C, N, P, Fe, and of the energy cost to make a cell in a photolithotroph. The corresponding cost for the heterotrophic (phagotrophic) apparatus is less than 10% (also see Jones 2000 for further discussion on these matters). The advantage of mixotrophy over obligate photolithotrophy and obligate photo-organotrophy emerges especially in the light (Jones 2000), although the large number of possible nutritional conditions makes it hard to provide an univocal outcome of competition between organisms with different trophisms. It should also be considered that chemo-organotrophy leads to greater C loss through respiration and thus can lead to an elemental stoichiometry with lower C relative to N, P, and Fe. The concomitant use of photosynthesis can compensate for such unbalance, to a variable degree (Beardall and Raven 2016). Osmomixotrophy may also be important to recapture leaked dissolved organic carbon (Raven and Beardall 2016).

6.4 Cell Composition Results from a Combination of Genotypic and Environmental Constraints

The composition of cells is the result of the interaction of the genome with the environment leading to the best suited structural and functional cell organization. The composition of the cell is thus strongly dependent on the condition in which cells live or are cultured. At the same time, because it depends on the genotype, cell composition is strongly species specific. Different genotypes will respond differently to the same kind of environmental perturbation, with different degrees of compositional and functional homeostasis (Giordano 2013). The attitude to homeostatically retain cell composition is possibly also a function of growth rate: Fanesi et al. (2014) showed that, other things being equal, fast-growing microalgae have a lower tendency to compositional homeostasis than slow-growing ones, because they have a higher probability of competitively taking advantage of the investment in reproduction, regardless of the duration of the perturbation (Giordano 2013). Depending on both the species and the type of environmental perturbations, cells can adjust their growth performance through (a) regulatory processes that do not require changes in the expressed proteome (Giordano 2013; Raven and Geider 2003); (b) the production of new protein and the degradation of protein present before the perturbation was applied (acclimation); and (c) by changes in the genotype (adaptation) (Giordano 2013; Raven and Geider 2003). Most species, if maintained for a prolonged time in the same conditions, will tend to change their expressed proteome and acclimate to the environmental condition (Giordano 2013), and if a sufficient genetic heterogeneity is present in the population, genotype selection is also likely (Venuleo et al. 2017).

The consequences of these considerations for the commercial cultivation of microalgae are that (a) a tight match between genotype and cultural conditions must be ensured to obtain the desired end product; (b) stability of culture conditions is required to ensure constancy in the quality of the product and in productivity; and (c) a control of the genetic stability of the population is important to prevent a shift in dominance that may lead to the prevalence of a strain with non optimal characteristics (from the commercial perspective).

6.4.1 Elemental Stoichiometry and Organic Cell Composition

Cell composition results from the availability of the various elements and their metabolic and structural requirements. Quigg et al. (2003, 2011) and Ho et al. (2010) reported the elemental composition for a large number of species cultured under presumably resource-replete conditions. The species-specificity of these cell stoichiometries emerges clearly; macronutrients (i.e., C, N, P, S) generally show a higher degree of similarity than micronutrients across species (Giordano 2013 and references therein). An obvious reciprocal relationship exists between cell stoichiometry and environmental chemistry; the oceanic "Redfield ratio" (Redfield 1934) is a typical example of this. A mechanistic basis for the Redfield ratio was proposed by Loladze and Elser (2011), who suggested that the fairly constant elemental stoichiometry of phytoplankton in the ocean, especially with respect to the N:P ratio, is imposed by the fact that rapid growth constrains the ratio between cell protein (main sink for N) and RNA (main sink for P) (e.g., Norici et al. 2011; Raven et al. 2012; Geider and La Roche 2002). The relative content of N:P can be used as a proxy for the protein to RNA ratio and this can be related to growth through the capacity for protein synthesis (growth rate hypothesis: Elser et al. 2011; Flynn et al. 2010; Loladze and Elser 2011; Giordano et al. 2015b) It is worthwhile noticing, however, that the growth rate hypothesis does not always apply to microalgae (Flynn et al. 2010; Nicklisch and Steinberg 2009; but also see Giordano et al. (2015b).

In a commercial cultivation system, most likely, growth conditions are resource replete, which may lead to the fact that cell elemental composition does not reflect the requirement to achieve the maximum possible growth rate but is influenced by luxury accumulation of some elements (Giordano 2013).

Elemental composition is connected to the organic cell composition also because an imbalance between C and N (or C and P, or C and S) can lead to different C allocation patterns. When, for instance, the C:N ratio is higher than the C:N ratio in protein and nucleic acids and the excess of C surpasses the requirement for structural non-N/P-containing pools, two options exist: the C in excess (relative to other nutrients) is not acquired (Kaffes et al. 2010) or C is assimilated beyond strict growth requirements with a consequent increase in the size of pools that do not contain N and P, such as carbohydrates and lipids (Giordano et al. 2015a; Giordano and Ratti 2013; Montechiaro and Giordano 2010; Palmucci et al. 2011). Whether the excess C is allocated to carbohydrates or lipids depends on genotypic, energetic, and size constraints (Palmucci et al. 2011). The genotypic constraints are associated with the preference for some metabolic pathways by a species/strain (Palmucci and Giordano 2012). The energetic constrains are associated with the different cost of allocating C to carbohydrate or lipid and mobilizing it (Montechiaro and Giordano 2010; Palmucci et al. 2011); this of course becomes relevant only when energy availability (i.e., light; as is often the case in dense commercial cultures) limits growth (Ruan and Giordano 2017; Ruan et al. 2017). The spatial constraints occur because the lower hydration of lipids makes it easier for them to accumulate when space is limited (Palmucci et al. 2011). The outcome of an imbalance between C and N is thus not easily forecasted and should be assessed case by case.

6.5 Genetic Modification of Algae: Tools and Aims

The enormous pool of metabolic possibilities constituted by microalgae translates into a very large and mostly unexplored potential for applications. It also minimizes the need for genetic manipulations, because many functional variants are present in nature (although they may not all have been discovered yet). Many applicative problems may be solved through the search of "natural" species/strains with the required metabolic capabilities. This notwithstanding, genetic manipulation is possible and, where allowed, it may offer the best solutions for specific problems (Gressel 2008). The generation of mutants is important for strain improvement for biotechnological applications, but it can also be used, in the secure space of a laboratory, for functional analyses of genes and proteins. Few genetically modified strains of microalgae are used commercially nowadays, partially because molecular tools (e.g., efficient nuclear transformation, availability of promoter and selectable maker genes, and stable expression of transgenes) are not available for some commercially important species (Amaro et al. 2011). However, recent developments of high-throughput technologies have enabled the profiling of mRNA, proteins, and metabolites, giving rise to the fields of transcriptomics, proteomics, and metabolomics, respectively (Lee et al. 2010); these methodologies allow the comprehension of the consequences of genetic manipulation at the whole cell level, thus facilitating their application in a productive context.

Despite the increasing number of sequenced microalgae genomes, precise and programmable genome editing has been reported for only a few eukaryotic microalgae, such as the eustigmatophyte *Nannochloropsis* sp. (Kilian et al. 2011), the green alga *Chlamydomonas reinhardtii* (Sizova et al. 2013), the diatoms *Phaeodactylum tricornutum* (Nymark et al. 2016) and *Thalassiosira pseudonana* (Poulsen et al. 2006). The genome editing methods used for these studies (Daboussi et al. 2014; Hopes et al. 2016; Nymark et al. 2016; Shin et al. 2016; Wang et al. 2016) (see following), together with the ever-growing number of tools for transgene expression, cloning, and transformation (e.g., Rasala et al. 2014; Scaife et al. 2015), open up very promising perspectives for the future of algal genetic manipulation.

The methods for targeted gene knockout and gene replacement based on homologous recombination (HR) have driven rapid progress in understanding many of the complex metabolic and regulatory networks in eukaryotic cells (Weeks 2011). The main obstacle for direct gene targeting is the low frequency of HR between nuclear genes and donor DNA. Recombination efficiency may be increased by the use of zinc-finger nucleases (ZFNs), which cut the genome at specific sites to facilitate HR. Sizova et al. (2013) published a nuclear gene targeting strategy for the green alga *Chlamydomonas reinhardtii* that is based on the application of ZFNs. In the case of *C. reinhardtii*, insertional mutagenesis to disrupt a gene of interest is commonly employed. For instance, by exploiting a collection of *C. reinhardtii* insertional mutants originally isolated for their insensitivity to ammonium, Emanuel et al. (2016) found a strain that, in addition to its ammonium-insensitive (AI) phenotype, was unable to correctly express nitrogen assimilation genes in response to NO signals. The difficulty of extending this approach to more species resides in the fact that it cannot prescind from the existence of large collections of mutants and from the screening of many thousands of clones.

MicroRNAs (miRNA) are 21- to 24-nucleotide RNAs present in many eukaryotes that guide the silencing effector Argonaute (AGO) protein to target mRNAs via a base-pairing process (Bartel 2009). Chung et al. (2016) showed that miRNAs in

C. reinhardtii regulate gene expression primarily by destabilizing mRNAs, using target sites that lie predominantly within coding regions.

Protein-based systems involving mega nucleases and "transcription activator-like effector nucleases" (TALENs) allow precisely targeted genome editing of eukaryotic microalgae genomes (Daboussi et al. 2014; Weyman et al. 2015). These systems bear great potential for research and generation of tailored strains, although they are labor intensive and rather costly. Recently, a much simpler and inexpensive method for genome editing in algae, CRISPR/Cas9, came about (Nymark et al. 2016). It was developed to generate stable targeted gene knockouts in the marine diatom Phaeodactylum *tricornutum*, but it should be easily adaptable for use in other microalgae. Shin et al. (2016) applied this system to *C. reinhardtii*; they directly delivered the Cas9 protein and the "single-chain guide RNAs" (sgRNAs) to three different genes, obtaining mutations at the Cas9 cut sites with a significantly improved targeted mutagenic efficiency. Wang et al. (2016) established a precise CRISPR/Cas9-based genome editing approach for the industrial oleaginous microalga *Nannochloropsis oceanica*, using the gene encoding nitrate reductase (NR; g7988) as an example. The isolated mutants, in which precise deletion of five bases caused a frameshift in NR translation, grew normally in the presence of NH_4^+ but failed to grow when N was supplied as NO_3^-. This demonstration of CRISPR/Cas9-based genome editing in industrial microalgae is very promising for microalgae-based biotechnological applications. Also, editing of the chloroplast genome is of interest for biotechnological applications, because it may allow transgene insertion via HR with expression that is not subject to nuclear gene-silencing mechanisms; furthermore, plastidial transformation may take advantage of the prokaryotic organization of chloroplast genomes to co-express multiple genes in operons (Wannathong et al. 2016). New simple and inexpensive protocols have recently been developed to this end (Wannathong et al. 2016), but their effectiveness for species different from *C. reinhardtii* is still to be demonstrated as is their applicability for biotechnological purposes.

Many cyanobacterial strains are amenable to transformation and homologous recombination. Cis genetic modification (through genome editing) is the most common approach for engineering cyanobacteria (Berla et al. 2013). Typically, chromosomal mutations are generated through the insertion of a plasmid that contains the gene(s) of interest, a selectable marker gene, and flanking sequences homologous to the targeted chromosomal sequence (homology arms). Numerous heterologous genes have been inserted through these methods in the model cyanobacteria *Synechocystis* sp. PCC 6803 and *Synechococcus elongatus* PCC 7942 (Savakis and Hellingwerf 2015). However, genome editing of cyanobacteria is more challenging than in model heterotrophic prokaryotes such as *Escherichia coli*, primarily because cyanobacteria often contain multiple genome copies per cell and long-term instability of the genes introduced (Kusakabe et al. 2013; Ramey et al. 2015). CRISPR interference is emerging as a promising method to repress expression of specific genes, with no need for gene knockout also for prokaryotes (Huang et al. 2016).

6.6 How Are Algae Cultured?

Large-scale microalgal cultivation can be attained by a number of culturing systems, the choice of which depends on cost, available technology, and desired quality of the biomass or added-value product. Consequently, no set recipe for a successful cultivation exists and physiological, engineering, and economic analyses must be conducted to ensure good results.

Algal commercial cultivation is often conducted empirically without a full understanding of the physiology behind it, which can lead to unsatisfactory results or to products that are highly variable in quality because of the lack of control on the biological processes controlling biomass production.

From a trophic point of view, microalgae can be grown photolithotrophically (phototrophy), chemo-organitrophically (heterotrophy), or mixotrophically (Chojnacka and Marquez-Rocha 2004; see above). Photolithotrophic growth is advantageous because it can use natural sunlight as the energy source and mineral media, which are relatively inexpensive; it is unavoidable when obligate photolithotrophs are used. Light must be in large supply to sustain photolithotrophic growth (Perez-Garcia et al. 2011), which makes this mode of mass cultivation economically convenient in area with high insolation. In intensive cultures, light penetration can be substantially attenuated and light can become the limiting factor for growth. This problem is usually addressed through careful design of culturing systems. In recent years, a molecular approach to the problem of light availability in intensive culture systems has also been taken, through the production of genetically modified strains that have antennae of smaller size and thus a decreased light attenuation across the culture (Mooij et al. 2015). These strains, however, have not yet found commercial application and still need to be tested for productivity at a usefully large scale.

In some cases, higher productivity can be attained through heterotrophic or mixotrophic cultivation methods (Chen et al. 2016). Heterotrophy can be maintained in total darkness by supplying organic compounds (e.g., glucose, acetate, glycerol) as both energy and carbon source, eliminating the need for illumination but adding a cost for the organic substrates. The cultivation of *Chlorella protothecoides* (now *Auxenochlorella protothecoides*) under mixotrophic condition was reported to increase the yield of biomass and lipid (Wang et al. 2013). The mixotrophic cultivation of the green alga *Chlorella* sp. C2 in a 5-L bioreactor resulted in a maximum biomass productivity of 9.87 g L^{-1} day^{-1}; this productivity declined to 7.93 g L^{-1} day^{-1} when the culture size was scaled up to 50 L (Chen et al. 2016), which is still a very small volume for commercial application. The change in productivity with increasing cultural volume warns us about the extrapolation of data obtained from small-scale tests to larger-scale cultivation. Cultivation on organic substrates currently is rarely utilized in the commercial production of algal biomass because the number of heterotrophic or mixotrophic algal strains that can be used is limited, the presence of organic carbon makes it very difficult to control bacterial proliferation, cases of growth inhibition by soluble organic compounds were reported, and because of the higher cost of growth media (Zhang et al. 2014a). There are however applications for which mixotrophy can be useful and almost unavoidable. One such application, for instance, is the use of algae in wastewater treatment, where the water affluent cannot be fully deprived of organic components and a concomitant utilization of the inorganic and organic components is desirable or necessary (Cai et al. 2013).

In terms of the engineering of the cultivation systems, both open ponds and photobioreactors are used for algal cultivation. When very large quantities of product must be generated at low cost, open ponds are the most common solution (Borowitzka 1999). However, not all species are amenable to effective cultivation in open ponds, and the susceptibility to weather (especially rain and temperature variations) and light availability makes an open pond suited mostly for tropical and subtropical regimes with low precipitation and cloud cover (Richmond 1986). Algae cultivation in open-pond production systems has been used since the 1950s (Chojnacka et al. 2004), in both natural (lakes, lagoons, ponds) and artificial basins. Shallow raceway ponds, in which the algal suspension is mixed with paddle wheels, are the most widely used systems because they are relatively easy and cheap to construct and operate (Doucha and Lívanský 2006). Currently, more than 90% of world microalgae biomass production is obtained in raceway ponds. Some ponds are built on non-arable lands adjacent to power plants to have access to CO_2 from flue gases or near wastewater treatment plants to easily access nutrient supplies. Although widespread, open ponds have their drawbacks such as relatively low culture density and biomass productivity, high evaporative losses, and susceptibility to weather and to contamination by bacteria or undesired algal strains (Chen et al. 2013; Richardson et al. 2012).

When the environmental conditions are not suitable for open ponds, or a high and verifiable quality of biomass is required or biomass is used for the production of added-value compounds, algae can be cultured in closed or nearly closed systems, the photobioreactors (PBR). The engineering of PBRs is very diverse (Behrebs 2005 and references therein), and there are various designs of PBRs for different uses and of different cost. We refer to the numerous reviews and research papers on these topics published in recent years for more details (e.g., Zittelli et al. 2013; Pires et al. 2017).

In a continuous culture system, such as most PBRs, the carrying capacity (i.e., maximum biomass that can be obtained) is determined by the concentration of the factor(s) that limit growth. If light is sufficient, the composition of the medium is therefore a crucial aspect in the planning of a successful production system. In most cases, care is taken to provide an excess of macronutrients but little attention is given to micronutrients and elemental stoichiometry. The consequence often is an imbalance in nutrient availability, resulting in unnecessary costs and lack of control on limiting factors (Giordano 2013). The rate of biomass production, instead, is a function of the rate by which limiting nutrients are supplied to the culture (dilution rate). A dilution rate that surpasses the genotypically fixed maximum growth potential of a strain will lead to the decrease of the cell number per unit of volume of culture. Dilution rates that are lower than this higher limit are sustainable. It should be considered that a suboptimal growth rate may impact cell metabolism and thus affect biomass quality (Fanesi et al. 2014).

6.6.1 Harvesting and Dehydration of Algal Biomass

After cultivation, biomass has to be separated from the growth medium and recovered for downstream processing. Harvesting usually involves two steps: (1) bulk harvesting (or primary harvesting), to separate the microalgae from their growth medium, usually done by sedimentation, flocculation, or flotation; (2) thickening (or secondary dewatering), to concentrate the microalgal slurry after bulk harvesting, typically by centrifugation or filtration (Lam and Lee 2012; Zhang et al. 2014b). Thickening by centrifugation and filtration use up considerable energy and, although often employed, represents one of the main costs for commercial algae (and algal products) production. Flocculation is used to increase the efficiency of gravity sedimentation (Brennan and Owende 2010). However, conventional flocculants are often toxic, whereas non toxic flocculants (e.g., organic polymers) are presently too expensive for large-scale applications (Lee et al. 2013). Autoflocculation, which can be induced by increasing the H^+ concentration in the medium, and electrolytic flocculation may be used to separate algae from the medium without the addition of chemicals; estimates suggest that these methods would be significantly more economical than other harvesting techniques (Beuckels et al. 2013; Coons et al. 2014; Lee et al. 2013). Bioflocculation is the process of flocculation induced by microorganisms or by compounds they produce; it is possibly the most environmentally friendly among the flocculation methods (Wan et al. 2015). In a study by Wang et al. (2015), co-culturing of *Chlorella* and bioflocculant-producing bacteria was optimized to decrease adverse effect of co-culturing and proved to be effective in facilitating harvesting; such an approach may be a good option for the collection of algal biomass in wastewater treatment plants, where the bacterial component is unavoidable. Electroflocculation is another option: Coons et al. (2014) reported that, in the production of algal lipids, the cost of electroflocculation with inert electrodes was appreciably lower than that of membrane filtration, which, in turn, was less costly than centrifugation. The same authors suggested that ultrasonic harvesting, which operates through a standing wave created by forward and reverse propagating pressure waves in the water, could afford substantial economic advantages in comparison to other harvesting methods.

Drying of biomass is among the most expensive steps in microalgal production, because the evaporation of large volumes of water drains large amounts of energy; yet, it is usually a necessary step, because the presence of water interferes with transport and processing (Kumar et al. 2010). Spray, drum, freeze, and solar drying are commonly applied methods. Solar drying is economical, but it requires large extensions of land and is not feasible in temperate climates, where sunlight is not always sufficient (Zhang et al. 2014a).

Another common problem in algal commercial cultures is the identification of the appropriate time for harvesting. A method that appears especially suited for such task is Fourier-transform infrared spectroscopy (FTIR) (Domenighini and Giordano 2009; Giordano et al. 2001). A number of papers have proved the reliability, rapidity, and low cost of this methodology, which affords a snapshot of cell composition (Jebsen et al. 2012; Montechiaro and Giordano 2010; Palmucci and Giordano 2012; Palmucci et al. 2011) and allows reliably following changes in biomass quality over time (Giordano et al. 2017; Giordano and Ratti 2013; Memmola et al. 2014). The advantage of FTIR is that various organic pools can be determined concomitantly, with no need for extractive procedures and in quasi-real time. The disadvantage is that, in complex whole-cell spectra, the identification of specific compounds or small pools may not be easy (and sometimes is not possible). In these cases, other methodologies may be better suited. For lipids, fluorescent probes such as the lipophilic Nile red and BODIPY 505/515 can be used; these compounds can detect neutral lipids in intact cells (Cooper et al. 2010). However, the relatively long time required for staining and detecting the fluorescent probe, the relatively high cost of the probe, and the potential errors caused by the different permeability of the probes into different microalgae cells represent drawbacks of the use of these molecular probes. Qiao et al. (2015) developed a method to determine the optimal harvest time in oil-producing microalgal cultivations by measuring maximal photosystem II quantum yield (Fv/Fm); although this method afforded good results, it must be considered that Fv/Fm is associated with a number of events occurring within the cells and is not highly specific.

6.7 Products and Applications

6.7.1 Biofuels from Algae

Microalgae can, in principle, be used for the production of several different types of biofuels: biodiesel can be obtained from algal oil (Ahmad et al. 2011), biomethane, also known as biogas, can be produced through anaerobic digestion of algal biomass (Frigon et al. 2013), hydrogen can be generated photobiologically (Zhang et al. 2012), and bioethanol can be produced in the dark by anaerobic fermentation (Bigelow et al. 2014). The production of these biofuels can be combined in the same process, because the residue of the oil extraction for biodiesel production can be further processed into ethanol, methane, or H_2 (Mata et al. 2010; Singh and Cu 2010) (Fig. 6.2). The literature on algal biofuels is vast (e.g., Demirbas 2009; Kapdan and Kargi 2006; Mata et al. 2010; Spolaore et al. 2006). We shall therefore not linger on details in this section of our review. In spite of the broad interest in algal biofuels, the actual commercial production of such forms of renewable energy is, to say the least, limited: the very large volumes of cultures required to obtain a meaningful quantity of biofuels, together with the still relatively low price of fuels

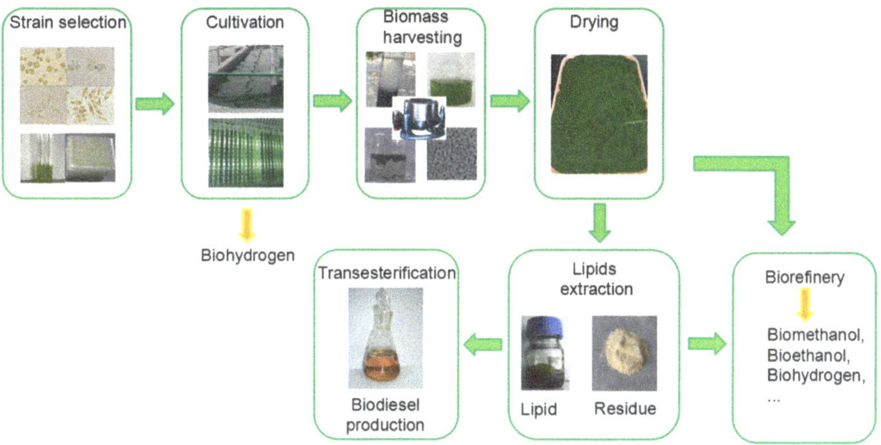

Fig. 6.2 The overall process flow for microalgal biofuel production

derived from fossil oil and other sources, is still in the way of a significant exploitation of algal fuels (Borowitzka 2016 and references therein).

6.7.1.1 Biodiesel

Biodiesel production comprises six steps: (i) strain selection; (ii) cultivation; (iii) biomass harvesting; (iv) biomass drying; (v) lipids extraction; and (vi) transesterification (Zhang et al. 2014b). We have already provided information on the first four steps; thus, the ensuing paragraphs focus on lipid extraction and transesterification.

Before lipid extraction, cells are usually lysed to facilitate access of the solvents to lipids within the cells. Various lysis procedures can be used: high pressure homogenization (HPH) (Samarasinghe et al. 2012), bead mills (Doucha and Lívanský 2008), ultrasonic disruption (Adam et al. 2012; Bigelow et al. 2014), and electroporation (Sheng et al. 2011). Ultrasonic disruption is possibly the procedure with the lowest energy requirement (Coons et al. 2014). Subsequently, lipids and fatty acids are extracted from the microalgal slurry mainly by two methods: the hexane Soxhlet method and the Bligh–Dyer method (Demirbas 2009; Kanda et al. 2013). Hexane-based oil extraction is more energy efficient and is therefore preferred for scaling-up efforts (Peralta-Ruiz et al. 2013). However, the use of chemical solvents has intrinsic problems associated with the toxicity of these compounds to humans and environment. Several supercritical fluids, especially supercritical CO_2, have been used for microalgal lipid extraction for the production of biodiesel. Although supercritical extraction is nontoxic and provides a nonoxidizing environment that avoids degradation of the extracts (Mouahid et al. 2013), it is expensive.

After lipid extraction, fatty acids transesterification is generally used to produce biodiesel (Lam et al. 2010). Lipid extraction and transesterification can be carried out simultaneously, simplifying the process and reducing the overall cost of microalgal biodiesel production (Lam and Lee 2012). Biodiesel recovery in in situ transesterification is negatively affected by excessive biomass moisture (>20% m/m) (Sathish et al. 2014). Given the aforementioned high cost of biomass dehydration, this excessive moisture has a nontrivial effect on the economic performance of the production system. An improved *in situ* transesterification process that directly converts wet oil-bearing microalgal biomass into biodiesel was recently proposed (Dang-Thuan et al. 2013).

6.7.1.2 Other Microalgal Biofuels

Bioethanol

Most microalgae do not contain lignin; this property of algal biomass facilitates the enzymatic hydrolysis necessary for bioethanol production (Sun and Cheng 2002). Furthermore, in appropriate culture conditions, many algal species can accumulate high amounts of ethanol if substrates can be fermented (Bibia et al. 2017; Farias Silva and Bertucco 2016). Green algae of the genera *Scenedesmus*, *Chlorella*, *Chlorococcum*, and *Tetraselmis* and cyanobacteria of the genus *Synechococcus* have been reported to be potentially good sources of bioethanol (Farias Silva and Bertucco 2016). Typically, bioethanol is produced by the hydrolysis of sugars and their subsequent fermentation in microaerobic or anaerobic conditions using yeasts (Farias Silva and Bertucco 2016). Algae can generate ethanol directly in the dark, by fermentative metabolism (Ueno et al. 1998); however, this process does not seem to be sufficiently efficient for commercial exploitation. Algae can also produce ethanol directly via photofermentation (photanol) (Hellingwerf and Mattos 2009). In photofermentation, the glyceraldehyde-3-phosphate generated in the Calvin cycle is converted to phosphoenolpyruvate and then to pyruvate; pyruvate is decarboxylated to acetaldehyde (by pyruvate decarboxylase), which is finally converted to ethanol by alcohol dehydrogenase. Some engineered cyanobacteria have been made able to directly produce ethanol (and other compounds) through photofermentation in amounts and rates that appear to be compatible with their commercial exploitation (Farias Silva and Bertucco 2016 and references therein).

Molecular Hydrogen

Some green microalgae are capable of H_2 generation, a clean fuel with H_2O as the only major by-product. As opposed to non biological production processes, bio-H_2 can be produced at ambient temperature and pressure and has no demand for metal catalysts. The matter has been excellently summarized by Eroglu and Melis (2016). We shall therefore not overly linger on this theme and simply mention that two light-dependent electron transport pathways leading to H_2 production have been identified in *Chlamydomonas reinhardtii*: one draws electrons from water lysis at photosystem II, and the other uses the reducing power allocated on quinones through their reduction by hydrogenase. Also, a light-independent fermentative pathway leading to H_2 production has been identified in *C. reinhardtii* (Eroglu and Melis 2016). Incompatibility of simultaneous O_2 and H_2 evolution from microalgae has so far hindered the development of large-scale bio-H_2 production (Rajvanshi and Sharma 2012). By using artificial miRNA (amiRNA) technology, a transgenic knockdown *C. reinhardtii* strain for the oxygen-evolving center (*OEE2* gene) was obtained; in this strain, O_2 is not released and the H_2 yield is about twofold higher than that of the wild type under similar growth conditions (Ngan et al. 2015).

Biogas
Microalgal biomass represents a potential alternative to biogas production from terrestrial crops (Dębowski et al. 2013). Photoautotrophically grown *Scenedesmus obliquus*, when used as biogas substrate, proved to produce more methane than maize silage (Wirth et al. 2015). However, a number of difficulties are associated with the use of algae for biogas production. For instance, the cell walls of some algae are resistant to anaerobic digestion and some algal strains generate compounds that are toxic to the bacteria that carry out anaerobic digestion; furthermore, in some cases, the C:N ratio of algae is unfavorable to anaerobic digestion (Dębowski et al. 2013 and references therein). This notwithstanding, the high turnover rate of algae and the possibility of selecting strains with suitable elemental stoichiometry make algae very interesting candidates for biogas production (Mussgnug et al. 2010). Some authors have also reported that when algae are mixed with traditional feedstocks they improve the efficiency of biogas production (Mussgnug et al. 2010; Zhong et al. 2012). Miao et al. (2014) showed that co-digestion of cyanobacteria with swine manure leads to an improved efficiency of both biodegradation and methane production as compared to the same processes without the addition of the algae. Zhao and Ruan (2013) also demonstrated the feasibility of adjusting the C/N ratio to increase biogas production by the addition of algae (mostly *Microcystis*) to kitchen wastes.

6.7.1.3 Challenges and Solutions for Algal Biofuel Production

Microalgal biofuel production is presently not conducted on a large scale because overwhelming investments in capital and operation are required (Chen et al. 2015b; Zhu 2015). In the case of biodiesel, for instance, the species that are known to be highly oleaginous often grow slowly. In such case, genetic manipulation may be advantageous and possibly necessary to obtain strains that can ensure sufficiently high productivity to make biodiesel production economically viable (Anandarajah et al. 2012; Iwai et al. 2014). It should also be considered that monocultures are susceptible to contamination, especially in conditions that intrinsically do not allow a tight control of the microbiota (e.g., wastewater); strains that grow slowly, such as the oleaginous ones, are especially likely to be outperformed by faster-growing competitors (Chen et al. 2015b). Mixed cultures of algae have been reported to persist in wastewater treatment systems and to be more stable and more resistant to exogenous invasion than monocultures (Chen et al. 2015b). This report would however need to be confirmed under a wider range of conditions; also, in-culture evolution (Borowitzka 2016) may have a stronger role in mixed cultures than in monospecific cultures because of the selective pressure exerted by interspecific interaction (Venuleo et al. 2017). Some algal species used in large-scale open-pond commercial production are restricted to geographic locations with warm climates and would be unable to grow at acceptable rates during the hot or cold seasons of certain geographic regions (Holbrook et al. 2014). One solution to this problem is to identify indigenous algae that are adapted to the local environment (Holbrook et al. 2014).

Lipid accumulation occurs within microalgal cells according to the general principles outlined in previous publications (Giordano 2013; Palmucci et al. 2011; Raven and Giordano 2016). Zhang et al. (2013) suggested a possible connection between the oxidative stress induced by N-shortage and neutral lipid accumulation; applications of N-limitation or starvation, however, are inefficient methods to increase lipid accumulation because they can also significantly lower biomass and lipid productivity (Borowitzka 2016). A two-stage cultivation strategy has often been proposed for the production of stress-induced algal compounds (Borowitzka (2016) and references therein), in which a full-strength medium is used to promote biomass buildup, followed by a stress treatment (e.g., N-starvation; Zhu et al. 2014) to trigger the accumulation of the target compound. Also, "mid-point" approaches (i.e., compromises between best condition for growth and production of the compound of interest) have been suggested to simplify processes and decrease production costs (Borowitzka 2016). Zhu et al. (2016) proposed a single-step approach for boosting lipid production: these authors showed that the addition of trace amounts of urea to the growth medium significantly stimulated the accumulation of neutral lipids without affecting growth rates.

6.8 Microalgae for Bioremediation

6.8.1 CO_2 Fixation and Flue Gas Treatment

Carbon is the main nutrient in microalgal cells (36–65% of dry matter). It is therefore extremely alluring to use algae to sequester CO_2, at least temporarily (Singh and Ahluwalia 2013). The frequent suggestions to utilize algae for this purpose have rarely considered the physiological nuances of the responses of algal cells to elevated CO_2. As explained earlier, the excess CO_2 may be not taken up by the cells (thus conferring no advantage and only increasing the cost of new biomass production) or it is assimilated and subsequently elicits a nutritional unbalance leading to a change in biomass quality (Beardall and Giordano 2002; Giordano and Ratti 2013; Lynn et al. 2010; Palmucci et al. 2011; Raven et al. 2011, 2012). Also, the impact of elevated CO_2 on growth rates is variable, mostly species specific, and depending on energy availability (Beardall and Giordano 2002; Raven et al. 2011, 2012; Wu et al. 2012). Nevertheless, the utilization of algal biomass for the mitigation of CO_2 emission has its merits, provided that appropriate strains and suitable culture conditions are selected and consideration is given to the fact that biomass will in the end release the CO_2 that it fixed. Liu et al. (2013) described a high-throughput screening method to rapidly identify microalgae strains that can tolerate high CO_2 condition or flue gases. Microalgae reported to tolerate high levels of CO_2 include *Chlorella* sp. (Qi et al. 2016), *Scenedesmus* sp. (Liu et al. 2013), and *Dunaliella tertiolecta* (Farrelly et al. 2013). Jacob et al. (2015) estimated that algal cultivation systems, whether they are tubular or flat photobioreactors or open ponds, can allow an effective and significant conversion of the CO_2 emitted by coal power plants into biomass. Once again, this work does not take into account the complex physiology associated with CO_2 fixation, but it does show that, at least in principle, algal cultivation can be coupled to industrial activity to minimize environmental impact. Experimental evidence (although in small-scale experiments) showed that the addition of flue gases to cultures of *Scenedesmus quadricauda* afforded a decrease by 85% v/v of CO_2, (and also 62% v/v of NO_x and 45% v/v of SO_x) in the flue gas; somewhat lower fixation capacities were obtained using *Botryococcus braunii* and *Chlorella vulgaris* (Kandimalla et al. 2016). Interestingly, the amount of fixed gases increased if the algae were cultured in mixotrophic conditions.

Flue gases contain different NO_x species, most of which are restricted by legislation and therefore must be removed (Van Den Hende et al. 2012). NO_x may serve as a nitrogen source for microalgae cultivation (Chen et al. 2016; Raven and Giordano 2016; Zhang et al. 2014a). Thus, "denoxification" (DeNOx) by microalgae (biodenox) may be a worthy contribution to flue gas treatment. As the efficiency of NO_x removal by microalgae varies dramatically among species, it is necessary to select or genetically modify suitable algal candidates for this purpose. Some strains of the genera *Chlorella*, *Scenedesmus*, and *Dunaliella* have been reported to significantly remove NO_x (Jin et al. 2008; Nagase et al. 2001; Santiago et al. 2010; Kandimalla et al. 2016), although high levels of NO_x tend to depress photosynthesis. In a typical flue gas from incineration processes, about 90–95% of the NO_x is given by NO (Fritz and Pitchon 1997). When NO dissolves in water, it is oxidized to nitrite and nitrate (Niu and Leung 2010). Nitrite has an inhibitory effect on algal growth, which is exerted through a retardation of electron transfer from Q_A to Q_B (Q_A is a bound quinone; Q_B is a quinone that binds and unbinds to photosystem II), and by interference with the donor side of PSII (Zhang et al. 2017). The screening of nitrite-tolerant microalgae species is therefore crucial for the use of algae in $DeNO_x$ approaches. Li et al. (2016) analyzed numerous *Chlorella* strains in this respect and found that the degree of nitrite tolerance was a strain-specific feature, although most *Chlorella* strains showed the ability to withstand high concentrations of nitrite. The nitrate and nitrite generated by the dissolution of NO in water can be directly assimilated by algae (Giordano and Raven 2014; Raven and Giordano 2016); NO dissolution is however rather slow and often limits the rate of combined nitrogen assimilation. Zhang et al. (2014a) reported on a two-step microalgal bio-$DeNO_x$ roadmap, in which NO_x-rich flue gases were first fixed, mostly as nitrite, to flue gas fixed salts (FGFS), and then used as nitrogen source for *Chlorella* sp. cultures. By using FGFS with NO_2^- equivalent to 5-fold that in the common culture medium BG11 (Stanier et al. 1971), up to 60% v/v of the NO_x was removed from the medium with an inoculated cell density of 0.07 g DW L^{-1}, together with the production of 33% algae lipids (Zhang et al. 2014a). The mixotrophic cultivation of *Chlorella* sp. with FGFSs and glucose achieved an overall $DeNO_x$ efficiency of 96%, demonstrating the feasibility and practicality of efficient biological $DeNO_x$ by microalgae (Chen et al. 2016).

In most incineration flue gases, SO_x are also present; they mainly consist of SO_2, with a minor contribution (2–4% v/v) by SO_3; both SO_2 and SO_3 are highly soluble in water; SO_2 tends to hydrate to H_2SO_3, which dissociates in protons and sulfite (at pH >6) and bisulfite (prominent between pH 2 and 6); SO_3 hydrates to H_2SO_4, which typically dissociates in protons and sulfate (SO_4^{2-}); SO_4^{2-} tends to prevail at pH >1.9; also the oxidation of H_2SO_3 can generate H_2SO_4 and SO_4^{2-} (Stumm and Morgan 1981; Van Den Hende et al. 2012 and references therein). The dissolution of SO_x, therefore, causes acidification of the medium, the extent of which depends on the SO_x content of the flue gas, which is a function of the combustion substrates from which it was generated. The consequence of SO_x dissolution in the growth medium can be such to limit the choice of algae to acidophilic and/or bisulfite-tolerant strains (see Van Den Hende et al. 2012 and references therein for details); in some cases, scrubbing SO_x from the flue gas may be a precondition for any microalgal treatment. If acidity and toxicity of SO_x-derived solutes do not prevent algal survival, algae can assimilate substantial amounts of SO_4^{2-} (Norici et al. 2005; Ratti et al. 2011; Giordano and Raven 2014; Prioretti and Giordano 2016), compatibly with elemental stoichiometry in the growth medium and stoichiometric constraints of cell growth (Giordano 2013).

6.8.2 Wastewater Treatment by Microalgae Cultivation

Large-scale microalgae culture may compete with crops and human activities with respect to water usage. Large amounts of nitrogen and phosphorus are also required, and their cost is high (Lardon et al. 2009). Both water and nutrients can be obtained from wastewaters; culturing algae in wastewaters also affords obvious environmental benefits. Microalgae are very effective at removing nitrogen, phosphorus, and toxic metals from wastewaters, producing cleaner effluents with high concentrations of dissolved oxygen (Gomez et al. 2013). Cabanelas et al. (2013) used *Chlorella vulgaris* for nitrogen and phosphorus removal from municipal wastewater with the highest removal rates of 9.8 (N) and 3.0 (P) mg l^{-1} $days^{-1}$. Some studies also reported on the cultivation of microalgae in sewage under mixotrophic conditions. Cheng et al. (2013) found that mixotrophic microalga–bacteria systems significantly promoted algal growth and nutrient removal efficiency; maximal biomass and lipid productivity was attained when the alga *Desmodesmus* sp. CHX1 was used to treat piggery wastewater. Moreover, the co-culture of microalgae and bacteria in wastewater was reported to obtain 50–60% and 68–81% dissolved organic carbon (DOC) removal efficiency from municipal and industrial wastewater mixtures, respectively (Nielsen 2015). Zhou et al. (2012) developed an effective organo-photolithotrophic system for improved wastewater nutrient removal, wastewater recycling, and enhanced algal lipid accumulation with *Auxenochlorella protothecoides* UMN280. Carbohydrate-rich and nitrogen-deficient solid wastes and some food industry wastewaters, such as olive mill wastewater, can be also used for hydrogen production (Keskin et al. 2011). It was reported that photosynthetic H_2 evolution from *C. reinhardtii* grown in advanced solid-state fermentation wastewater was increased by more than 700% compared to the cells grown in TAP medium (Chen et al. 2014). A study was also carried out to evaluate the potential of the green alga *Scenedesmus obliquus* grown in different concentrations of wastewater to produce biomass rich in sugar to produce bioethanol by fermentation processes; it was found that the highest removal efficiency of biological oxygen demand (BOD) and chemical oxygen demand (COD) were 18% for *S. obliquus* grown under aeration conditions and that the highest ethanol efficiency of biomass hydrolysate was 20.33% (Hamouda et al. 2016). Also, biomethane production in digesters could be improved by the addition of microalgae biomass harvested from algae-based swine wastewater digestate (Perazzoli et al. 2016). Because of the complex nature of wastewaters, issues such as contamination, inconsistent wastewater components, and unstable biomass production hinder efforts to use wastewater for large-scale algal cultivation (Cai et al. 2013).

The combination of CO_2 and/or NO_x fixation from flue gases and nutrient removal from wastewaters may provide a very promising alternative to current bioremediation strategies; the concomitant supply of nutrients from the gas and the liquid phase synergistically increases the effectiveness of depuration by algae (Chen et al. 2015b) and also stimulates algal growth and accumulation of added-value metabolic products (e.g., lipids) within the cells (Devi and Mohan 2012). Chinnasamy et al. (2010) cultured *Chlamydomonas globosa*, *Chlorella minutissima*, and *Scenedesmus bijuga* in untreated wastewater from the carpet industry to which a gas stream containing 5–6% v/v CO_2 was added; biomass productivity reached 5.9–21.1 g m^{-2} day^{-1}. The cyanobacterium *Aphanothece microscopica* Nägeli cultivated in a photobioreactor using supplemented wastewater from an oil refinery was found to assimilate CO_2 when light was present; the capacity for CO_2 sequestration was lowered by one fourth when the algae were cultured in a light/dark photoperiod rather than under continuous light (Jacob-Lopes et al. 2010). The other important finding of this study was that only a small portion (about 3% v/v) of the CO_2 sequestered during cultivation was in the end effectively fixed in algal biomass, whereas the rest was probably released as biopolymers or volatile organic compounds. This finding is a warning about the direct extrapolation to commercial application of physiological studies that do not include a thorough analysis of biomass.

Recently, the use of microalgae for the concomitant remediation of environmental pollution and biofuel production has also been proposed; this would allow decreasing energy, nutrients, water cost, and also CO_2 emissions (Chen et al. 2015b; Sun et al. 2013), making biofuel production from microalgae more environmentally sustainable, cost-effective, and profitable (Chen et al. 2015b; Nayak et al. 2016) (Fig. 6.3).

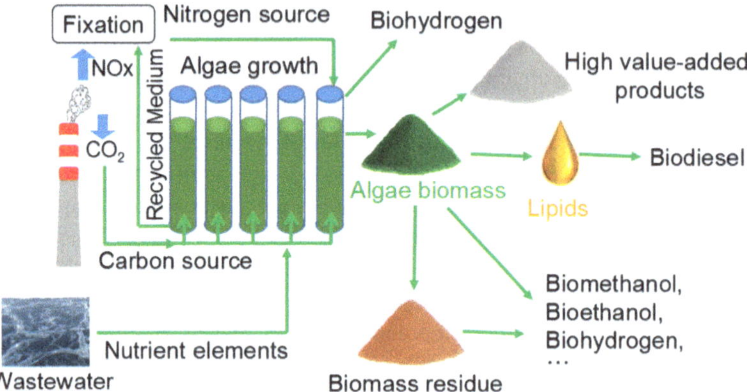

Fig. 6.3 Flowchart of the combination of environmental pollution control and biofuel production

Chlorogonium sp. showed good potential in the simultaneous purification of saline sewage effluent and CO_2 sequestration while delivering feedstock for potential biofuel production in a waste-recycling manner, achieving high removal efficiencies of NH_3-N, NO_3^--N, TN, and PO_4^{3-}-P, at a CO_2 consumption rate of 58.96 mg l^{-1} day^{-1}, and lipid content of 24.26% m/m of the algal biomass (Lee et al. 2015). An economically viable algal biofuel-based $DeNO_x$ process using *Chlorella* was evaluated and verified in actual industrial flue gas condition by Zhang et al. (2014a). To reduce the mismatch between the large amount of NO_x contained in flue gases and the relatively low capacity for its assimilation in photolithotrophic algal growth, the possibility of managing NO_x by culturing oil-producing *Chlorella* strains mixotrophically was tested (Chen et al. 2016). After a stepwise optimization of mixotrophic cultivation of *Chlorella* using FGFS, an impressive $DeNO_x$ efficiency of more than 96%, with a biomass productivity of 9.87 g l^{-1} day^{-1} and a high lipid productivity of 1.83 g l^{-1} day^{-1}, were obtained.

6.9 Microalgal Cultivation for Food or Feed Production

The large and increasing demand for animal feed exerts a tremendous pressure on food crops, because, on a purely economic basis, the conversion of land use from crops (for humans) to animal feed production is more profitable; this trend is, however, in obvious conflict with the need to support the increasing human population on our planet. Microalgae can be effectively and conveniently used as animal feed (Norambuena et al. 2015; Packer et al. 2016; Tibbetts et al. 2017; Vidyashankar et al. 2015); furthermore, their cultivation poses minimal or no threat to crop production (Vidyashankar et al. 2015). The nutritional and bioactive effects of microalgal biomass have been assessed in a variety of studies (Benemann 2013; Wells et al. 2017). The composition of algae, whose cells are rich in carotenoids and other antioxidants, essential polyunsaturated fatty acids, minerals, and protein with a balanced amino acid profile, makes them an excellent alternative to conventional feedstocks such as corn, soya, barley, and skimmed milk (Shields and Lupatsch 2012). There are several examples of the utilization of algae for animal feed: *Chlorella vulgaris* has been used for the development of pet and fish feed (Groza et al. 1966; Li et al. 2015); *Dunaliella* can be used directly as a nutritional additive for fish or for secondary biological baits (such as rotifers and *Artemia*) and other aquacultured animals (Del Campo et al. 2007; Elbermawi 2009); *Spirulina maxima* has been used in swine feed (Saeid et al. 2013). Several studies have suggested that small amounts (2.5–10% of the diet) of algae in fish diets result in higher growth rates, feed utilization efficiency, carcass quality, physiological activity, intestinal microbiota, disease resistance, stress response, modulation of lipid metabolism, and protein retention during periods of reduced feed intake, and also lead to a higher palatability in sea urchin formulated feed (Cyrus et al. 2015; Valente et al. 2006; Nakagawa 1997; Norambuena et al. 2015; Wassef et al. 2005).

The widespread and growing interest in algae as food and food complements for humans emerges clearly in the recent literature (Cottin et al. 2011; Hafting et al. 2015; Harnedy and Fitzgerald 2011; Knies 2017; Packer et al. 2016; Pangestuti and Kim 2011; Sinéad et al. 2011; Wells et al. 2017). Limiting our *excursus* to microalgae (see Packer et al. 2016 for a panorama on macroalgae used for food), numerous species have been traditionally grown or have been collected as food: *Nostoc sphaeroides*, for instance, is a edible cyanobacteria widely cultivated in Hubei Province, China (Yi et al. 2016); there is evidence that in Central America the Aztecs were already eating cyanobacteria (*Spirulina*) collected from lakes in the fifteenth century; populations inhabiting the banks of Chad Lake, in Africa, have also traditionally used the cyanobacterium *Arthrospira* (formerly *Spirulina*) (Reed et al. 1985). *Arthrospira* was possibly the first microalga that spread widely across the shelves of supermarkets and "natural food" shops; this species encountered the favor of consumers for its rich protein, linolenic acid, and phycocyanin content. Nowadays, China is the main producer of *Spirulina* in the world (Lu et al. 2011). Also, *Dunaliella salina* encountered substantial success by its high content of β-carotene, an antioxidant in its own right and a precursor of vitamin A, with the first large production plants becoming operative in the 1980s in Israel, Australia, and the USA (Borowitzka 2016 and references therein). In more recent years, the fad of natural nutritional complements has facilitated the expansion of the market for nutritional products from algae, to which, with variably sound scientific bases, antioxidant, antibacterial, antiinflammatory, antiviral, and anti-cancer functions, for example, have been attributed (Wells et al. 2017). The consumption of microalgae as food has also, to some extent, been driven by the producers, who with the decline of profitability in algal biofuels have looked for alternative uses of their biomass (Packer et al. 2016). More species are now cultured in large-scale plants for the production and commercialization of β-carotene, astaxanthin, phycocyanin, some fatty acids (including Ω-3 and Ω-6), and other bioactive substances (Borowitzka 2016). *Chlorella* has been marketed as a health food because of its alleged ability to stimulate the human immune system; and its production is mainly distributed in

Japan, China, France, Portugal, and South Korea (An et al. 2008; Liu and Hu 2013; Saad et al. 2006). The green alga *Haematococcus pluvialis* can accumulate carotenoids, mostly astaxanthin and its ester derivatives, when subject to nutrient limitation, high temperature, or excessive light (Borowitzka 2016); these compounds have a high value on the market as antioxidants (Bagchi et al. 2001; Hagen and Grunewald 2001). The mass culture of *H. pluvialis* is mainly concentrated in Japan, Israel, and the USA (Gómez et al. 2013). Also, some heterotrophic species, such as the dinoflagellate *Crypthecodinium cohnii* and the labyrinthulid *Ulkenia*, have been used for production of docosahexaenoic acid (DHA), which has been proposed as a baby food additive (Ganuza et al. 2008; Lee Chang et al. 2014).

6.10 Conclusions

In the light of all these facts, it seems fair to conclude that, although some applications of microalgal cultivation appear not to be economically sustainable, at this point in time (e.g., the still fashionable use of these organisms for the sole production of biofuels), the use of microalgae in large-scale multifunctional plants is feasible and promising. Also, the direct use of algal biomass for human and animal nutrition or for the production of nutritional complements appears to have a positive outlook in terms of market demand and economic sustainability. However, further studies must be conducted, both on the engineering aspects of large-scale algal culturing systems and, possibly more importantly, on the specific challenges that industrial applications pose to algal physiology (e.g., responses to high CO_2, NO_x, and SO_x concentration, temperature, and low light penetration; C allocation under different growth regimes) and on the functional diversity of algae, which has been only marginally explored.

Acknowledgments MG applicative research has been funded by Fondi di Ricerca di Anteneo 2013–2017 and Fondi Strategici di Ateneo 2017.

References

Adam F, Abert-Vian M, Peltier G, Chemat F (2012) "Solvent-free" ultrasound-assisted extraction of lipids from fresh microalgae cells: a green, clean and scalable process. Bioresour Technol 114:457–465

Ahmad AL, Yasin NHM, Derek CJC, Lim JK (2011) Microalgae as a sustainable energy source for biodiesel production: a review. Renew Sust Energ Rev 15(1):584–593. https://doi.org/10.1016/j.rser.2010.09.018

Amaro HM, Guedes AC, Malcata FX (2011) Advances and perspectives in using microalgae to produce biodiesel. Appl Energy 88(10):3402–3410. https://doi.org/10.1016/j.apenergy.2010.12.014

An HJ, Rim HK, Lee JH, Seo MJ, Hong JW, Kim NH, Myung NY, Moon PD, Choi IY, Na HJ (2008) Effect of *Chlorella vulgaris* on immune-enhancement and cytokine production in vivo and in vitro. Food Sci Biotechnol 17(5):953–958

Anandarajah K, Mahendraperumal G, Sommerfeld M, Hu Q (2012) Characterization of microalga *Nannochloropsis* sp. mutants for improved production of biofuels. Appl Energy 96(0):371–377. https://doi.org/10.1016/j.apenergy.2012.02.057

Bagchi D, Garg A, Krohn RL, Bagchi M, Tran MX, Stohs SJ (2001) Oxygen free radical scavenging abilities of vitamins C, E, β-carotene, pycnogenol, grape seed proanthocyanidin extract and astaxanthins in vitro. Res Commun Mol Pathol Pharmacol 95(2):179–189

Bartel DP (2009) MicroRNAs: target recognition and regulatory functions. Cell 136(2):215

Beardall J, Giordano M (2002) Ecological implications of microalgal and cyanobacterial CO_2 concentrating mechanisms, and their regulation. Funct Plant Biol 29(2-3):335–347

Beardall J, Raven JA (2016) Carbon acquisition by microalgae. In: The physiology of microalgae. Springer, Cham

Behrebs P (2005) Photobioreactors and fermenters: the light and dark sides of growing algae. In: Andersen RA (ed) Algal culturing techniques. Academic Press, Burlington

Ben-Amotz A (2004) Industrial production of microalgal cell-mass and secondary products major industrial species. In: Richmond A (ed) Handbook of microalgal culture biotechnology and applied phycology. Blackwell, Oxford, pp 273–280

Benemann JR (2013) Microalgae for biofuels and animal feeds. Biotechnol Bioeng 110(9):2319–2321

Berla BM, Rajib S, Immethun CM, Maranas CD, Seok MT, Pakrasi HB (2013) Synthetic biology of cyanobacteria: unique challenges and opportunities. Front Microbiol 4:246

Beuckels A, Depraetere O, Vandamme D, Foubert I, Smolders E, Muylaert K (2013) Influence of organic matter on flocculation of *Chlorella vulgaris* by calcium phosphate precipitation. Biomass Bioenergy 54:107–114

Bigelow TA, Xu J, Stessman DJ, Yao L, Spalding MH, Wang T (2014) Lysis of *Chlamydomonas reinhardtii* by high-intensity focused ultrasound as a function of exposure time. Ultrason Sonochem 21(3):1258–1264

Bibia R, Ahmad Z, Imran M, Hussain S, Ditta A, Mahmood S, Khalid A (2017) Algal bioethanol production technology: a trend towards sustainable development. Renew Sustain Energy Rev 7:976–985

Blank CE, Sanchez-Baracaldo P (2010) Timing of morphological and ecological innovations in the cyanobacteria–a key to understanding the rise in atmospheric oxygen. Geobiology 8(1):1–23

Borowitzka MA (1999) Commercial production of microalgae: ponds, tanks, tubes and fermenters. J Biotechnol 70(1-3):313–321

Borowitzka MA (2016) Algal physiology and large-scale outdoor cultures of microalgae. In: The physiology of microalgae. Springer, Cham

Bowes G, Ogren WL, Hageman RH (1971) Phosphoglycolate production catalyzed by ribulose diphosphate carboxylase. Biochem Biophys Res Commun 45(3):716–722

Brennan L, Owende P (2010) Biofuels from microalgae—a review of technologies for production, processing, and extractions of biofuels and co-products. Renew Sust Energ Rev 14(2):557–577

Brodie J, Chan CX, De Clerck O, Cock JM, Coelho SM, Gachon C, Grossman AR, Mock T, Raven JA, Smith AG, Yoon HS, Bhattacharya D (2017) The algal revolution. Trends Plant Sci. https://doi.org/10.1016/j.tplants.2017.05.005

Cabanelas ITD, Ruiz J, Arbib Z, Chinalia FA, Garrido-Perez C, Rogalla F, Nascimento IA, Perales JA (2013) Comparing the use of different domestic wastewaters for coupling microalgal production and nutrient removal. Bioresour Technol 131:429–436. https://doi.org/10.1016/j.biortech.2012.12.152

Cai T, Park SY, Li Y (2013) Nutrient recovery from wastewater streams by microalgae: status and prospects. Renew Sust Energ Rev 19(0):360–369. https://doi.org/10.1016/j.rser.2012.11.030

Chen Y, Wang J, Zhang W, Chen L, Gao L, Liu T (2013) Forced light/dark circulation operation of open pond for microalgae cultivation. Biomass Bioenergy 56(56):464–470

Chen M, Zhang L, Li S, Chang S, Wang W, Zhang Z, Wang J, Gang Z, Qi L, Xu W (2014) Characterization of cell growth and photobiological H_2 production of *Chlamydomonas reinhardtii* in ASSF industry wastewater. Int J Hydrog Energy 39(25):13462–13467

Chen H, Hu J, Qiao Y, Chen W, Rong J, Zhang Y, He C, Wang Q (2015a) Ca^{2+}-regulated cyclic electron flow supplies ATP for nitrogen starvation-induced lipid biosynthesis in green alga. Sci Rep 5:15117. https://doi.org/10.1038/srep15117. http://www.nature.com/articles/srep15117#supplementary-information

Chen H, Qiu T, Rong J, He C, Wang Q (2015b) Microalgal biofuel revisited: an informatics-based analysis of developments to date and future prospects. Appl Energy 155:585–598. https://doi.org/10.1016/j.apenergy.2015.06.055

Chen WX, Zhang SS, Rong JF, Li X, Chen H, He CL, Wang Q (2016) Effective biological $DeNO_x$ of industrial flue gas by the mixotrophic cultivation of an oil-producing green alga *Chlorella* sp. C2. Environ Sci Technol 50(3):1620–1627. https://doi.org/10.1021/acs.est.5b04696

Cheng HX, Tian GM, Liu JZ (2013) Enhancement of biomass productivity and nutrients removal from pretreated piggery wastewater by mixotrophic cultivation of *Desmodesmus* sp. CHX1. Desalin Water Treat 51(37-39):7004–7011. https://doi.org/10.1080/19443994.2013.769917

Chinnasamy S, Bhatnagar A, Hunt RW, Das KC (2010) Microalgae cultivation in a wastewater dominated by carpet mill effluents for biofuel applications. Bioresour Technol 101(9):3097–3105

Chojnacka K, Marquez-Rocha FJ (2004) Kinetic and stoichiometric relationships of the energy and carbon metabolism in the culture of microalgae. Biotechnology 3:21–34

Chojnacka K, Chojnacki A, Górecka H (2004) Trace element removal by *Spirulina* sp. from copper smelter and refinery effluents. Hydrometallurgy 73(1):147–153

Chung BY, Deery MJ, Groen AJ, Howard J, Baulcombe DC (2016) mRNA turnover through CDS-targeting is the primary role of miRNA in the green alga *Chlamydomonas*. bioRxiv. https://doi.org/10.1101/088807

Coons JE, Kalb DM, Dale T, Marrone BL (2014) Getting to low-cost algal biofuels: a monograph on conventional and cutting-edge harvesting and extraction technologies. Algal Res 6:250–270

Cooper MS, Hardin WR, Petersen TW, Cattolico RA (2010) Visualizing "green oil" in live algal cells. J Biosci Bioeng 109(2):198–201. https://doi.org/10.1016/j.jbiosc.2009.08.004

Cottin SC, Sanders TA, Hall WL (2011) The differential effects of EPA and DHA on cardiovascular risk factors. Proc Nutr Soc 70(2):215

Cyrus MD, Bolton JJ, Scholtz R, Macey BM (2015) The advantages of *Ulva* (Chlorophyta) as an additive in sea urchin formulated feeds: effects on palatability, consumption and digestibility. Aquacult Nutr 21(5):578–591

Daboussi F, Leduc S, Marechal A, Dubois G, Guyot V, Perez-Michaut C, Amato A, Falciatore A, Juillerat A, Beurdeley M, Voytas DF, Cavarec L, Duchateau P (2014) Genome engineering empowers the diatom *Phaeodactylum tricornutum* for biotechnology. Nat Commun 5:3831

Dang-Thuan T, Chen CL, Chang JS (2013) Effect of solvents and oil content on direct transesterification of wet oil-bearing microalgal biomass of *Chlorella vulgaris* ESP-31 for biodiesel synthesis using immobilized lipase as the biocatalyst. Bioresour Technol 135:213–221. https://doi.org/10.1016/j.biortech.2012.09.101

de Vargas C, Audic S, Henry N, Decelle J, Mahe F, Logares R, Lara E, Berney C, Le Bescot N, Probert I, Carmichael M, Poulain J, Romac S, Colin S, Aury JM, Bittner L, Chaffron S, Dunthorn M, Engelen S, Flegontova O, Guidi L, Horak A, Jaillon O, Lima-Mendez G, Lukes J, Malviya S, Morard R, Mulot M, Scalco E, Siano R, Vincent F, Zingone A, Dimier C, Picheral M, Searson S, Kandels-Lewis S, Acinas SG, Bork P, Bowler C, Gorsky G, Grimsley N, Hingamp P, Iudicone D, Not F, Ogata H, Pesant S, Raes J, Sieracki ME, Speich S, Stemmann L, Sunagawa S, Weissenbach J, Wincker P, Karsenti E, Coordinators TO (2015) Eukaryotic plankton diversity in the sunlit ocean. Science 348(6237). doi:https://doi.org/10.1126/science.1261605

Dębowski M, Zieliński M, Grala A, Dudek M (2013) Algae biomass as an alternative substrate in biogas production technologies: review. Renew Sust Energ Rev 27:596–604

Del Campo JA, García-González M, Guerrero MG (2007) Outdoor cultivation of microalgae for carotenoid production: current state and perspectives. Appl Microbiol Biotechnol 74(6):1163–1174

Demirbas A (2009) Progress and recent trends in biodiesel fuels. Energy Convers Manag 50(1):14–34. https://doi.org/10.1016/j.enconman.2008.09.001
Devi MP, Mohan SV (2012) CO_2 supplementation to domestic wastewater enhances microalgae lipid accumulation under mixotrophic microenvironment: effect of sparging period and interval. Bioresour Technol 112:116–123. https://doi.org/10.1016/j.biortech.2012.02.095
Domenighini A, Giordano M (2009) Fourier transform infrared spectroscopy of microalgae as a novel tool for biodiversity studies, species identification, and the assessment of water quality. J Phycol 45(2):522–531
Doucha J, Lívanský K (2006) Productivity, CO_2/O_2 exchange and hydraulics in outdoor open high density microalgal (*Chlorella* sp.) photobioreactors operated in a middle and southern European climate. J Appl Phycol 18(6):811–826
Doucha J, Lívanský K (2008) Influence of processing parameters on disintegration of *Chlorella* cells in various types of homogenizers. Appl Microbiol Biotechnol 81(3):431–440
Elbermawi NM (2009) Using *Dunaliella salina* extract to improve survival, stress tolerance and growth performance of freshwater prawn *Macrobrachium rosenbergii*. Egypt J Nutr Feeds 12(1):113–125
Elser JJ, Acquisti C, Kumar S (2011) Stoichiogenomics: the evolutionary ecology of macromolecular elemental composition. Trends Ecol Evol 26(1):38–44
Emanuel SL, Francisco OC, Aurora G, Emilio F, Amaury DM (2016) Characterization of a mutant deficient for ammonium and nitric oxide signalling in the model system *Chlamydomonas reinhardtii*. PLoS One 11(5):e0155128
Enzing C, Ploeg M, Barbosa M, Sijtsma L (2014) Microalgae-based products for the food and feed sector: an outlook for Europe. JRC Scientific and Policy Reports. Report EUR 26255 EN. European Union 2014. doi:https://doi.org/10.2791/3339
Eroglu E, Melis A (2016) Microalgal hydrogen production research. Int J Hydrog Energy 41(30):12772–12798. https://doi.org/10.1016/j.ijhydene.2016.05.115
Fanesi A, Raven JA, Giordano M (2014) Growth rate affects the responses of the green alga *Tetraselmis suecica* to external perturbations. Plant Cell Environ 37(2):512–519
Farrelly DJ, Brennan L, Everard CD, McDonnell KP (2014) Carbon dioxide utilisation of *Dunaliella tertiolecta* for carbon bio-mitigation in a semicontinuous photobioreactor. Appl Microbiol Biotechnol. https://doi.org/10.1007/s00253-013-5322-y
Farias Silva CE, Bertucco A (2016) Bioethanol from microalgae and cyanobacteria: a review and technological outlook. Process Biochem 51:1833–1842
Flynn KJ, Raven JA, Rees TAV, Finkel Z, Quigg A, Beardall J (2010) Is the growth rate hypothesis applicable to microalgae? J Phycol 46(1):1–12
Frigon J-C, Matteau-Lebrun F, Hamani Abdou R, McGinn PJ, O'Leary SJB, Guiot SR (2013) Screening microalgae strains for their productivity in methane following anaerobic digestion. Appl Energy 108(0):100–107. https://doi.org/10.1016/j.apenergy.2013.02.051
Fritz A, Pitchon V (1997) The current state of research on automotive lean NO_x catalysis. Appl Catal B Environ 13(1):1–25
Ganuza E, Benítez-Santana T, Atalah E, Vega-Orellana O, Ganga R, Izquierdo MS (2008) *Crypthecodinium cohnii* and *Schizochytrium* sp. as potential substitutes to fisheries-derived oils from seabream (*Sparus aurata*) microdiets. Aquaculture 277(1-2):109–116
Geider RJ, La Roche J (2002) Redfield revisited: variability of C:N:P in marine microalgae and its biochemical basis. Eur J Phycol 37(1):1–17
Gentil J, Hempel F, Moog D, Zauner S, Maier UG (2017) Origin of complex algae by secondary endosymbiosis: a journey through time. Protoplasma 254:1835–1843
Giordano M (2013) Homeostasis: an underestimated focal point of ecology and evolution. Plant Sci 211:92–101
Giordano M, Ratti S (2013) The biomass quality of algae used for CO_2 sequestration is highly species-specific and may vary over time. J Appl Phycol 25(5):1431–1434
Giordano M, Kansiz M, Heraud P, Beardall J, Wood B, McNaughton D (2001) Fourier transform infrared spectroscopy as a novel tool to investigate changes in intracellular macromolecular pools in the marine microalga *Chaetoceros muellerii* (Bacillariophyceae). J Phycol 37(2):271–279

Giordano M, Beardall J, Raven JA (2005) CO_2 concentrating mechanisms in algae: mechanisms, environmental modulation, and evolution. Annu Rev Plant Biol 56:99–131
Giordano M, Palmucci M, Norici A (2015a) Taxonomy and growth conditions concur to determine the energetic suitability of algal fatty acid complements. J Appl Phycol 27(4):1401–1413
Giordano M, Palmucci M, Raven JA (2015b) Growth rate hypothesis and efficiency of protein synthesis under different sulphate concentrations in two green algae. Plant Cell Environ 38(11):2313–2317
Giordano M, Raven JA (2014) Nitrogen and sulfur assimilation in plants and algae. Aquat Bot 118:45–61. https://doi.org/10.1016/j.aquabot.2014.06.012
Giordano M, Norici A, Beardall J (2017) Impact of inhibitors of amino acid, protein, and RNA synthesis on C allocation in the diatom *Chaetoceros muellerii*: a FTIR approach. Algae 32(2):161–170. https://doi.org/10.4490/algae.2017.32.6.6
Gómez PI, Inostroza I, Pizarro M, Pérez J (2013) From genetic improvement to commercial-scale mass culture of a Chilean strain of the green microalga *Haematococcus pluvialis* with enhanced productivity of the red ketocarotenoid astaxanthin. AoB Plants 5(1):plt026
Gomez C, Escudero R, Morales MM, Figueroa FL, Fernandez-Sevilla JM, Acien FG (2013) Use of secondary-treated wastewater for the production of *Muriellopsis* sp. Appl Microbiol Biotechnol 97(5):2239–2249. https://doi.org/10.1007/s00253-012-4634-7
Gressel J (2008) Transgenics are imperative for biofuel crops. Plant Sci 174(3):246–263. https://doi.org/10.1016/j.plantsci.2007.11.009
Grobbelaar JU (2009) From laboratory to commercial production: a case study of a *Spirulina* (*Arthrospira*) facility in Musina, South Africa. J Appl Phycol 21(5):523–527
Groza I, Boldijar A, Vlad G (1966) *Chlorella vulgaris* an important source of protein and vitamins in animal feeding. Revista Zooteh Med Vet 7:24–26
Hafting JT, Craigie JS, Stengel DB, Loureiro RR, Buschmann AH, Yarish C, Edwards MD, Critchley AT (2015) Prospects and challenges for industrial production of seaweed bioactives. J Phycol 51(5):821
Hagen C, Grunewald K (2001) Compartmentation of astaxanthin biosynthesis in *Haematococcus pluvialis*. Sci Access 3(1). https://doi.org/10.1071/SA0403037
Hamouda RA, Yeheia DS, Hussein MH, Hamzah HA (2016) Removal of heavy metals and production of bioethanol by green alga *Scenedesmus obliquus* grown in different concentrations of wastewater. Sains Malays 45(3):467–476
Harnedy PA, Fitzgerald RJ (2011) Bioactive proteins, peptides, and amino acids from macroalgae. J Phycol 47(2):218
Hellingwerf KJ, De Mattos MJT (2009) Alternative routes to biofuels: light-driven biofuel formation from CO2 and water based on the 'photanol' approach. J Biotechnol 142:87–90
Ho TY, Quigg A, Finkel ZV, Milligan AJ, Wyman K, Falkowski PG, Morel FMM (2010) The elemental composition of some marine phytoplankton. J Phycol 39(6):1145–1159
Holbrook GP, Davidson Z, Tatara RA, Ziemer NL, Rosentrater KA, Scott Grayburn W (2014) Use of the microalga *Monoraphidium* sp. grown in wastewater as a feedstock for biodiesel: cultivation and fuel characteristics. Appl Energy 131(0):386–393. https://doi.org/10.1016/j.apenergy.2014.06.043
Hopes A, Nekrasov V, Kamoun S, Mock T (2016) Editing of the urease gene by CRISPR-Cas in the diatom *Thalassiosira pseudonana*. Plant Methods 12:49
Huang CH, Shen CR, Li H, Sung LY, Wu MY, Hu YC (2016) CRISPR interference (CRISPRi) for gene regulation and succinate production in cyanobacterium *S. elongatus* PCC 7942. Microb Cell Factories 15(1):196
Iwai M, Ikeda K, Shimojima M, Ohta H (2014) Enhancement of extraplastidic oil synthesis in *Chlamydomonas reinhardtii* using a type-2 diacylglycerol acyltransferase with a phosphorus starvation-inducible promoter. Plant Biotechnol J 12(6):808–819. https://doi.org/10.1111/Pbi.12210
Jacob A, Xia A, Murphy JD (2015) ChemInform abstract: a perspective on gaseous biofuel production from micro-algae generated from CO_2 from a coal-fired power plant. ChemInform 148(51):396–402

Jacob-Lopes E, Scoparo CHG, Queiroz MI, Franco TT (2010) Biotransformations of carbon dioxide in photobioreactors. Energy Convers Manag 51(5):894–900

Jebsen C, Norici A, Wagner H, Palmucci M, Giordano M, Wilhelm C (2012) FTIR spectra of algal species can be used as physiological fingerprints to assess their actual growth potential. Physiol Plant 146(4):427–438

Jin H-F, Santiago DEO, Park J, Lee K (2008) Enhancement of nitric oxide solubility using Fe(II)-EDTA and its removal by green algae *Scenedesmus* sp. Biotechnol Bioprocess Eng 13(1):48–52. https://doi.org/10.1007/s12257-007-0164-z

Jones RI (2000) Mixotrophy in planktonic protists: an overview. Freshw Biol 45(2):219–226

Kaffes N, Thoms S, Trimborn S, Rost B, Richter K-U, Köhler A, Norici A, Giordano M (2010) Carbon and nitrogen fluxes in the marine coccolithophore Emiliania huxleyi grown under different nitrate concentrations. J Exp Mar Biol Ecol 393:1–8. https://doi.org/10.1016/j.jembe.2010.06.004

Kanda H, Li P, Yoshimura T, Okada S (2013) Wet extraction of hydrocarbons from *Botryococcus braunii* by dimethyl ether as compared with dry extraction by hexane. Fuel 105:535–539. https://doi.org/10.1016/j.fuel.2012.08.032

Kandimalla P, Desi S, Vurimindi H (2016) Mixotrophic cultivation of microalgae using industrial flue gases for biodiesel production. Environ Sci Pollut Res 23(10):9345–9354

Kapdan IK, Kargi F (2006) Bio-hydrogen production from waste materials. Enzyme Microb Technol 38(5):569–582. https://doi.org/10.1016/j.enzmictec.2005.09.015

Keskin T, Abo-Hashesh M, Hallenbeck PC (2011) Photofermentative hydrogen production from wastes. Bioresour Technol 102(18):8557–8568. https://doi.org/10.1016/j.biortech.2011.04.004

Kilian O, Benemann CSE, Niyogi KK, Vick B (2011) High-efficiency homologous recombination in the oil-producing alga *Nannochloropsis* sp. Proc Natl Acad Sci USA 108(52):21265–21269

Knies JM (2017) Algae and algal products as novel foods. Ernahr Umsch 64(2):M84–M93

Knoll AH (2003) Biomineralization and evolutionary history. Rev Mineral Geochem 54(1):329–356. https://doi.org/10.2113/054032

Kumar A, Ergas S, Yuan X, Sahu A, Zhang Q, Dewulf J, Malcata FX, van Langenhove H (2010) Enhanced CO_2 fixation and biofuel production via microalgae: recent developments and future directions. Trends Biotechnol 28(7):371–380. https://doi.org/10.1016/j.tibtech.2010.04.004

Kusakabe T, Tatsuke T, Tsuruno K, Hirokawa Y, Atsumi S, Liao JC, Hanai T (2013) Engineering a synthetic pathway in cyanobacteria for isopropanol production directly from carbon dioxide and light. Metab Eng 20(5):101

Lam MK, Lee KT (2012) Microalgae biofuels: a critical review of issues, problems and the way forward. Biotechnol Adv 30(3):673–690

Lam MK, Lee KT, Mohamed AR (2010) Homogeneous, heterogeneous and enzymatic catalysis for transesterification of high free fatty acid oil (waste cooking oil) to biodiesel: a review. Biotechnol Adv 28(4):500–518

Lane N (2017) Serial endosymbiosis or singular event at the origin of eukaryotes? J Theor Biol 434:58–67

Lardon L, Hélias A, Sialve B, Steyer J-P, Bernard O (2009) Life-cycle assessment of biodiesel production from microalgae. Environ Sci Technol 43(17):6475–6481

Lee Chang KJ, Nichols CM, Blackburn SI, Dunstan GA, Koutoulis A, Nichols PD (2014) Comparison of thraustochytrids *Aurantiochytrium* sp., *Schizochytrium* sp., *Thraustochytrium* sp., and *Ulkenia* sp. for production of biodiesel, long-chain omega-3 oils, and exopolysaccharide. Mar Biotechnol 16(4):396–411

Lee WNP, Wahjudi PN, Xu J, Go VL (2010) Tracer-based metabolomics: concepts and practices. Clin Biochem 43(16-17):1269–1277

Lee YC, Kim B, Farooq W, Chung J, Han JI, Shin HJ, Jeong SH, Park JY, Lee JS, YK O (2013) Harvesting of oleaginous *Chlorella* sp. by organoclays. Bioresour Technol 132:440–445. https://doi.org/10.1016/j.biortech.2013.01.102

Lee KY, Ng TW, Li GY, An TC, Kwan KK, Chan KM, Huang GC, Yip HY, Wong PK (2015) Simultaneous nutrient removal, optimised CO_2 mitigation and biofuel feedstock production by *Chlorogonium* sp. grown in secondary treated non-sterile saline sewage effluent. J Hazard Mater 297:241–250

Li J, Liu Y, Cheng JJ, Mos M, Daroch M (2015) Biological potential of microalgae in China for biorefinery-based production of biofuels and high value compounds. New Biotechnol 32(6):588

Li T, Xu G, Rong J, Chen H, He C, Giordano M, Wang Q (2016) The acclimation of *Chlorella* to high-level nitrite for potential application in biological NO_x removal from industrial flue gases. J Plant Physiol 195:73–79

Liu J, Hu Q (2013) *Chlorella*: industrial production of cell mass and chemicals. In: Handbook of microalgal culture: applied phycology and biotechnology. Wiley, Chichester

Liu Z, Zhang F, Chen F (2013) High throughput screening of CO_2-tolerating microalgae using GasPak bags. Aquat Biosyst 9(1):23. https://doi.org/10.1186/2046-9063-9-23

Loladze I, Elser JJ (2011) The origins of the Redfield nitrogen-to-phosphorus ratio are in a homoeostatic protein-to-rRNA ratio. Ecol Lett 14(3):244–250

Lu YM, Xiang WZ, Wen YH (2011) Spirulina (*Arthrospira*) industry in Inner Mongolia of China: current status and prospects. J Appl Phycol 23(2):265

Lynn SG, Kilham SS, Kreeger DA, Interlandi SJ (2010) Effect of nutrient availability on the biochemical and elemental stoichiometry in the freshwater diatom *Stephanodiscus minutulus* (Bacillariophyceae). J Phycol 36(3):510–522

Marcus Y, Altmangueta H, Wolff Y, Gurevitz M (2011) Rubisco mutagenesis provides new insight into limitations on photosynthesis and growth in *Synechocystis* PCC6803. J Exp Bot 62(12):4173–4182

Mata TM, Martins AA, Caetano NS (2010) Microalgae for biodiesel production and other applications: a review. Renew Sust Energ Rev 14(1):217–232. https://doi.org/10.1016/j.rser.2009.07.020

McFadden GI (2001) Primary and secondary endosymbiosis and the origin of plastids. J Phycol 37:951–959

Memmola F, Mukherjee B, Moroney JV, Giordano M (2014) Carbon allocation and metal use in four *Chlamydomonas* mutants defective in CCM-related genes. Photosynth Res 121:111–124

Miao HF, Wang SQ, Zhao MX, Huang ZX, Ren HY, Yan Q, Ruan WQ (2014) Codigestion of Taihu blue algae with swine manure for biogas production. Energy Convers Manag 77:643–649

Mitchell P (1961) Coupling of phosphorylation to electron and hydrogen transfer by a chemiosmotic type of mechanism. Nature (Lond) 191(478):144–148

Mitra A, Flynn KJ, Tillmann U, Raven JA, Caron D, Stoecker DK, Not F, Hansen PJ, Hallegraeff G, Sanders R, Wilken S, McManus G, Johnson M, Pitta P, Vage S, Berge T, Calbet A, Thingstad F, Jeong HJ, Burkholder J, Glibert PM, Graneli E, Lundgren V (2016) Defining planktonic protist functional groups on mechanisms for energy and nutrient acquisition: incorporation of diverse mixotrophic strategies. Protist 167(2):106–120

Montechiaro F, Giordano M (2010) Compositional homeostasis of the dinoflagellate *Protoceratium reticulatum* grown at three different pCO_2. J Plant Physiol 167(2):110–113

Mooij TD, Janssen M, Cerezo-Chinarro O, Mussgnug JH, Kruse O, Ballottari M, Bassi R, Bujaldon S, Wollman FA, Wijffels RH (2015) Antenna size reduction as a strategy to increase biomass productivity: a great potential not yet realized. J Appl Phycol 27(3):1063–1077

Mouahid A, Crampon C, Toudji SAA, Badens E (2013) Supercritical CO_2 extraction of neutral lipids from microalgae: experiments and modelling. J Supercrit Fluids 77:7–16. https://doi.org/10.1016/j.supflu.2013.01.024

Mussgnug JH, Klassen V, Schluter A, Kruse O (2010) Microalgae as substrates for fermentative biogas production in a combined biorefinery concept. J Biotechnol 150(1):51–56

Nagase H, Yoshihara K, Eguchi K, Okamoto Y, Murasaki S, Yamashita R, Hirata K, Miyamoto K (2001) Uptake pathway and continuous removal of nitric oxide from flue gas using microalgae. Biochem Eng J 7(3):241–246. https://doi.org/10.1016/S1369-703x(00)00122-4

Nakagawa H (1997) Effect of dietary algae on improvement of lipid metabolism in fish. Biomed Pharmacother 51(8):345–348
Nayak M, Karemore A, Sen R (2016) Sustainable valorization of flue gas CO_2 and wastewater for the production of microalgal biomass as a biofuel feedstock in closed and open reactor systems. RSC Adv 6(94):91111–91120
Ngan CY, Wong CH, Choi C, Yoshinaga Y, Louie K, Jia J, Chen C, Bowen B, Cheng HY, Leonelli L, Kuo R, Baran R, Garcia-Cerdan JG, Pratap A, Wang M, Lim J, Tice H, Daum C, Xu J, Northen T, Visel A, Bristow J, Niyogi KK, Wei CL (2015) Lineage-specific chromatin signatures reveal a regulator of lipid metabolism in microalgae. Nat Plants 1(8):15107
Nicklisch A, Steinberg CEW (2009) RNA/protein and RNA/DNA ratios determined by flow cytometry and their relationship to growth limitation of selected planktonic algae in culture. Eur J Phycol 44:297–308. https://doi.org/10.1080/09670260802578518
Nielsen A (2015) Treatment of wastewater with microalgae under mixotrophic growth: a focus on removal of DOC from municipal and industrial wastewater. Degree thesis in Geoecology, Umeå University
Niu HJY, Leung DYC (2010) A review on the removal of nitrogen oxides from polluted flow by bioreactors. Environ Rev 18:175–189. https://doi.org/10.1139/A10-007
Norambuena F, Hermon K, Skrzypczyk V, Emery JA, Sharon Y, Beard A, Turchini GM (2015) Algae in fish feed: performances and fatty acid metabolism in juvenile Atlantic salmon. PLoS One 10(4):e0124042
Norici A, Bazzoni AM, Pugnetti A, Raven JA, Giordano M (2011) Impact of irradiance on the C allocation in the coastal marine diatom *Skeletonema marinoi* Sarno and Zingone. Plant Cell Environ 34(10):1666–1677
Norici A, Hell R, Giordano M (2005) Sulfur and primary production in aquatic environments: an ecophysiological perspective. Photosynth Res 85:409–417. https://doi.org/10.1007/s11120-005-3250-0
Nymark M, Sharma AK, Sparstad T, Bones AM, Winge P (2016) A CRISPR/Cas9 system adapted for gene editing in marine algae. Sci Rep 6:24951
Packer MA, Harris GC, Adams SL (2016) Food and feed applications of algae. In: Bux F, Chisti Y (eds) Algae biotechnology, Green Energy and Technology. Springer, Cham
Palmucci M, Giordano M (2012) Is cell composition related to the phylogenesis of microalgae? An investigation using hierarchical cluster analysis of Fourier transform infrared spectra of whole cells. Environ Exp Bot 75:220–224
Palmucci M, Ratti S, Giordano M (2011) Ecological and evolutionary implications of carbon allocation in marine phytoplankton as a function of nitrogen availability: a Fourier transform infrared spectroscopy approach. J Phycol 47(2):313–323
Pangestuti R, Kim SK (2011) Biological activities and health benefit effects of natural pigments derived from marine algae. J Funct Foods 3(4):255–266
Peralta-Ruiz Y, González-Delgado AD, Kafarov V (2013) Evaluation of alternatives for microalgae oil extraction based on exergy analysis. Appl Energy 101(0):226–236. https://doi.org/10.1016/j.apenergy.2012.06.065
Perazzoli S, Bruchez BM, Michelon W, Steinmetz RLR, Mezzari MP, Nunes EO, da Silva MLB (2016) Optimizing biomethane production from anaerobic degradation of *Scenedesmus* spp. biomass harvested from algae-based swine digestate treatment. Int Biodeter Biodegr 109:23–28
Perez-Garcia O, Escalante FM, de-Bashan LE, Bashan Y (2011) Heterotrophic cultures of microalgae: metabolism and potential products. Water Res 45(1):11–36
Pires JCM, Alvim-Ferraz MCM, Martins FG (2017) Photobioreactor design for microalgae production through computational fluid dynamics: a review. Renew Sust Energ Rev 79:248–254
Poulsen N, Chesley PM, Kröger N (2006) Molecular genetic manipulation of the diatom *Thalassiosira pseudonana* (Bacillariophyceae). J Phycol 42(5):1059–1065
Prioretti L, Giordano M (2016) Direct and indirect influence of sulfur availability on phytoplankton evolutionary trajectories. J Phycol 52:1094–1102

Qi F, Pei HY, Hu WR, Mu RM, Zhang S (2016) Characterization of a microalgal mutant for CO_2 biofixation and biofuel production. Energy Convers Manag 122:344–349
Qiao Y, Rong J, Chen H, He C, Wang Q (2015) Non-invasive rapid harvest time determination of oil-producing microalgae cultivations for biodiesel production by using chlorophyll fluorescence. Front Energy Res 3:44
Quigg A, Finkel ZV, Irwin AJ, Rosenthal Y, Ho TY, Reinfelder JR, Schofield O, Morel FMM, Falkowski PG (2003) The evolutionary inheritance of elemental stoichiometry in marine phytoplankton. Nature (Lond) 425(6955):291–294
Quigg A, Irwin AJ, Finkel ZV (2011) Evolutionary inheritance of elemental stoichiometry in phytoplankton. Proc R Soc Lond B Biol Sci 278(1705):526–534
Rajvanshi S, Sharma MP (2012) Micro algae: a potential source of biodiesel. J Sustain Bioenergy Syst 2(3):49–59
Ramey CJ, Barón-Sola Á, Aucoin HR, Boyle NR (2015) Genome engineering in cyanobacteria: where we are and where we need to go. ACS Synth Biol 4(11):1186
Rasala BA, Chao SS, Pier M, Barrera DJ, Mayfield SP (2014) Enhanced genetic tools for engineering multigene traits into green algae. PLoS One 9(4):e94028
Ratti S, Knoll AH, Giordano M (2011) Did sulfate availability facilitate the evolutionary expansion of chlorophyll a+c phytoplankton in the oceans? Geobiology 9(4):301–312. https://doi.org/10.1111/j.1472-4669.2011.00284.x
Raven JA (1995) Costs and benefits of low intracellular osmolarity in cells of fresh-water algae. Funct Ecol 9(5):701–707
Raven JA (1997) Inorganic carbon acquisition by marine autotrophs. Adv Bot Res 27:85–209
Raven JA, Beardall J (2016) Dark respiration and organic carbon loss. In: The physiology of microalgae. Springer, Cham
Raven JA, Geider RJ (2003) Adaptation, acclimation and regulation in algal photosynthesis. In: Larkum AWD, Douglas SE, Raven JA (eds) Photosynthesis in algae, Advances in photosynthesis and respiration, vol 14. Springer, Dordrecht
Raven JA, Giordano M (2014) Algae. Curr Biol 24(13):R590–R595. https://doi.org/10.1016/j.cub.2014.05.039
Raven JA, Giordano M (2016) Combined nitrogen. In: Borowitzka M, Beardall J, Raven JA (eds) The physiology of microalgae. Springer, Cham, pp 143–154
Raven JA, Ball LA, Beardall J, Giordano M, Maberly SC (2005) Algae lacking carbon-concentrating mechanisms. Can J Bot 83(7):879–890
Raven JA, Giordano M, Beardall J (2008) Insights into the evolution of CCMs from comparisons with other resource acquisition and assimilation processes. Physiol Plant 133(1):4–14
Raven JA, Beardall J, Flynn KJ, Maberly SC (2009) Phagotrophy in the origins of photosynthesis in eukaryotes and as a complementary mode of nutrition in phototrophs: relation to Darwin's insectivorous plants. J Exp Bot 60(14):3975–3987
Raven JA, Giordano M, Beardall J, Maberly SC (2011) Algal and aquatic plant carbon concentrating mechanisms in relation to environmental change. Photosynth Res 109(1-3):281–296
Raven JA, Giordano M, Beardall J, Maberly SC (2012) Algal evolution in relation to atmospheric CO_2: carboxylases, carbon-concentrating mechanisms and carbon oxidation cycles. Philos Trans R Soc Lond B Biol Sci 367(1588):493–507
Raven JA, Beardall J, Larkum AWD, Sanchez-Baracaldo P (2013) Interactions of photosynthesis with genome size and function. Philos Trans R Soc Lond B Biol Sci 368(1622):20120264
Raven JA, Beardall J, Giordano M (2014) Energy costs of carbon dioxide concentrating mechanisms in aquatic organisms. Photosynth Res 121(2-3):111–124
Redfield AC (1934) The haemocyanins. Biol Rev Camb Philos Soc 9(2):175–212
Reed RH, Warr SRC, Richardson DL, Moore DJ, Stewart WDP (1985) Blue-green algae (cyanobacteria): prospects and perspectives. Plant Soil 89(1):97–106
Richardson JW, Johnson MD, Outlaw JL (2012) Economic comparison of open pond raceways to photo bio-reactors for profitable production of algae for transportation fuels in the southwest. Algal Res 1(1):93–100

Richmond AE (1986) Microalgaculture. Crit Rev Biotechnol 4:369–438. https://doi.org/10.3109/07388558609150800

Ruan Z, Giordano M (2017) The use of NH_4^+ rather than NO_3^- affects cell stoichiometry, C allocation, photosynthesis and growth in the cyanobacterium *Synechococcus* sp. UTEX LB 2380, only when energy is limiting. Plant Cell Environ 40:227–236

Ruan Z, Raven JA, Giordano M (2017) In *Synechococcus* sp. competition for energy between assimilation and acquisition of C and those of N only occurs when growth is light limited. J Exp Bot 68(14):3829–3839

Saad SM, Yusof YAM, Ngah WZW (2006) Comparison between locally produced *Chlorella vulgaris* and *Chlorella vulgaris* from Japan on proliferation and apoptosis of liver cancer cell line, HepG2. Malays J Biochem Mol Biol 13:32–36

Saeid A, Chojnacka K, Korczyński M, Korniewicz D, Dobrzański Z (2013) Biomass of *Spirulina maxima* enriched by biosorption process as a new feed supplement for swine. J Appl Phycol 25(2):667–675

Samarasinghe N, Fernando S, Lacey R, Faulkner WB (2012) Algal cell rupture using high pressure homogenization as a prelude to oil extraction. Renew Energy 48:300–308

Santiago DE, Jin H-F, Lee K (2010) The influence of ferrous-complexed EDTA as a solubilization agent and its auto-regeneration on the removal of nitric oxide gas through the culture of green alga *Scenedesmus* sp. Process Biochem 45(12):1949–1953

Sathish A, Smith BR, Sims RC (2014) Effect of moisture on in situ transesterification of microalgae for biodiesel production. J Chem Technol Biotechnol 89(1):137–142. https://doi.org/10.1002/jctb.4125

Savakis P, Hellingwerf KJ (2015) Engineering cyanobacteria for direct biofuel production from CO_2. Curr Opin Biotechnol 33(33):8–14

Scaife MA, Nguyen GTDT, Rico J, Lambert D, Helliwell KE, Smith AG (2015) Establishing *Chlamydomonas reinhardtii* as an industrial biotechnology host. Plant J 82:532–546

Schmollinger S, Merchant SS (2014) Nitrogen-sparing mechanisms in *Chlamydomonas* affect the transcriptome, the proteome, and photosynthetic metabolism. Plant Cell 26(4):1410–1435

Sheng J, Vannela R, Rittmann BE (2011) Evaluation of cell-disruption effects of pulsed-electric-field treatment of *Synechocystis* PCC 6803. Environ Sci Technol 45(8):3795–3802

Shields RJ, Lupatsch I (2012) Algae for aquaculture and animal feeds. Technikfolgenabschätzung – Theorie Und Praxis 21(1):23–37

Shin SE, Lim JM, Koh HG, Kim EK, Kang NK, Jeon S, Kwon S, Shin WS, Lee B, Hwangbo K, Kim J, Ye SH, Yun JY, Seo H, Oh HM, Kim KJ, Kim JS, Jeong WJ, Chang YK, Jeong BR (2016) CRISPR/Cas9-induced knockout and knock-in mutations in *Chlamydomonas reinhardtii*. Sci Rep 6:27810

Sinéad L, Paul RR, Catherine S (2011) Marine bioactives as functional food ingredients: potential to reduce the incidence of chronic diseases. Mar Drugs 9(6):1056–1100

Singh J, Cu S (2010) Commercialization potential of microalgae for biofuels production. Renew Sust Energ Rev 14(9):2596–2610

Singh UB, Ahluwalia AS (2013) Microalgae: a promising tool for carbon sequestration. Mitig Adapt Strat Glob Chang 18:73–79

Sizova I, Greiner A, Awasthi M, Kateriya S, Hegemann P (2013) Nuclear gene targeting in *Chlamydomonas* using engineered zinc-finger nucleases. Plant J 73(5):873–882

Spolaore P, Joannis-Cassan C, Duran E, Isambert A (2006) Commercial applications of microalgae. J Biosci Bioeng 101(2):87–96. https://doi.org/10.1263/Jbb.101.87

Stanier RY, Kunisawa R, Mandel M, Cohen-Bazire G (1971) Purification and properties of unicellular blue-green algae (order Chroococcales). Bacteriol Rev 35:171–205

Stumm W, Morgan JJ (1981) Aquatic chemistry: an introduction emphasizing chemical equilibria in natural waters, 2nd edn. Wiley-Interscience, New York. 780 pp

Sun T, Cheng J (2002) Hydrolysis of lignocellulosic materials for ethanol production: a review. Bioresour Technol 83:1–11

Sun X, Wang C, Li Z, Wang W, Tong Y, Wei J (2013) Microalgal cultivation in wastewater from the fermentation effluent in riboflavin (B2) manufacturing for biodiesel production. Bioresour Technol 143:499–504

Tibbetts SM, Yasumaru F, Lemos D (2017) In vitro prediction of digestible protein content of marine microalgae (*Nannochloropsis granulata*) meals for Pacific white shrimp (*Litopenaeus vannamei*) and rainbow trout (*Oncorhynchus mykiss*). Algal Res 21:76–80

Ueno Y, Kurano K, Miyachi S (1998) Ethanol production by dark fermentation in the marine green alga, *Chlorococcum littorale*. J Ferment Bioeng 86:38–43

Valente LMP, Gouveia A, Rema P, Matos J, Gomes FF, Pinto IS (2006) Evaluation of three seaweeds *Gracilaria bursa-pastoris*, *Ulva rigida* and *Gracilaria cornea* as dietary ingredients in European sea bass (*Dicentrarchus labrax*) juveniles. Aquaculture 252(1):85–91

Van Den Hende S, Vervaeren H, Boon N (2012) Flue gas compounds and microalgae: (Bio-)chemical interactions leading to biotechnological opportunities. Biotechnol Adv 30:1405–1424. https://doi.org/10.1016/j.biotechadv.2012.02.015

Venuleo M, Raven JA, Giordano M (2017) Intraspecific chemical communication in microalgae. New Phytol 215:516–530

Vidyashankar S, VenuGopal KS, Chauhan VS, Muthukumar SP, Sarada R (2015) Characterisation of defatted *Scenedesmus dimorphus* algal biomass as animal feed. J Appl Phycol 27(5):1871–1879

Wan C, Alam MA, Zhao XQ, Zhang XY, Guo SL, Ho SH, Chang JS, Bai FW (2015) Current progress and future prospect of microalgal biomass harvest using various flocculation technologies. Bioresour Technol 184:251–257. https://doi.org/10.1016/j.biortech.2014.11.081

Wang Y, Rischer H, Eriksen NT, Wiebe MG (2013) Mixotrophic continuous flow cultivation of *Chlorella protothecoides* for lipids. Bioresour Technol 144:608–614. https://doi.org/10.1016/j.biortech.2013.07.027

Wang Y, Yang Y, Ma F, Xuan L, Xu Y, Huo H, Zhou D, Dong S (2015) Optimization of *Chlorella vulgaris* and bioflocculant-producing bacteria co-culture: enhancing microalgae harvesting and lipid content. Lett Appl Microbiol 60(5):497–503. https://doi.org/10.1111/Lam.12403

Wang Q, Lu Y, Xin Y, Wei L, Huang S, Xu J (2016) Genome editing of model oleaginous microalgae *Nannochloropsis* spp. by CRISPR/Cas9. Plant J 88(6):1071

Wannathong T, Waterhouse JC, Young REB, Economou CK, Purton S (2016) New tools for chloroplast genetic engineering allow the synthesis of human growth hormone in the green alga *Chlamydomonas reinhardtii*. Appl Microbiol Biotechnol 100(12):5467–5477

Wassef EA, El-Sayed AFM, Kandeel KM, Sakr EM (2005) Evaluation of *Pterocla dia* and *Ulva* meals as additives to gilthead seabream *Sparus aurata* diets. Egypt J Aquat Res 31:321–332

Weeks DP (2011) Homologous recombination in *Nannochloropsis*: a powerful tool in an industrially relevant alga. Proc Natl Acad Sci USA 108(52):20859–20860

Wells ML, Potin P, Craigie JS, Raven JA, Merchant SS, Helliwell KE, Smith AG, Camire ME, Brawley SH (2017) Algae as nutritional and functional food sources: revisiting our understanding. J Appl Phycol 29(2):949–982

Weyman PD, Beeri K, Lefebvre SC, Rivera J, Mccarthy JK, Heuberger AL, Peers G, Allen AE, Dupont CL (2015) Inactivation of *Phaeodactylum tricornutum* urease gene using transcription activator-like effector nuclease-based targeted mutagenesis. Plant Biotechnol J 13(4):460

Wirth R, Lakatos G, Bojti T, Maroti G, Bagi Z, Kis M, Kovacs A, Acs N, Rakhely G, Kovacs KL (2015) Metagenome changes in the mesophilic biogas-producing community during fermentation of the green alga *Scenedesmus obliquus*. J Biotechnol 215:52–61

Wu XJ, Gao G, Giordano M, Gao KS (2012) Growth and photosynthesis of a diatom grown under elevated CO_2 in the presence of solar UV radiation. Fundam Appl Limnol 180(4):279–290

Yi YU, Wang H, Yang Y, Pei G, Qin Z, Jiujun JI (2016) Study on preparation of nutritional *Nostoc sphaeroides* jelly. Amino Acids Biot Resour 03:34–39

Zhang YM, Chen H, He CL, Wang Q (2013) Nitrogen starvation induced oxidative stress in an oil-producing green alga *Chlorella sorokiniana* C3. PLoS One 8(7):e69225. https://doi.org/10.1371/journal.pone.0069225

Zhang Y, Fan X, Yang Z, Wang H, Yang D, Guo R (2012) Characterization of H2 photoproduction by a new marine green alga, *Platymonas helgolandica* var. tsingtaoensis. Appl Energy 92:38–43

Zhang X, Chen H, Chen W, Qiao Y, He C, Wang Q (2014a) Evaluation of an oil-producing green alga *Chlorella* sp. C2 for biological $DeNO_x$ of industrial flue gases. Environ Sci Technol 48(17):10497–10504. https://doi.org/10.1021/es5013824

Zhang X, Rong J, Chen H, He C, Wang Q (2014b) Current status and outlook in the application of microalgae in biodiesel production and environmental protection. Energy Res. https://doi.org/10.3389/fenrg.2014.00032

Zhang X, Ma F, Zhu X, Zhu J, Rong J, Zhan J, Chen H, He C, Wang Q (2017) The acceptor side of photosystem II is the initial target of nitrite stress in *Synechocystis* sp. strain PCC 6803. Appl Environ Microbiol 83(3):e02952–e03016

Zhao MX, Ruan WQ (2013) Biogas performance from co-digestion of Taihu algae and kitchen wastes. Energy Convers Manag 75:21–24

Zhong W, Zhang Z, Luo Y, Qiao M, Xiao W (2012) Biogas productivity by co-digesting Taihu blue algae with corn straw as an external carbon source. Bioresource 114:281–286

Zhou W, Min M, Li Y, Bing H, Ma X, Cheng Y, Liu Y, Chen P, Ruan R (2012) A hetero-photoautotrophic two-stage cultivation process to improve wastewater nutrient removal and enhance algal lipid accumulation. Bioresour Technol 110(1):448

Zhu LD (2015) Biorefinery as a promising approach to promote microalgae industry: an innovative framework. Renew Sust Energ Rev 41:1376–1384. https://doi.org/10.1016/j.rser.2014.09.040

Zhu S, Wang Y, Huang W, Xu J, Wang Z, Xu J, Yuan Z (2014) Enhanced accumulation of carbohydrate and starch in *Chlorella zofingiensis* induced by nitrogen starvation. Appl Biochem Biotechnol 174(7):2435–2445

Zhu J, Chen W, Chen H, Zhang X, He C, Rong J, Wang Q (2016) Improved productivity of neutral lipids in *Chlorella* sp. A2 by minimal nitrogen supply. Front Microbiol 7(557). https://doi.org/10.3389/fmicb.2016.00557

Zittelli GC, Rodolfi L, Bassi N, Biondi N, Tredici MR (2013) Photobioreactors for microalgal biofuel production. In: Algae for biofuels and energy. Springer, Dordrecht

Chapter 7
Enzymatic Conversion of First- and Second-Generation Sugars

Roger A. Sheldon

Abstract Processes for the hydrolytic conversion of polysaccharides to fermentable sugars as feedstocks for biofuels and commodity chemicals are discussed. The production of first-generation biofuels, for example, bioethanol, involves the conversion of sucrose or starch; the latter requires initial enzymatic hydrolysis of the starch to glucose in a two-step process catalyzed by α-amylase and glucoamylase. These methods are established industrial processes that are conducted on an enormous scale. Although the enzymes involved are relatively inexpensive, they are used on a single-use, throw-away basis, and substantial cost savings can be achieved by immobilization of the enzymes to enable their recycling. In particular, immobilization of the enzymes as magnetic cross-linked enzyme aggregates (mCLEAs), in combination with magnetic separation using commercially available equipment, offers possibilities for achieving substantial cost reductions.

The production of second-generation biofuels involves, in the long term, more sustainable conversion of waste lignocellulose to fermentable sugars, a much more complicated process requiring multiple enzymes. The hydrolytic step is preceded by a pretreatment step that opens the structure of the recalcitrant lignocellulose to make it accessible for the hydrolytic enzymes. This step is usually conducted in water, in which the lignocellulose is insoluble, but there is currently much interest in the use of ionic liquids or deep eutectic solvents in combination with water. Subsequent hydrolysis of the cellulose and hemicellulose to fermentable sugars involves a complex cocktail of enzymes referred to as "cellulase." In this case the percentage cost contribution of the enzymes to the biofuel is even higher than with first-generation biofuels. Consequently, it is even more important to reduce the costs of enzyme usage by immobilization, and magnetic separation of magnetic immobilized enzymes, such as magnetic CLEAs, is a potentially attractive way to achieve this.

R.A. Sheldon (✉)
Molecular Science Institute, School of Chemistry, University of the Witwatersrand, Johannesburg, South Africa

Department of Biotechnology, Delft University of Technology, Delft, Netherlands
e-mail: roger.sheldon@wits.ac.za

S. Vaz Jr. (ed.), *Biomass and Green Chemistry*,
https://doi.org/10.1007/978-3-319-66736-2_7

Keywords Biocatalysis • Lignocellulose • Enzymatic hydrolysis • Cellulase • Enzyme immobilization • Magnetic separation • Ionic liquids • Deep eutectic solvents

7.1 Introduction

One of the grand challenges in chemistry and biology, motivated by the pressing need to mitigate climate change, is to devise green and sustainable technologies for the conversion of renewable biomass to fuels, commodity chemicals, and bio-based materials such as bioplastics in integrated biorefineries (Imhof and van der Waal 2013; Yang et al. 2013; Gallezot 2012; De Jong et al. 2012) The majority (60–80%) of all biomass consists of carbohydrates, which can be divided into storage carbohydrates—starch, inulin, and sucrose—and structural polysaccharides, such as cellulose, hemicelluloses, and chitin. In particular, lignocellulose, a fibrous material that constitutes the cell walls of plants, is available in very large quantities. In addition, aquatic carbohydrates derived from micro- and macroalgae consist of a variety of polysaccharides that differ in structure from those of terrestrial biomass. The remainder of biomass comprises triglycerides (from fats and oils), proteins, and terpene hydrocarbons.

The use of first-generation (1G) biomass feedstocks, comprising sucrose, starch, and triglycerides from edible oil seeds, is not perceived as a sustainable option in the long term because it competes, directly or indirectly, with food production. Second-generation (2G) feedstocks, in contrast, comprise lignocellulose and triglycerides produced by the deliberate cultivation of fast-growing, nonedible crops or, in a more attractive option, by the valorization of waste triglycerides (oils and fats) and, in particular, the enormous amounts of waste lignocellulose generated in the harvesting, processing, and use of agricultural products, including foods and beverages. Pertinent examples include sugarcane bagasse, corn stover, wheat straw, rice husks, and orange peel (Tuck et al. 2012). Indeed, the so-called bio-based economy is founded on the full utilization of agricultural biomass by using green and sustainable chemistry.

The first step in the conversion of polysaccharide feedstocks to, for example, biofuels and commodity chemicals is generally hydrolytic depolymerization to fermentable sugars, mainly glucose. The preferred method, from both economic and environmental aspects, is enzymatic hydrolysis. The hydrolysis of starch, for example, consists of two steps: liquefaction and saccharification. The former is conducted at 90 °C and pH 7 and is catalyzed by α-amylase (E.C. 3.2.1.1), which hydrolyzes α-(1-4) glycosidic bonds, affording a mixture of glucose oligomers (maltodextrins). The latter is conducted at 60 °C and pH 5 and involves hydrolysis of both α-(1-4) and α-(1-6) glycosidic bonds catalyzed by glucoamylase (E.C. 3.2.1.3).

Lignocellulose is much more difficult to process than starch. It consists of three major polymeric components: lignin (~20%), cellulose (~40%), and hemicellu-

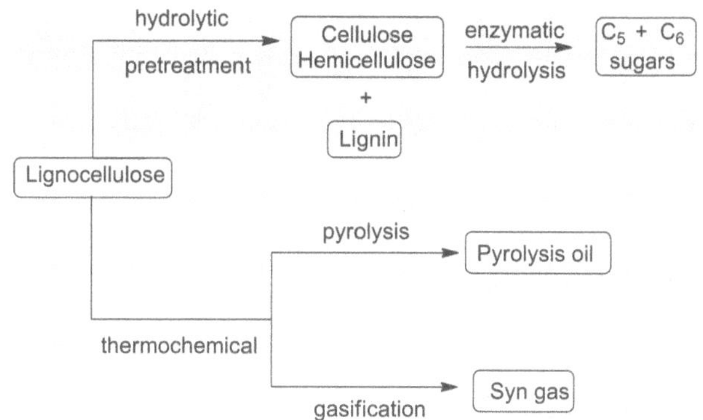

Fig. 7.1 Methods for depolymerization of lignocellulosic biomass

lose (~25%) in weight composition. Lignin is a three-dimensional polyphenolic biopolymer having a nonuniform structure that imparts rigidity and recalcitrance to plant cell walls. In volume, it is the second largest biopolymer after cellulose and the only one composed entirely of aromatic subunits. Basically, depolymerization and (partial) deoxygenation of lignocellulose can be performed in two ways: hydrolytic and thermochemical (Fig. 7.1). The latter involves pyrolysis to a mixture of charcoal and pyrolysis oil or gasification to afford syngas (a mixture of carbon monoxide and hydrogen), analogous to syngas from coal gasification (Sheldon 1983). The syngas can be subsequently converted to liquid fuels or platform chemicals using established technologies such as the well-known Fischer–Tropsch process or methanol synthesis, respectively. Alternatively, it can be used as a feedstock for the microbial production of biofuels and platform chemicals (Munasinghe and Khanal 2007; Henstra et al. 2007). In this chapter, we are concerned with the hydrolytic conversion of the polysaccharide feedstock, starch or lignocellulose, to monosaccharides (primarily glucose) for subsequent conversion to 1G and 2G biofuels and commodity chemicals, respectively.

7.2 Enzymatic Hydrolysis of Starch to Glucose

Although lignocellulosic biofuels are seen as the long-term option, in the short term biofuels consist primarily of corn- or sugar-based bioethanol. The United States (USA) is the largest producer, with a production of 14.7 billion gallons in 2015 (Renewable Fuels Association 2016), from cornstarch as the feedstock. This process includes the enzymatic hydrolysis of the starch to glucose (Fig. 7.2), followed by fermentation of the latter to ethanol; it can be conducted in a separate hydrolysis and fermentation (SHF) mode or, to be more cost-effective, in a combined simultaneous saccharification and fermentation (SSF) process (Olofsson et al. 2008). An SSF process has the added advantage that the glucose is immediately consumed by the fermenting organism, thus circumventing possible inhibition by increasing concentrations of glucose.

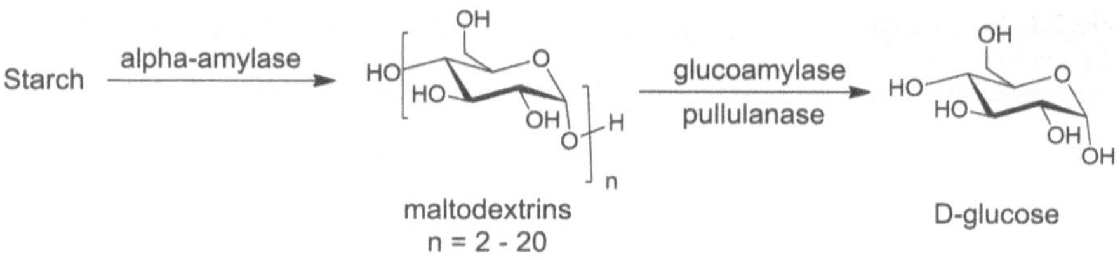

Fig. 7.2 Enzymatic hydrolysis of starch

7.3 Enzyme Immobilization

The enzyme costs per kilogram of product are a crucial factor in determining the economic viability of these processes, particularly in the case of lignocellulosic feedstocks (see later). Enzyme manufacturers have achieved remarkable results, in recent years, in reducing enzyme costs by optimizing the production of the enzymes involved, but there is still room for improving the enzyme usage. The enzyme(s) used in the hydrolysis step are dissolved in the aqueous reaction mixture and, consequently, are discarded with the wash water; that is, they are employed on a single-use, throw-away basis. Multiple recycling of the enzymes represents a clear opportunity to substantially reduce the enzyme costs and environmental footprint to drive competitiveness and sustainability. It can be achieved by using the enzyme in a solid, immobilized form, which can be easily recovered and reused. A further benefit of immobilization is that the resulting decrease in flexibility leads to increasing operational stability by suppressing the propensity of enzymes to unfold (denature) under the influence of heat or organic solvents.

Immobilization typically involves binding the enzyme to a prefabricated carrier (support), such as an organic resin or silica, or entrapment in a polymeric inorganic or organic matrix that is formed in the presence of the enzyme. Binding to a prefabricated carrier can involve simple adsorption, such as via hydrophobic or ionic interactions or the formation of covalent bonds (Cao 2005). Typical supports are synthetic resins (Cantone et al. 2013), biopolymers such as polysaccharides, or inorganic solids such as (mesoporous) silicas (Hartmann and Kostrov 2013; Zhou and Hartmann 2012; Magner 2013). Entrapment involves inclusion of an enzyme in an organic or inorganic polymer matrix, such as polyacrylamide and silica sol-gel, respectively, or a membrane device such as a hollow fiber or a microcapsule (Reetz 2013). The use of a carrier inevitably leads to 'dilution of activity,' owing to the introduction of a large portion of noncatalytic ballast, ranging from 90% to more than 99%, giving rise to lower space–time yields and catalyst productivities (Cao et al. 2003).

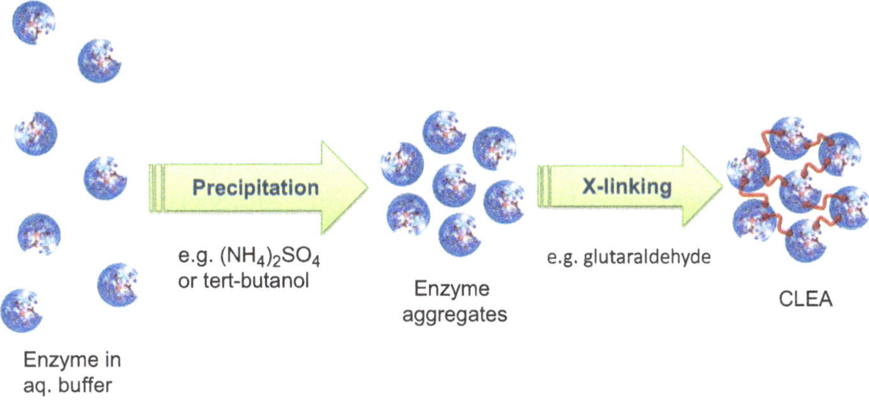

Fig. 7.3 Enzyme immobilization as cross-linked enzyme aggregates (CLEAs)

7.3.1 Immobilization Via Cross-Linking: CLEAs

In contrast, immobilization by cross-linking of enzyme molecules affords carrier-free immobilized enzymes with high productivities and avoids the extra costs of a carrier. For example, cross-linked enzyme aggregates (CLEAs) are formed by precipitation of the enzyme from aqueous buffer, as physical aggregates held together by noncovalent bonding without perturbation of their tertiary structure, followed by cross-linking with a bifunctional reagent, such as glutaraldehyde (Fig. 7.3) (Cao et al. 2000). The method is simple and inexpensive as it does not require highly pure enzymes. Indeed, because selective precipitation with ammonium sulfate is commonly used to purify enzymes, the CLEA methodology essentially combines two processes, purification and immobilization, into a single unit operation. The CLEA technology has subsequently been applied to the immobilization of a broad spectrum of enzymes and forms the subject of several reviews (Sheldon 2011, 2013; Talekar et al. 2013a; Cui and Ja 2015; Valesco-Lozano et al. 2015; Sheldon et al. 2013).

Multipurpose CLEAs can be prepared from crude enzyme extracts consisting of multiple enzymes (Dalal et al. 2007a, 2007b). Interestingly, enzymes that would not be compatible in solution can function effectively when co-immobilized. For example, conducting reactions with a mixture of a protease and a lipase is not feasible with the free enzymes as the protease would break down the lipase, but it is possible when the two enzymes are co-immobilized in a novel lipase/protease multi-CLEA (Mahmod et al. 2015).

The term combi-CLEA is generally used when two or more enzymes are deliberately co-immobilized in a single CLEA for the purpose of performing two or more sequential biotransformations in a multi-enzyme cascade process. Catalytic cascade processes have several advantages compared with classical multistep syntheses: fewer unit operations, less solvent, smaller reactor volume, shorter cycle times, higher volumetric and space–time yields, and less waste. Furthermore, coupling of reactions can be used to drive equilibria toward the product, thus avoiding the need for excess reagents. Because biocatalytic processes generally proceed under roughly the same conditions—in water at ambient temperature and pressure—they can be readily integrated into cascade processes. These processes have become a focus of attention in recent years, largely motivated by these envisaged environmental and economic benefits (Lopez-Gallego and Schmidt-Dannert 2010; Schrittweiser et al. 2011; Santacoloma et al. 2011; Xue and Woodley 2012; Muschiol et al. 2015; Land et al. 2016). However, different enzymes can be incompatible, and Nature solves this problem by compartmentalizing enzymes in different parts of the cell. Hence, compartmentalization via immobilization is a possible solution in enzymatic cascade processes.

The rates of sequential biocatalytic cascades can be substantially increased by simulating the close proximity of the enzymes extant in microbial cells by co-immobilization of the respective enzymes in, for example, combi-CLEAs. For example, combi-CLEAs of glucose oxidase or galactose oxidase with catalase exhibited significantly better activities and stabilities than corresponding mixtures of the two separate CLEAs (Schoevaart et al. 2004). Moreover, the combi-CLEAs could be recycled without significant loss of activity. Oxidases catalyze the aerobic oxidations of various substrates, generally with concomitant production of an equivalent of hydrogen peroxide. The latter can cause oxidative degradation of the enzyme and, in vivo, oxidases generally occur together with catalase, which catalyzes the spontaneous decomposition of hydrogen peroxide to oxygen and water.

Combi-CLEAs have also been widely used in carbohydrate conversions. As already mentioned, glucoamylase catalyzes the hydrolysis (saccharification) of α-(1,4)- and α-(1,6)-glycosidic bonds in starch. Because the hydrolysis of the α-(1,6) branches is relatively slow, a second enzyme, pullulanase (E.C. 3.2.1.41), is sometimes added to facilitate this hydrolysis. Co-immobilization of the two enzymes in a combi-CLEA results in a shift in optimum pH values (from 5 to 7) and temperature (from 60 °C to 70 °C) (Talekar et al. 2013b). In a batch mode hydrolysis of starch, the combi-CLEA gave 100% conversion after 3 h compared with 30% with the free enzyme. A mixture of the two separate CLEAs gave 80% conversion. The combi-CLEA had good stability, retaining 90% and 85% of the glucoamylase and pullulanase activity, respectively, after eight recycles.

More recently, the same group prepared a tri-enzyme combi-CLEA containing α-amylase, glucoamylase, and pullulanase from commercially available enzyme preparations (Talekar et al. 2013c). In a one-pot starch hydrolysis (Fig. 7.4) in batch mode, 100%, 60%, and 40% conversions were observed with the combi-CLEA, a mixture of the separate CLEAs, and a mixture of the free enzymes, respectively. Co-immobilization increased the thermal stability of all three enzymes, and the catalytic performance was maintained for up to five cycles.

Similarly, combi-CLEAs have been used with success in the hydrolysis of lignocellulose, which has become the focus of attention in connection with second-generation biofuels production from waste lignocellulose streams (see Sect. 7.5).

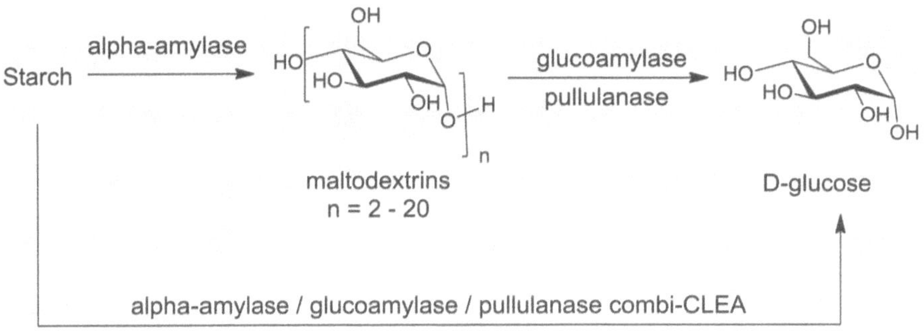

Fig. 7.4 One-pot starch hydrolysis with a combi-CLEA

7.3.2 Advantages and Limitations of CLEAs

Immobilization of enzymes as cross-linked enzyme aggregates can lead to dramatic increases in storage and operational stability (van Pelt et al. 2008). Moreover, because the enzyme molecules in CLEAs are bound by covalent bonds, no leaching of enzyme is observed in aqueous media, even under drastic conditions such as in the presence of surfactants.

An important advantage from a cost-effectiveness aspect is their ease of recovery and reuse, as heterogeneous catalysts, by filtration, centrifugation, or, alternatively, in a fixed-bed reactor. In practice this translates to simpler, less expensive downstream processing. Another important cost advantage is the fact that CLEAs can be prepared from enzyme samples of low purity, including crude cell lysate obtained from fermentation broth. Furthermore, because they consist mainly of active enzyme, CLEAs exhibit high catalyst productivities (kilograms per product per kilograms of enzyme) compared to carrier-bound enzymes, and the costs of the carrier ballast are avoided.

A limitation of the technique is that, because every enzyme is a different molecule, the protocol has to be optimized for every enzyme. However, in practice optimizing the aggregation plus cross-linking protocol is not a lengthy procedure and it can be readily automated. Another limitation is the relatively small particle size, as already mentioned, which can be an issue for processes conducted in a fixed-bed reactor, for example. The problem can be alleviated by mixing the CLEA with an inert, less compressible solid, such as controlled microporous glass or perlite (Hickey et al. 2007).

7.4 Magnetically Separable Immobilized Enzymes

Applications in processes involving suspensions of other water-insoluble solids, such as fibers or yeasts in SSF conversions of 1G and 2G biomass, is a challenge for standard immobilized enzymes. However, industrially viable separation on a large

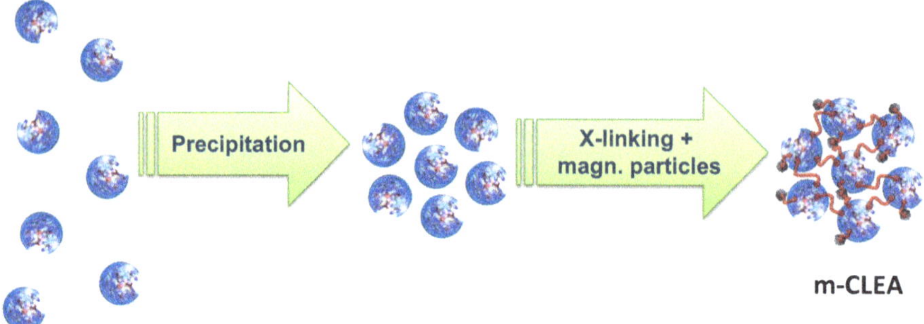

Fig. 7.5 Formation of magnetic cross-linked enzyme aggregates (mCLEAs)

scale, using standard commercial equipment, can be readily achieved using enzyme-ferromagnetic particle composites, affording novel combinations of biocatalysis with downstream processing. Indeed, recently increasing attention has been devoted to the design of magnetically recoverable catalysts, based on ferromagnetic magnetite (Fe_3O_4) or maghemite (γ-Fe_2O_3) (nano)particles, in chemocatalytic (Rossi et al. 2014; Ranganath and Glorius 2011; Gawande et al. 2013; Zamani and Hosseini 2014) and biocatalytic (Ansari and Husain 2012; Johnson et al. 2008; Liu et al. 2011; Yiu and Keane 2012; Netto et al. 2013) processes. The magnetic separation of ferromagnetic heterogeneous catalysts, such as Raney nickel, has long been known (Whitesides et al. 1976; Lindley 1982) and is practiced on an industrial scale. In combination with commercially available magnetic separation equipment (e.g., see www.eclipsemagnetics.com), very high recoveries can be obtained at industrially acceptable flow rates. Magnetic separation can also be used to alleviate problems encountered with separation of relatively small particles by filtration or centrifugation. Ideally, one would like to have the high activity of small particles while maintaining the ease of processing of large particles, and this goal can be achieved with magnetically separable immobilized enzymes.

Similarly, the CLEA technology has been raised to a new level of sophistication and industrial relevance with the invention of robust, cost-effective ferromagnetic CLEAs (mCLEAs), produced by conducting the cross-linking in the presence of ferromagnetic (nano)particles (Fig. 7.5) (Sheldon et al. 2012). Using the latest methodology (Janssen et al. 2016), conversion to mCLEAs adds little cost to regular CLEAs and it does not require many recycles to achieve cost reductions. Hence, mCLEAs are expected to find applications in a variety of processes, including 1G and 2G biofuels.

mCLEAs have been prepared from a variety of enzymes, including lipases (Cruz-Izquierdo et al. 2014; Zhang et al. 2015a; Cui et al. 2016; Tudorache et al. 2016), penicillin G amidase (Kopp et al. 2014), phenyl ammonia lyase (PAL, EC. 4.3.1.24) (Cui et al. 2014), laccase (Kumar et al. 2014,) and even a combi-mCLEA of horseradish peroxidase and glucose oxidase for use in dye decolorization (Zhou et al. 2016). However, industrial interest is primarily in carbohydrate conversions, particularly in the conversion of polysaccharides such as starch and lignocellulose in connection with 1G and 2G biofuels and food and beverage processing.

For example, Talekar and coworkers (Talekar et al. 2012) described the preparation of mCLEAs of α-amylase, with an activity recovery of 100%, for use in starch hydrolysis. Thermal and storage stability was improved compared to the free enzyme, and the mCLEA retained 100% of its activity after six recycles. A mCLEA of α-amylase was also prepared with pectin dialdehyde as the cross-linker and exhibited 95% activity recovery compared to 85% using glutaraldehyde as the cross-linker (Nadar and Rathod 2016). The authors attributed the higher activity recovery to better mass transfer with macromolecular substrates in the more open porous structure. mCLEAs of glucoamylase from *Aspergillus niger* were prepared by Gupta and coworkers (Gupta et al. 2013) with 93% activity recovery and showed enhanced thermal and storage stability and reusability.

Probably the most exciting and challenging application of the mCLEA technology is in the complex chemistry of lignocellulose conversion, which is discussed in the following section.

7.5 Enzymatic Depolymerization of Lignocellulose

Conversion of lignocellulose to biofuels and commodity chemicals basically involves three steps: pretreatment, enzymatic hydrolysis to fermentable sugars (saccharification), and fermentation, for example, to ethanol, or chemocatalytic conversion. Alternatively, so-called consolidated bioprocessing (CBP) involves the use of cellulolytic enzyme-producing microbes—bacteria, fungi, or yeasts—in a single-step process in which enzyme production, enzymatic hydrolysis, and fermentation proceed simultaneously (Jouzani and Taherzadeh 2015). Although CBP is potentially very attractive, it is still in its infancy, and a detailed discussion falls outside the scope of this review, which focuses on the more established classical approach of using separately produced enzyme cocktails.

7.5.1 Pretreatment of Lignocellulose

Some form of pretreatment, such as a steam explosion, ammonia fiber expansion (AFEX), or lime treatment, is generally necessary to open up the recalcitrant lignocellulose structure and render it accessible to the enzyme cocktail (Rabemanolontsoa and Saka 2016; Kumar et al. 2009; Alvira et al. 2010). The pretreatment generally accounts for a large fraction of the total energy requirements of lignocellulose processing (Menon and Rao 2012.) It is generally conducted in water in which the cellulose, hemicellulose, and lignin are present as suspended solids. The use of alternative reaction media, which (partially) dissolve these polymeric substrates, could have processing advantages. However, to be economically and

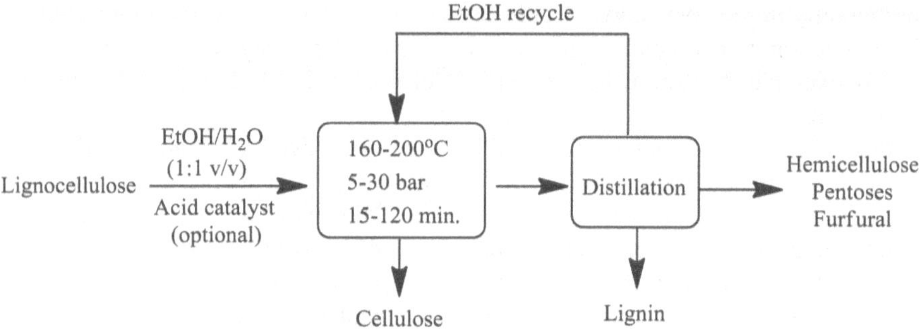

Fig. 7.6 The Organosolv process with aqueous ethanol

environmentally viable, the solvent should be inexpensive, nontoxic, biodegradable, recyclable, and preferably derived from renewable resources.

In the Organosolv process (Fig. 7.6) (Wildschut et al. 2013), for example, lignocellulose is subjected to elevated temperatures (185–210°C) in water/organic solvent (e.g., 50% v/v aqueous ethanol) mixtures, in the presence (Zhao et al. 2009) or absence of an acid catalyst (Viell et al. 2013a). This reaction results in hydrolysis of the hemicellulose and dissolution of the lignin. The remaining cellulose is separated, and the dissolved lignin is precipitated by water addition or ethanol evaporation. Overall the products comprise cellulose, solid lignin, and an aqueous stream containing hemicellulose, C_5 sugars, and derivatives thereof, such as furfural.

In a variation on this theme, ionic liquids (ILs), which are known to dissolve polysaccharides, are being considered as potential alternatives for the deconstruction of lignocellulosic biomass in a so-called Ionosolv process (Brandt et al. 2013). Rogers and coworkers (Swatloski et al. 2002) showed that the room temperature IL, [bmim] [Cl], can dissolve 100 g l^{-1} cellulose at 100 °C. Currently, much attention is focused on the use of ILs (Bogel-Lukasic 2016; Fang et al. 2014; Dominguez de Maria 2014; Vancov et al. 2012; Zavrel et al. 2009) as reaction media for the saccharification of lignocellulose. A variety of waste lignocellulosic biomass, such as wood chips (Sun et al. 2009; Xu et al. 2014) and wheat straw (Li et al. 2009; Magalhaes da Silva et al. 2013), has been shown to dissolve in ILs, and the latter could be used as reaction media for subsequent chemocatalytic (Rinaldi 2014; Morales-de la Rosa et al. 2012; Long et al. 2012) or enzymatic hydrolysis. For example, [emim] [OAc] was used as the solvent for the fractionation of wheat straw into lignin and carbohydrate fractions (da Costa Lopes et al. 2013) and for the combined pretreatment and enzymatic hydrolysis of wood chips (Viell et al. 2013b), switchgrass (*Miscanthus giganteus*) (Shi et al. 2013), and sugarcane bagasse (Qiu et al. 2012). Cholinium salts of amino acids, such as cholinium lysinate [Ch] [Lys], are examples of ILs derived from renewable resources. A 20:80 mixture (v/v) of [Ch] [Lys] and water was used for the pretreatment and subsequent enzymatic hydrolysis of rice straw (Hou et al. 2013) and switchgrass (Sun et al. 2014).

Ultimately, the goal is to develop an integrated process for IL pretreatment of lignocellulose and enzymatic hydrolysis, with efficient recycling of both the IL and the (immobilized) enzyme (Ungerean et al. 2014). The commercial viability of ILs in lignocellulose pretreatment and saccharification depends on the biomass loading, the cost of the IL, its environmental acceptability and stability under operating conditions, and its recycling efficiency. Protic ionic liquids (PILs) are interesting from a cost and availability aspect as they are readily prepared by mixing commodity amines with inexpensive acids. Welton and coworkers (Verdía et al. 2014; Brandt et al. 2011) used a mixture of the PIL, 1-butylimidazolium hydrogen sulfate, [bhim][HSO_4], and water (80:20 v/v) for dissolution and subsequent cellulase-catalyzed saccharification of switchgrass. In another study (George et al. 2015), a range of PILs containing the hydrogen sulfate anion, the cost of which is primarily determined by the price of the amine, was prepared by simply mixing the corresponding amines with sulfuric acid. The best results were obtained with triethylammonium hydrogen sulfate, [Et_3NH] [HSO_4]. Similarly, Henderson and coworkers (Achinivu et al. 2014) used pyrrolidinium acetate for extracting lignin, with little or no cellulose extraction, from lignocellulosic biomass. The PIL was recovered by vacuum distillation.

Deep eutectic solvents (DESs), formed by mixing a salt with a hydrogen bond donor and gently heating, constitute inexpensive alternatives to ILs and PILs for lignocellulose pretreatment (see Fig. 7.7 for structures of Ils, PILs, and DESs). An example of a DES derived from renewable raw materials is the 1:2 mixture of choline chloride and glycerol (Ch/Gly). Aqueous ChCl/Gly was effective as the reaction medium for the pretreatment and saccharification of lignocellulose from switchgrass (Xia et al. 2014). Similarly, ChCl-based DESs were successfully used in the pretreatment and saccharification of lignocellulose from empty fruit bunches of the oil palm (Nor et al. 2016), corncob residues (Procentese et al. 2015), and wheat straw (Jablonski et al. 2015). Kroon and coworkers (Francisco et al. 2012; Kroon et al. 2013) screened a range of natural deep eutectic solvents (NADES), derived from mixtures of CHCl or natural amino acids as hydrogen bond acceptor and natural carboxylic acids as hydrogen bond donor, as solvents for biopolymers (lignin, cellulose, and starch). Most combinations exhibited high lignin solubilities and negligible solubility for cellulose, and, hence, are potential candidates for lignocellulose fractionation. Similarly, Kumar and coworkers (Kumar et al. 2016) recently reported the use of NADES, such as CHCl/lactic acid, in the pretreatment of rice straw lignocellulose, with separation of high-quality lignin and holocellulose in a single step. Chitin was also shown to dissolve in DESs such as CHCl/Gly (Sharma et al. 2013).

7.5.2 Enzymatic Hydrolysis of Cellulose and Hemicellulose

Hydrolysis of cellulose and hemicellulose to fermentable sugars requires the involvement of a complex cocktail of cellulolytic and hemicellulolytic enzymes (Bhattachariya et al. 2015) in a process referred to as saccharification (Bornscheuer et al. 2014). The hydrolysis of cellulose involves catalysis by at least five enzymes:

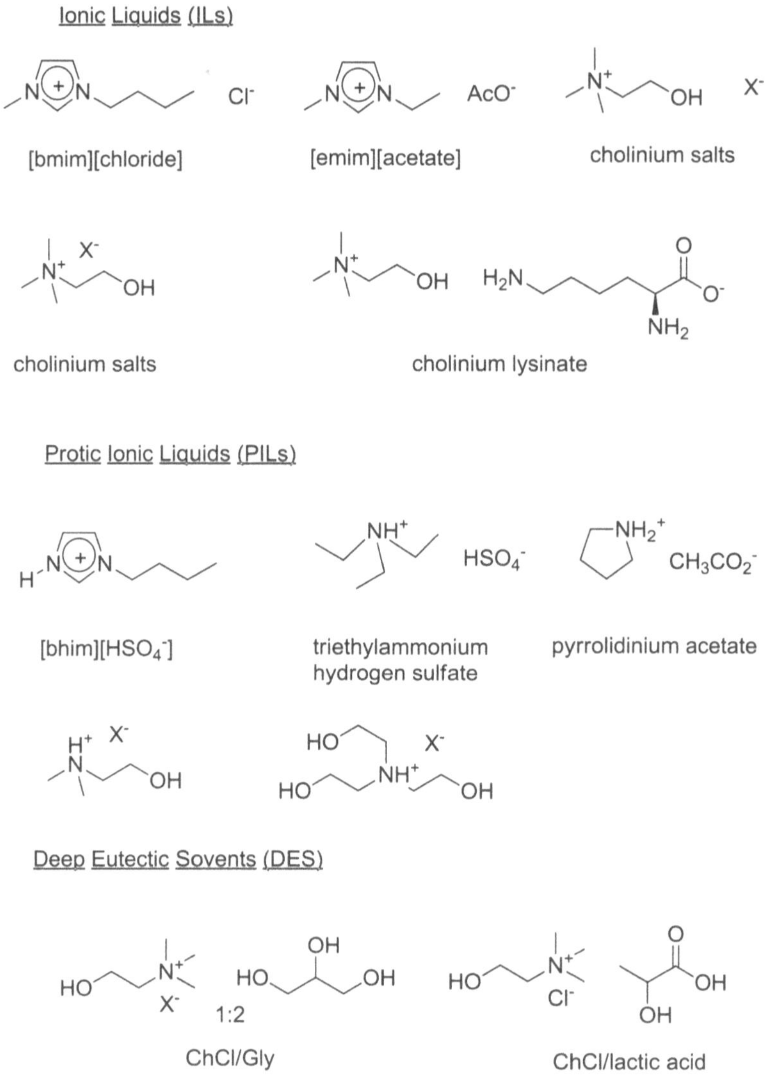
Ionic Liquids (ILs)
Cl-
AcO-
X-
[bmim][chloride]
[emim][acetate]
cholinium salts
X-
cholinium salts
cholinium lysinate
Protic Ionic Liquids (PILs)
HSO4-
CH3CO2-
[bhim][HSO4-]
triethylammonium hydrogen sulfate
pyrrolidinium acetate
X-
X-
Deep Eutectic Sovents (DES)
X-
1:2
ChCl/Gly
Cl-
ChCl/lactic acid

Fig. 7.7 Structures of ionic liquids (ILs), protic ionic liquids (PILs), and deep eutectic solvents (DESs)

exo-1,4-β-glucanase (EC 3.2.1.91), *endo*-1,4-β-glucanase (EC 3.2.1.4), cellobiohydrolase (EC 3.2.1.176), and β-glucosidase (EC 3.2.1.21), and the more recently discovered, copper-dependent lytic polysaccharide monooxygenases (LPMO) (Johansen 2016; Hemsworth et al. 2015; Walton and Davies 2016), which catalyze the oxidative cleavage of polysaccharides. Hemicellulose has a more complicated structure than cellulose and requires a diverse suite of enzymes to effect its hydrolysis to its constituent sugars, mainly xylose and mannose. These enzymes can be divided into two groups: core enzymes that catalyze cleavage of the polysaccharide backbone and ancillary enzymes that perform the removal of functional groups. Examples of the core enzymes are *endo*-β-1,4-xylanase (EC 3.2.1.8), xylan-1,4-β-xylosidase (EC 3.2.1.37), *endo*-1,4-β-mannanase (EC 3.2.1.78), and β-1,4-mannosidase (EC 3.2.1.25). Ancillary enzymes include β-glucuronidase (EC 3.2.1.139), acetyl xylan esterase (EC 3.2.1.55), ferulic acid esterase (EC 3.1.1.73), and *p*-coumaroyl acid esterase (EC 3.1.1.B10).

In vivo these enzymes are contained in multi-enzyme complexes, so-called cellulosomes (Artzi et al. 2017) produced by many cellulolytic fungi and bacteria. Cellulosomes have a distinct advantage compared to simple mixtures of the free enzymes owing to the close proximity of the enzymes in the former. This advantageous close proximity of the individual enzymes can be mimicked in combi-CLEAs without many of the disadvantages of cellulosomes.

For example, a xylanase-mannanase combi-CLEA was prepared and successfully applied in the conversion of lime-pretreated sugarcane bagasse and milled corn stover (Bhattacharya and Pletschke 2015). The authors concluded that the efficiency of combi-CLEAs makes them ideal candidates for achieving cost-effective application of lignocellulytic enzymes. Similarly, co-immobilization of xylanase, β-1,3-glucanase, and cellulase gave a combi-CLEA that was more thermally stable than the free enzymes and retained more than 97% of its activity on storing at 4 °C for 11 weeks, compared to 65% for the free enzymes (Periyasamy et al. 2016). The combi-CLEA was successfully used in the hydrolysis of ammonia-cooked sugarcane bagasse and could be recycled six times. Various groups have reported the successful immobilization of a cellulase cocktail as cross-linked enzyme aggregates (CLEAs) (Dalal et al. 2007b; Perzon et al. 2017; Jones and Vasudevan 2010; Li et al. 2012; Jamwal et al. 2016). Interestingly, hybrid cellulase CLEAs containing a silica core, prepared by physical adsorption of cellulase CLEAs on a highly porous silica support, exhibited twice as much activity as the regular CLEA, and settled better after the hydrolysis, thus facilitating its separation (Sutarlie and Yang 2013).

7.5.3 *Magnetic Immobilized Enzymes in Lignocellulose Conversion*

Probably the most exciting and challenging development in the processing of lignocellulosic biomass is the use of magnetic immobilized enzymes, to provide the possibility of achieving substantial cost reductions by magnetic separation and recycling of the enzymes (Asgher et al. 2014). Immobilization of the cellulase enzyme cocktail on magnetic particles has been extensively studied, either on prefabricated magnetic (nano)carriers (Alftren and Hobley 2014; Abraham et al. 2014; Zhang et al. 2015b, 2016; Jordan et al. 2011; Khoshnevisan et al. 2011; Zang et al. 2014; Roth et al. 2016) or as magnetic CLEAs (Khorshidi et al. 2016; Xie et al. 2012; Jia et al. 2017), but activities were generally measured only in the hydrolysis of the water-soluble carboxymethyl cellulose as a model for the complex mixture derived from lignocellulose. Immobilization of β-glucosidase, one of the enzymes contained in the cellulase cocktail, on magnetic silica-based particles has also been described (Alftren and Hobley 2013). It is also worth noting that it may not be essential to immobilize all the enzymes in the cellulase cocktail. Immobilization and recycling of a selection of the enzymes present could lead to significant cost reductions.

Bhattachariya and Pletschke (2014) prepared mCLEAs of a bacterial xylanase and observed that incorporation of Ca^{2+} ions in the CLEA led to increased thermal stability. The Ca-mCLEA exhibited 35% more activity than the free enzyme and a ninefold higher sugar release from ammonia pretreated sugarcane bagasse. Similarly, Illias and coworkers (Shaarani et al. 2016) prepared mCLEAs of a recombinant xylanase from *Trichoderma reesei* using maghemite (γ-Fe_2O_3) rather than the more usual magnetite (Fe_3O_4) nanoparticles.

As discussed in Sect. 7.5.1, much attention is currently being devoted to lignocellulose pretreatment in ILs and DESs or their mixtures with water. Obviously, it would be attractive to conduct subsequent enzymatic hydrolysis in the same medium, which requires that the cellulase enzyme cocktail be active and stable in the presence of ILs and/or DESs, which has been demonstrated with [emim] [OAc] (Datta et al. 2010). However, to our knowledge, such magnetic cellulase CLEAs have not yet been used in IL- or DES-containing media. Xu and coworkers (Xu et al. 2015) used a cellulase from *Trichoderma aureoviride*, encapsulated in alginate beads, in the enzymatic in situ saccharification of rice straw pretreated in aqueous [emim] [$(MeO)_2PO_2$]. Interestingly, Ogino and coworkers (Nakashima et al. 2011) combined pretreatment and saccharification with fermentation in a direct SSF production of bioethanol by a cellulase-displaying, ionic liquid-resistant yeast. The cellulase enzyme cocktail is known to be active and stable in designer ILs and DESs (Nakashima et al. 2011). For example, cellulase was engineered to improve its stability in mixtures of ChCl/Gly and concentrated seawater as a low-cost reaction medium for enzymatic hydrolysis of cellulose (Lehmann et al. 2012).

7.6 Conclusions and Future Outlook

The costs of the hydrolytic conversion of first- and second-generation biomass to fermentable sugars can be optimized by reducing the costs of the enzymes involved, which are currently used on a single-use, throw-away basis. This change can be achieved by enabling the recycling of the enzymes via immobilization. However, this move can be challenging because of the heterogeneous nature of the reaction mixture, which contains various suspended solids. A potentially attractive methodology is to produce magnetic immobilized enzymes, such as mCLEAs, which can be separated from other suspended solids on an industrial scale using commercially available equipment. Second-generation biofuels are still at the beginning of the learning curve, and significant future cost reductions are still needed to achieve commercial viability. In this context, cost-effective immobilization and recycling of the complex enzyme cocktail involved can make an important contribution. In addition, we expect that such technologies will, in the future, be widely applied to conversions of polysaccharides in other areas such as food and beverage processing (Sojitra et al. 2016; Dal Magro et al. 2016).

References

Abraham RE, Verma ML, Barrow CJ, Puri M (2014) Suitability of magnetic nanoparticle immobilised cellulases in enhancing enzymatic saccharification of pretreated hemp biomass. Biotechnol Biofuels 7:90

Achinivu EC, Howard RM, Li G, Gracz H, Henderson WA (2014) Lignin extraction from biomass with protic ionic liquids. Green Chem 16:1114–1119

Alftren J, Hobley TJ (2013) Covalent immobilization of β-glucosidase on magnetic particles for lignocellulose hydrolysis. Appl Biochem Biotechnol 169:2076–2089

Alftren J, Hobley TJ (2014) Immobilization of cellulase mixtures on magnetic particles for hydrolysis of lignocellulose and ease of recycling. Biomass Bioenergy 65:72–78

Alvira P, Tomas-Pejo E, Ballesteros M, Negro MJ (2010) Pretreatment technologies for an efficient bioethanol production process based on enzymatic hydrolysis: a review. Bioresour Technol 101:4851–4861

Ansari SA, Husain Q (2012) Potential applications of enzymes immobilized on/in nanomaterials: a review. Biotechnol Adv 30:512–523

Artzi L, Bayer EA, Moraïs S (2017) Cellulosomes: bacterial nanomachines for dismantling plant polysaccharides. Nat Rev Microbiol 15:83–95

Asgher M, Shahid M, Kamal S, Iqbal HMN (2014) Recent trends and valorization of immobilization strategies and ligninolytic enzymes by industrial biotechnology. J Mol Catal B Enzym 101:56–66

Bhattacharya A, Pletschke BI (2014) Magnetic cross-linked enzyme aggregates (CLEAs): a novel concept towards carrier free immobilization of lignocellulolytic enzymes. Enzyme Microb Technol 61–62:17–27

Bhattacharya A, Pletschke BI (2015) Strategic optimization of xylanase-mannanase combi-CLEAs for synergistic and efficient hydrolysis of complex lignocellulosic substrates. J Mol Catal B Enzym 115:140–150

Bhattacharya AS, Bhattacharya A, Pletschke BI (2015) Synergism of fungal and bacterial cellulases and hemicellulases: a novel perspective for enhanced bio-ethanol production. Biotechnol Lett 37:1117–1129

Bogel-Lukasic R (ed) (2016) Ionic liquids in the biorefinery concept: challenges and perspectives. RSC Green Chemistry Series No. 36. Royal Society of Chemistry, London

Bornscheuer U, Buchholz K, Seibel J (2014) Enzymatic degradation of (ligno)cellulose. Angew Chem Int Ed 53:10876–10893

Brandt A, Ray MJ, To TQ, Leak DJ, Murphy RJ, Welton T (2011) Ionic liquid pretreatment of lignocellulosic biomass with ionic liquid–water mixtures. Green Chem 13:2489–2499

Brandt A, Gräsvik J, Hallett JP, Welton T (2013) Deconstruction of lignocellulosic biomass with ionic liquids. Green Chem 15:550–583

Cantone S, Ferrario V, Corici L, Ebert C, Fattor D, Spizzo P, Gardossi L (2013) Efficient immobilisation of industrial biocatalysts: criteria and constraints for the selection of organic polymeric carriers and immobilisation methods. Chem Soc Rev 42:6262–6278

Cao L (2005) Carrier-bound immobilized enzymes, principles, applications and design. Wiley-VCH, Weinheim

Cao L, van Rantwijk F, Sheldon RA (2000) Cross-linked enzyme aggregates: a simple and effective method for the immobilization of penicillin acylase. Org Lett 2:1361–1364

Cao L, van Langen L, Sheldon RA (2003) Immobilised enzymes: carrier-bound or carrier-free? Curr Opin Biotechnol 14:387–394

da Costa Lopes AM, Joao KG, Rubik DF, Bogel-Lukasik E, Duarte LC, Andreaus J, Bogel-Lukasik R (2013) Pre-treatment of lignocellulosic biomass using ionic liquids: wheat straw fractionation. Bioresour Technol 142:198–208

Cruz-Izquierdo A, Pico EA, Lopez C, Serra JL, Lama MJ (2014) Magnetic cross-linked enzyme aggregates (mCLEAs) of *Candida antarctica* lipase: an efficient and stable biocatalyst for biodiesel synthesis. PLoS One. https://doi.org/10.1371/journal.pone.0115202

Cui JD, Ja SR (2015) Optimization protocols and improved strategies of cross-linked enzyme aggregates technology: current development and future challenges. Crit Rev Biotechnol 35(1):15–28

Cui JD, Cui LL, Zhang SP, Zhang YF, Su ZG, Ma GH (2014) Hybrid magnetic cross-linked enzyme aggregates of phenylalanine ammonia lyase from *Rhodotorula glutinis*. Plos One 9(5):e97221

Cui J, Cui L, Jia S, Su Z, Zhang S (2016) Hybrid cross-linked lipase aggregates with magnetic nanoparticles: a robust and recyclable biocatalysis for the epoxidation of oleic acid. J Agric Food Chem 64:7179–7187

Dal Magro L, Hertz PF, Fernandez-Lafuenta R, Klein MP, Rodrigues RC (2016) Preparation and characterization of a Combi-CLEAs from pectinases and cellulases: a potential biocatalyst for grape juice clarification. RSC Adv 6:27242–27251

Dalal S, Kapoor M, Gupta MN (2007a) Preparation and characterization of combi-CLEAs catalyzing multiple non-cascade reactions. J Mol Catal B Enzym 44:128–132

Dalal S, Sharma A, Gupta MN (2007b) A multipurpose immobilized biocatalyst with pectinase, xylanase and cellulase activities. Chem Cent J 1:16

Datta S, Holmes B, Park JI, Chen Z, Dibble DC, Hadi M, Blanch HW, Simmons BA, Sapra R (2010) Ionic liquid tolerant hyperthermophilic cellulases for biomass pretreatment and hydrolysis. Green Chem 12:338–345

De Jong E, Higson A, Walsh P, Wellisch M (2012) Product developments in the bio-based chemicals arena. Biofuels Bioprod Bioref 6:606–624

Dominguez de Maria P (2014) Recent trends in (ligno)cellulose dissolution using neoteric solvents: switchable, distillable and bio-based ionic liquids. J Chem Technol Biotechnol 89:11–18

Fang Z, Smith RL, Qi X (eds) (2014) Production of biofuels and chemicals with ionic liquids. Springer, Dordrecht

Francisco M, van den Bruinhorst A, Kroon MC (2012) New natural and renewable low transition temperature mixtures (LTTMs): screening as solvents for lignocellulosic biomass processing. Green Chem 14:2153–2157

Gallezot P (2012) Conversion of biomass to selected chemical products. Chem Soc Rev 41:1538–1558

Gawande MB, Branco PS, Varma RS (2013) Nano-magnetite (Fe_3O_4) as a support for recyclable catalysts in the development of sustainable methodologies. Chem Soc Rev 42:3371–3393

George A, Brandt A, Tran K, Nizan SMS, Zahari S, Klein-Marcuschamer D, Sun N, Sathitsuksanoh N, Shi J, Stavila V, Parthasari R, Singh S, Holmes BM, Welton T, Simmons BA, Hallett JP (2015) Design of low cost ionic liquids for lignocellulosic biomass pretreatment. Green Chem 17:1728–1734

Gupta K, Kumar Jana A, Kumar S, Maiti M (2013) Immobilization of amyloglucosidase from SSF of *Aspergillus niger* by crosslinked enzyme aggregate onto magnetic nanoparticles using minimum amount of carrier and characterizations. J Mol Catal B Enzym 98:30–36

Hartmann M, Kostrov X (2013) Immobilization of enzymes on porous silicas: benefits and challenges. Chem Soc Rev 42:6277–6289

Hemsworth GR, Johnston EM, Davies GJ, Walton PH (2015) Lytic polysaccharide monooxygenases in biomass conversion. Trends Biotechnol 33(12):747–761

Henstra AM, Sipma J, Rinzema A, Stams AJM (2007) Microbiology of synthesis gas fermentation for biofuel production. Curr Opin Biotechnol 18:200–206

Hickey AM, Marle L, McCreedy T, Watts P, Greenway GM, Littlechild JA (2007) Immobilization of thermophilic enzymes in minitaturized flow reactors. Biochem Soc Trans 35:1621–1623

Hou XD, Li N, Zong MH (2013) Significantly enhancing enzymatic hydrolysis of rice straw after pretreatment using renewable ionic liquid–water mixtures. Bioresour Technol 136:469–474

Imhof P, van der Waal JC (eds) (2013) Catalytic process development for renewable materials. Wiley-VCH, Weinheim

Jablonski M, Skulcova A, Kamenska L, Vrska M, Sima J (2015) Deep eutectic solvents: fractionation of wheat straw. Bioresources 10:8039–8047

Jamwal GS, Chauhan JH, Ahn N, Reddy S (2016) Cellulase stabilization by crosslinking with ethylene glycol dimethacrylate and evaluation of its activity including in a water-ionic liquid mixture. RSC Adv 6:25485–25491

Janssen MHA, van Pelt S, Rasmussen J, Sheldon RA, Koning P (2016) Patent submitted

Jia J, Zhang W, Yang Z, Yang X, Wang N, Yu X (2017) Novel magnetic cross-linked cellulose aggregates with a potential application in lignocellulosic biomass bioconversion. Molecules 22(2):269/1–269/14

Johansen KS (2016) Discovery and industrial applications of lytic polysaccharide monooxygenases. Biochem Soc Trans 44(1):143–149

Johnson AK, Zawadzka AM, Deobald LA, Crawford RL, Paszczynski AJ (2008) Novel method for immobilization of enzymes to magnetic nanoparticles. J Nanopart Res 10:1009–1025

Jones PO, Vasudevan PT (2010) Cellulose hydrolysis by immobilized *Trichoderma reesei* cellulase. Biotechnol Lett 32:103–106

Jordan J, Kumar CSSR, Theegala C (2011) Preparation and characterization of cellulase-bound magnetite nanoparticles. J Mol Catal B Enzym 68:139–146

Jouzani GS, Taherzadeh TJ (2015) Advances in consolidated bioprocessing systems for bioethanol and biobutanol production from biomass: a comprehensive review. Biofuel Res J 5:152–195

Khorshidi KJ, Lenjannezhadian H, Jamalan M, Zeinali M (2016) Preparation and characterization of nanomagnetic cross-linked cellulase aggregates for cellulose bioconversion. J Chem Technol Biotechnol 91:539–546

Khoshnevisan K, Bordbar AK, Zare D, Davoodi D, Noruzi M, Barkhi M, Tabatabaei M (2011) Immobilization of cellulase enzyme on superparamagnetic nanoparticles and determination of its activity and stability. Chem Eng J 171:669–673

Kopp W, Da Costa TP, Pereira SC, Jafelicci M, Giordano RC, Marques RFC, Araujo-Moreira FM, Giordano RLC (2014) Easily handling penicillin G acylase magnetic cross-linked enzymes aggregates: catalytic and morphological studies. Process Biochem 49:38–46

Kroon MC, Casal MF, Van den Bruinhorst A (2013) Pretreatment of lignocellulosic biomass and recovery of substituents using natural deep eutectic solvents/compound mixtures with low transition temperatures. WO 2013/153203 A1. Technische Universiteit, Eindhoven

Kumar P, Barrett DM, Delwiche MJ, Stroeve P (2009) Methods for pretreatment of lignocellulosic biomass for efficient hydrolysis and biofuel production. Ind Eng Chem Res 48:3713–3729

Kumar VV, Sivanesan S, Cabana H (2014) Magnetic cross-linked laccase aggregates: bioremediation tool for decolorization of distinct classes of recalcitrant dyes. Sci Total Environ 487:830–839

Kumar AK, Parikh BS, Pravakar M (2016) Natural deep eutectic solvent mediated pretreatment of rice straw: bioanalytical characterization of lignin extract and enzymatic hydrolysis of pretreated biomass residue. Environ Sci Pollut Res 23(10):9265–9275

Land H, Hendit-Forssell P, Martinelle M, Berglund P (2016) One-pot biocatalytic amine transaminase/acyl transferase cascade for aqueous formation of amides from aldehydes or ketones. Catal Sci Technol 6:2897–2900

Lehmann C, Sibilla F, Maugeri Z, Streit WR, Domínguez de María P, Martinez R, Schwaneberg U (2012) Reengineering CelA2 cellulase for hydrolysis in aqueous solutions of deep eutectic solvents and concentrated seawater. Green Chem 14:2719–2726

Li Q, He Y, Xian M, Jun G, Ju X, Yang J, Li L (2009) Improving enzymatic hydrolysis of wheat straw using ionic liquid 1-ethyl-3-methylimidazolium diethyl phosphate pretreatment. Bioresour Technol 100:3570–3575

Li B, Dong S, Xie X, Xu Z, Li L (2012) Preparation and properties of cross-linked enzyme aggregates of cellulase. Adv Mater Res 581–582:257–260

Lindley J (1982) The use of magnetic techniques in the development of a hydrogenation process. IEEE Trans Magnetics 18:836–840

Liu Y, Jia S, Wu Q, Ran J, Zhang W, Wu S (2011) Studies of Fe_3O_4-chitosan nanoparticles prepared by co-precipitation under the magnetic field for lipase immobilization. Catal Commun 12:717–720

Long J, Li X, Guo G, Wang F, Yu Y, Wang L (2012) Simultaneous delignification and selective catalytic transformation of agricultural lignocellulose in cooperative ionic liquid pairs. Green Chem 14:1935–1941

Lopez-Gallego F, Schmidt-Dannert C (2010) Multi-enzymatic synthesis. Curr Opin Chem Biol 14:174–183

Magalhaes da Silva SP, da Costa Lopes AM, Roseiro LB, Bogel-Lukasik R (2013) Novel pretreatment and fractionation method for lignocellulosic biomass using ionic liquids. RSC Adv 3:16040–16050

Magner E (2013) Immobilisation of enzymes on mesoporous silicate materials. Chem Soc Rev 42:6213–6222

Mahmod SS, Yusof F, Saedi Jami M, Khanahmadi S, Shah H (2015) Development of an immobilized biocatalyst with lipase and protease activities as a multipurpose cross-linked enzyme aggregate. Process Biochem 50:2144–2157

Menon V, Rao M (2012) Trends in bioconversion of lignocellulose: biofuels, platform chemicals and biorefinery concept. Prog Energy Combust Sci 38:522–550

Morales-de la Rosa M, Campos-Martin JM, Fierro JLG (2012) High glucose yields from the hydrolysis of cellulose dissolved in ionic liquids. Chem Eng J 181–182:538–541

Munasinghe PC, Khanal SK (2007) Biomass derived syn gas fermentation into biofuels. Bioresource Technol 101:5013–5022

Muschiol J, Peters C, Oberleitner N, Mihovilovic MD, Bornscheuer UT, Rudroff F (2015) Cascade catalysis: strategies and challenges en route to preparative synthetic biology. Chem Commun 51:5798–5811

Nadar SS, Rathod VK (2016) Magnetic macromolecular cross-linked enzyme aggregates (CLEAs) of glucoamylase. Enzyme Microb Technol 83:78–87

Nakashima K, Yamaguchi K, Tanguchi N, Arai S, Yamada R, Katahira S, Ishida N, Takahashi H, Ogino C, Kondo A (2011) Direct bioethanol production from cellulose by the combination of cellulase-displaying yeast and ionic liquid pretreatment. Green Chem 13:2948–2953

Netto CGCM, Toma HE, Andrade LH (2013) Superparamagnetic nanoparticles as versatile carriers and supporting materials for enzymes. J Mol Catal B Enzym 85–86:71–92

Nor NAM, Mustapha WAW, Hassa O (2016) Deep eutectic solvent (DES) as a pretreatment for oil palm empty fruit bunch (OPEFB) in sugar production. Proc Chem 18:147–154

Olofsson K, Bertilsson M, Liden G (2008) A short review on SSF: an interesting process option from lignocellulosic feedstocks. Biotechnol Biofuels 1:7

van Pelt S, Quignard S, Kubac S, Sorokin DY, van Rantwijk F, Sheldon RA (2008) Nitrile hydratase CLEAs: the immobilization and stabilization of an industrially important enzyme. Green Chem 10:395–400

Periyasamy K, Santhalembi L, Mortha G, Aurousseau M, Subramanian S (2016) Carrier-free co-immobilization of xylanase, cellulase and β-1,3-glucanase as combined cross-linked enzyme aggregates (combi-CLEAs) for one-pot saccharification of sugarcane bagasse. RSC Adv 6:32849–32857

Perzon A, Dicko C, Cobanoglu O, Yuekselen O, Eryilmaz J, Dey ES (2017) Cellulase cross-linked enzyme aggregates (CLEA) activities can be modulated and enhanced by precipitant selection. J Chem Technol Biotechnol 92(7):1645–1649

Procentese A, Johnson E, Orr V, Garruto Campanile A, Wood JA, Marzocchella A, Rehmann L (2015) Deep eutectic solvent pretreatment and subsequent saccharification of corncob. Bioresour Technol 192:31–36

Qiu Z, Aita GM, Walker MS (2012) Effect of ionic liquid pretreatment on the chemical composition, structure and enzymatic hydrolysis of energy cane bagasse. Bioresour Technol 117:251–256

Rabemanolontsoa H, Saka S (2016) Various pretreatments of lignocellulosics. Bioresour Technol 199:83–91

Ranganath KVS, Glorius F (2011) Superparamagnetic nanoparticles for symmetric catalysis: a perfect match. Catal Sci Technol 1:13–22

Reetz MT (2013) Practical protocols for lipase immobilization via sol-gel techniques. In: Guisan JM (ed) Immobilization of enzymes and cells, 3rd edn. Springer, New York, pp 241–254

Renewable Fuels Association (2016) Ethanol Industry Outlook: see http://www.ethanolrfa.org/wp-content/uploads/2016/02/Ethanol-Industry-Outlook-2016

Rinaldi R (2014) Plant biomass fractionation meets catalysis. Angew Chem Int Ed 53:8559–8560

Rossi LM, Costa NJS, Silva FP, Wojcieszak R (2014) Magnetic nanomaterials in catalysis: advanced catalysts for magnetic separation and beyond. Green Chem 16:2906–2933

Roth HC, Schwaminger SP, Peng F, Berensmeier S (2016) Immobilization of cellulase on magnetic nanocarriers. Chemistry Open 5:183–187

Santacoloma PAD, Sin G, Gernacy KV, Woodley JM (2011) Multienzyme-catalyzed processes: next-generation biocatalysis. Org Proc Res Dev 15:203–212

Schoevaart R, Wolbers MW, Golubovic M, Ottens M, Kieboom APG, van Rantwijk F, van der Wielen LAM, Sheldon RA (2004) Preparation, optimization and structure of cross-linked enzyme aggregates. Biotechnol Bioeng 87:754–762

Schrittweiser JH, Sattler J, Resch V, Mutti FG, Kroutil W (2011) Recent biocatalytic oxidation-reduction cascades. Curr Opin Chem Biol 15:249

Shaarani SM, Jahim JM, Rahman RA, Idris A, Murad AMA, Illias RM (2016) Silanized maghemite for cross-linked enzyme aggregates of recombinant xylanase from *Trichoderma reesei*. J Mol Catal B Enzym 133:65–76

Sharma M, Mukesh M, Mondal D, Prasad K (2013) Dissolution of α-chitin in deep eutectic solvents. RSC Adv 3:18149–185155

Sheldon RA (1983) Chemicals from synthesis gas. Reidel, Dordrecht

Sheldon RA (2011) Cross-linked enzyme aggregates as industrial biocatalysts. Org Proc Res Dev 15:213–223

Sheldon RA (2013) Characteristic features and biotechnological applications of cross-linked enzyme aggregates (CLEAs). Appl Microbiol Biotechnol 92:467–477

Sheldon RA, Sorgedrager MJ, Kondor B (2012) Non-leachable magnetic cross-linked enzyme aggregate. PCT Int Appl WO 2012/023847 A2; US Patent Appl., US 2013/0196407 A1, 2013, to CLEA Technologies B.V. (NL)

Sheldon RA, van Pelt S, Kanbak-Adsu S, Rasmussen J, Janssen MHA (2013) Cross-linked enzyme aggregates (CLEAs) in organic synthesis. Aldrichim Acta 46(3):81–93

Shi J, Gladden JM, Sathitsuksanoh N, Kambam P, Sandoval L, Mitra D, Zhang S, George A, Singer SW, Simmons BA, Singh S (2013) One-pot ionic liquid pretreatment and saccharification of switchgrass. Green Chem 15:2579–2589

Sojitra UV, Nadar SS, Rathod VK (2016) A magnetic tri-enzyme nanobiocatalyst for fruit juice clarification. Food Chem 213:296–305

Sun N, Rahman M, Qin Y, Maxim ML, Rodriguez H, Rogers RD (2009) Complete dissolution and partial delignification of wood in the ionic liquid 1-ethyl-3-methylimidazolium acetate. Green Chem 11:646–655

Sun N, Parthasarathi R, Socha AM, Shi J, Zhang S, Stavila V, Sale KL, Simmons BA, Singh S (2014) Understanding pretreatment efficacy of four cholinium and imidazolium ionic liquids by chemistry and computation. Green Chem 16:2546–2557

Sutarlie L, Yang KL (2013) Hybrid cellulase aggregate with a silica core for hydrolysis of cellulose and biomass. J Colloid Interface Sci 411:76–81
Swatloski RP, Spear SK, Holbrey JD, Rogers RD (2002) Dissolution of cellulose with ionic liquids. J Am Chem Soc 124:4974–4975
Talekar S, Ghodake V, Ghotage T, Rathod P, Deshmukh P, Nadar S, Mulla M, Ladole M (2012) Novel magnetic cross-linked enzyme aggregates (magnetic CLEAs) of alpha amylase. Bioresour Technol 123:542–547
Talekar S, Joshi A, Joshi G, Kamat P, Haripurkar R, Kambale S (2013a) Parameters in preparation and characterization of cross linked enzyme aggregates (CLEAs). RSC Adv 3:12485–12511
Talekar S, Desai S, Pillai M, Nagavekar N, Ambarkar S, Surnis S, Ladole M, Nadar S, Mulla M (2013b) Carrier free co-immobilization of glucoamylase and pullulanase as combi-cross linked enzyme aggregates (combi-CLEAs). RSC Adv 3:2265–2271
Talekar S, Pandharbale A, Ladole M, Nadar S, Mulla M, Japhalekar K, Pattankude K, Arage D (2013c) Carrier free co-immobilization of alpha amylase, glucoamylase and pullulanase as combined cross-linked enzyme aggregates (combi-CLEAs): a tri-enzyme biocatalyst with one pot starch hydrolytic activity. Bioresour Technol 147:269–275
Tuck CO, Perez E, Horvath IT, Sheldon RA, Poliakoff M (2012) Valorization of biomass: deriving more value from waste. Science 337:695–699
Tudorache M, Gheorghe A, Viana AS, Parvulescu VI (2016) Biocatalytic epoxidation of α-pinene to oxy-derivatives over cross-linked lipase aggregates. J Mol Catal B Enzym 134:9–15
Ungerean M, Csanádi Z, Gubicza L, Peter F (2014) An integrated process of ionic liquid pretreatment and enzymatic hydrolysis of lignocellulosic biomass with immobilised "cellulase". Bioresources 9(4):6100–6106
Valesco-Lozano S, Lopez-Gallego F, Mateoes-Diaz JC, Favela-Torres E (2015) Cross-linked enzyme aggregates (CLEA) in enzyme improvement: a review. Biocatalysis 1:166–177
Vancov T, Alston AS, Brown T, McIntosh S (2012) Use of ionic liquids in converting lignocellulosic material to biofuels. Renew Energy 45:1–6
Verdía P, Brandt A, Hallett JP, Ray MJ, Welton T (2014) Fractionation of lignocellulosic biomass with the ionic liquid 1-butylimidazolium hydrogen sulfate. Green Chem 16:1617–1627
Viell J, Harwardt A, Seiler J, Marquardt W (2013a) Is biomass fractionation by organosolv-like processes economically viable? A conceptual design study. Bioresour Technol 150:89–97
Viell J, Wulfhorst H, Schmidt T, Commandeur U, Fischer R, Spiess A, Marquardt W (2013b) An efficient process for the saccharification of wood chips by combined ionic liquid pretreatment and enzymatic hydrolysis. Bioresour Technol 146:144–151
Walton PH, Davies GJ (2016) On the catalytic mechanisms of lytic polysaccharide monooxygenases. Curr Opin Chem Biol 31:195–207
Whitesides GM, Hill CL, Brunie JC (1976) Magnetic filtration of small heterogeneous catalyst particles. preparation of ferrimagnetic catalyst supports. I&EC Process Des Dev 15:226
Wildschut J, Smit AT, Reith JH, Huijgen WJJ (2013) Ethanol-based organosolv fractionation of wheat straw for the production of lignin and enzymatically digestible cellulose. Bioresour Technol 135:58–66
Xia S, Baker GA, Ravula S, Zhao H (2014) Aqueous ionic liquids and deep eutectic solvents for cellulosic biomass pretreatment and saccharification. RSC Adv 4:10586–10596
Xie X, Li B, Wu Z, Dong S, Li L (2012) Preparation of cross-linked cellulase aggregates onto magnetic chitosan microspheres. Adv Mater Res 550-553:1566–1571
Xu JK, Sun YC, Sun RC (2014) Ionic liquid pretreatment of woody biomass to facilitate biorefinery: structural elucidation of alkali-soluble hemicelluloses. ACS Sustain Chem Eng 2:1035–1042
Xu J, Liu J, He L, Hu B, Dai B, Wu B (2015) Enzymatic in situ saccharification of rice straw in aqueous-ionic liquid media using encapsulated *Trichoderma aureoviride* cellulase. J Chem Technol Biotechnol 90:57–63
Xue R, Woodley JM (2012) Process technology for multi-enzymatic reaction systems. Bioresour Technol 115: 183–195

Yang ST, El-Enshasy HA, Thongchul N (2013) Bioprocessing technologies biorefinery for sustainable production of fuels, chemicals and polymers. Wiley, Hoboken

Yiu HHP, Keane MA (2012) Enzyme–magnetic nanoparticle hybrids: new effective catalysts for the production of high value chemicals. J Chem Technol Biotechnol 87:583–594

Zamani F, Hosseini SM (2014) Palladium nanoparticles supported on Fe_3O_4/amino acid nanocomposite: highly active magnetic catalyst for solvent-free aerobic oxidation of alcohols. Catal Commun 43:164–168

Zang L, Qiu J, Wu X, Zhang W, Sakai E, Wei Y (2014) Preparation of magnetic chitosan nanoparticles as support for cellulase immobilization. Ind Eng Chem Res 53:3448–3454

Zavrel M, Bross D, Funke M, Büchs J, Spiess AC (2009) High-throughput screening for ionic liquids dissolving (ligno-)cellulose. Bioresour Technol 100:2580–2587

Zhang WW, Yang XL, Jia JQ, Wang N, Hu CJ, Yu XQ (2015a) Surfactant-activated magnetic cross-linked enzyme aggregates (magnetic CLEAs) of *Thermomyces lanuginosus* lipase for biodiesel production. J Mol Catal B Enzym 115:83–89

Zhang W, Qiu J, Feng H, Zang L, Sakai E (2015b) Increase in stability of cellulase immobilized on functionalized magnetic nanospheres. J Magn Magn Mater 375:117–123

Zhang Q, Kang JK, Yang B, Zhao L, Hou Z, Tang B (2016) Immobilized cellulase on Fe_3O_4 nanoparticles as a magnetically recoverable biocatalyst for the decomposition of corncob. Chin J Catal 37:389–397

Zhao X, Cheng K, Liu D (2009) Organosolv pretreatment of lignocellulosic biomass for enzymatic hydrolysis. Appl Microbiol Biotechnol 82:815–827

Zhou Z, Hartmann M (2012) Recent progress in biocatalysis with enzymes immobilized on mesoporous hosts. Top Catal 55:1081–1100

Zhou L, Tang W, Jiang Y, Ma L, He Y, Gao J (2016) Magnetic combined cross-linked enzyme aggregates of horseradish peroxidase and glucose oxidase: an efficient biocatalyst for dye decolourization. RSC Adv 6:90061–90068

Chapter 8
Sustainability of Biomass

Arnaldo Walter, Joaquim E.A. Seabra, Pedro Gerber Machado, Bruna de Barros Correia, and Camila Ortolan Fernandes de Oliveira

Abstract The bio-based economy is considered one of the options for mitigating greenhouse gas (GHG) emissions and is pursued by many countries seeking not only emissions reductions but also greater independency and security. In this context, biofuels production has expanded in the first decade of this century, and the same increase can occur with biomaterials in the years to come. However, despite the large appeal of biofuel, various concerns regarding its sustainability have been raised, constraining production and imposing the necessity to attest compliance with some principles and criteria. As a result of interest group advocacy, a diversity of sustainability initiatives has emerged in recent years in the bioenergy context, which may soon be extended to chemicals and biomaterials as well. This chapter presents the main technical regulations and standards for bioenergy currently in place and discusses the social, economic, and environmental issues these address. Guided by the set principles and criteria, there is evidence supporting that, if implemented correctly, the bio-based economy can indeed offer significant contributions toward sustainable development.

Keywords Bioeconomy • Climate change mitigation • Socioeconomic development

8.1 Introduction

Bioenergy has been promoted recently in different countries with the aims of reducing dependence on nonrenewable resources, supporting local economies, and offering a better quality of life in rural areas. The same arguments would be valid for large-scale production of renewable chemicals and biomaterials: the development of a bioeconomy is considered one of the options for mitigating greenhouse gas (GHG) emissions and is pursued by many countries seeking not only emissions reductions but also greater independency and security.

A. Walter • J.E. A. Seabra (✉) • P.G. Machado • B. de Barros Correia • C.O.F. de Oliveira
Faculdade de Engenharia Mecânica, Universidade Estadual de Campinas (Unicamp), Campinas, SP, Brazil
e-mail: jseabra@fem.unicamp.br

S. Vaz Jr. (ed.), *Biomass and Green Chemistry*,
https://doi.org/10.1007/978-3-319-66736-2_8

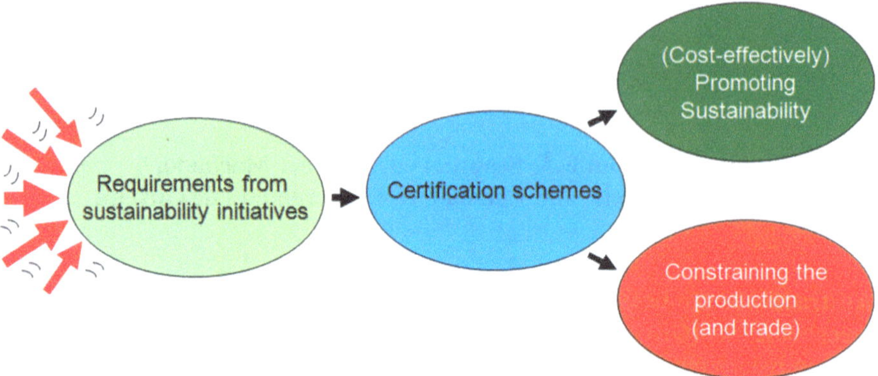

Fig. 8.1 Certification schemes as the consequence of imposed sustainability requirements, and the main related concerns

A bio-based economy relies on biotechnology to guarantee the supply of goods without compromising nonrenewable resources. The bioeconomy concept was born in the 1990s from the search for a new economic paradigm (Richardson 2012), but it is still immature.

In the case of bioenergy, despite its large appeal, various concerns regarding its sustainability have been raised, constraining production and imposing the necessity to attest whether some principles and criteria have been fulfilled. Again, it seems that the same can occur with chemicals and biomaterials, defining limits for production.

Criticisms regarding the benefits of liquid biofuels followed the very fast growth of production in the first decade of this century (REN21 2016), such as no effective reduction of GHG emissions in comparison to the fossil fuels they displace, negative impacts on the environment and on natural resources (e.g., on biodiversity, on water resources), negative impacts of biofuels production over food supply, undesirable working conditions, land ownership considerations, and disrespect for land and water use rights.

The pressures that resulted in different sustainability initiatives came first from nongovernmental organizations (NGOs), mostly European NGOs, supported by the outcomes of academics that put priority on studying potential impacts of large-scale biofuel production. Pilgrim and Harvey (2010) state that "…. a consortium of NGOs has played a significant role in shaping the market for, and restricting the use of, biofuels as an alternative to conventional fuels for road transport in Europe," and that they (e.g., Greenpeace, Oxfam, WWF, Friends of Earth, …) "have developed policy, agreed a political campaign, and exercised political influence" that resulted in strict regulations.

Hence, as a result of interest group advocacy (Endres et al. 2015), policymakers decided to implement sustainability initiatives that set conditions for commercializing liquid biofuels in the most important consumer markets (i.e., Europe and the United States). In Europe, to verify compliance with principles and criteria, it was determined that certified production is required. Figure 8.1 schematically represents the process and the concerns regarding the adoption of sustainability initiatives: Are these schemes able to cost-effectively promote sustainability? Could sustainability initiatives foster or constrain bioenergy production (and trade)?

This chapter focuses on the main sustainability initiatives for bioenergy currently in place, as well as the social, economic, and environmental issues these initiatives address. Bioenergy has been the focus of these initiatives, but renewable chemicals and biomaterials may also be included in the future. The chapter first presents the main concerns raised with the expansion of biofuels development and the technical regulations, standards, and certification schemes that have emerged. The requirements of these initiatives are then discussed, with a separate section on the overarching issue of land use change.

8.2 The Scientific Arguments

In the early 2000s, accompanying the growth of biofuels production came criticisms of the real benefits of biofuels, which resulted in different sustainability initiatives. Some examples of the academic publications that illuminated concerns regarding liquid biofuels are presented next. Some of these papers addressed extreme or very specific cases and thus their conclusions cannot be generalized, whereas the conclusions of others were minimized, or even denied, by subsequent publications on the same subject.

Among these papers, those published by Fargione et al. (2008) and by Searchinger et al. (2008) were crucial; both consider the land use problem, which presented the idea that indirect impacts of land use change (i.e., deforestation) could void the benefits of biofuels for reducing GHG emissions. Also related to the avoidance of GHG emissions, Crutzen et al. (2007) concluded that nitrous oxide emissions from fertilizer use could nullify the contribution of biofuels.

In addition, Delucchi (2010) concluded that biofuels production from crops, using conventional agricultural practices, will likely exacerbate stress on water resources (considering water quality availability); this conclusion is basically the same as that presented by Dominguez-Faus et al. (2009).

As for the potential impacts of biofuels production on biodiversity, Koh and Ghazoul (2008) highlighted threats to ecosystems, mainly to forests, and Hennenberg et al. (2010) advocated that policies should preserve areas of significant biodiversity value and promote adequate agricultural practices. Concerns regarding the impacts on biodiversity have been exacerbated by the growing production of palm oil in Southeast Asia (Koh and Wilcove 2007; Venter et al. 2008), mainly as related to threats to orangutans (Nantha and Tisdell 2009).

In 2007–2008, during the crisis of food commodities prices, some international organizations and experts quickly stated that biofuels production was responsible for a significant part of the price rise (e.g., the World Bank and the Monetary Fund), thus causing a food shortage for poor people (UN FAO) and a "crime against humanity" (Jean Ziegler, an independent UN expert, in 2007). Examples of papers published at that time, stressing the impact of biofuels on food supply, are those by Pimentel et al. (2009) and Kullander (2010), both concluding that biofuels production should not be based on food crops. However, later papers demonstrated that the conclusions at that time were an overestimation of the impacts of biofuels production (Hochman et al. 2010; Zilberman et al. 2012).

A final example concerns the potential adverse social impacts of biofuels production, a subject that has been addressed by many authors: by Martinelli and Filoso (2008), in the case of ethanol production from sugarcane in Brazil; by Semino et al. (2009), regarding biodiesel production from soy, in Argentina; and by Phalan (2009), considering biofuels production in Asia. A special concern was the deterioration of basic rights (Eriksen and Watson 2009) and land grabbing (Robertson and Pinstrup-Andersen 2010) in new production regions, such as Africa (see further information in Sect. 8.5.3.2).

8.3 Sustainability Initiatives

The combination of the foregoing concerns, voiced by NGOs and by scientists, has led to diverse sustainability initiatives for bioenergy. These sustainability initiatives are, from a legal point of view, technical barriers to trade. They are classified as "soft law" and are accepted by the World Trade Organization (WTO) so long as their targets are legitimate. In the case of the initiatives for liquid biofuels, they aim at (1) promoting the sustainability of biomass production, trade, and consumption; (2) preventing biofuels that are harmful (to society and to the environment) from entering the international trade; (3) reducing GHG emissions related to the transportation sector; and (4) minimizing the dependence on fossil fuels (Correia 2011). In the case of bioenergy (and, possibly, in the future, for renewable chemicals and biomaterials), they bring technical conditions to products and services related to the cultivation of biomass.

These initiatives can be classified as (1) technical regulations, technical standards, or conformity assessment procedures, according to their scope; (2) public, private, or mixed, considering their nature; (3) voluntary or mandatory, according to flexibility; and (4) aiming at guidance, verification, or certification, considering the purpose.

Examples of technical regulations are the Renewable Energy Directive of the European Union (EU-RED), the Renewable Fuel Standard (RFS-2) in the United States, and the Californian Low Carbon Fuel Standard (LCFS). These rules are public by nature, and mandatory, in the sense that all these three regulations define conditions for the commercialization of liquid biofuels in their markets.

The Global Bioenergy Partnership (GBEP) and the standard ISO 13065 (Sustainability Criteria for Bioenergy) are examples of technical standards that aim at guiding governments, producers, and consumers about how production occurs, its impacts, and important needs for verifying product sustainability. In both cases, adoption of these technical standards is voluntary.

Conformity assessment procedures aim at verification or certification, and these schemes are mostly private and always voluntary. Examples are the certification schemes recognized by the European Commission, in the context of the EU-RED, such as the RSB standard (Roundtable on Sustainable Biomaterials), the Bonsucro Production Standard, and the ISCC (International Sustainability & Carbon Certification) standard. Currently, 19 schemes are recognized by the European Commission (European Commission 2017).

A brief description of these initiatives is presented next.

8.3.1 Technical Regulations

EU-RED is a legislative act that intends to achieve, by 2020, a 20% share of renewable energy sources in the EU's final consumption (energy basis), as well as a 10% share of renewable sources in the transport sector (also energy basis). Each Member State shall accomplish these targets and all are free to define their own strategies. EU countries have already agreed on a new renewable energy target, that is, at least 27% of final energy consumption in the EU as a whole, by 2030 (Goovaerts et al. 2013).

The RED has set either specific minimum sustainability standards for biofuels or requirements for their verification; only biofuels of which the production fulfills these standards can be accounted toward the mandatory national renewable energy targets. Compliance with the sustainability requirements needs to be checked by Member States or through voluntary schemes that have been approved by the European Commission (the most used solution).

The original EU-RED sustainability requirements include the following:

1. Minimum GHG savings of 35% compared to the fossil fuel displaced, a level that will be increased to 50% from 2017 onward, and to 60% for new installations (from 2018)
2. Raw material cannot be extracted/produced from land with high biodiversity value (i.e., primary forest, highly biodiverse grasslands, and protected areas, except when biomass extraction is part of management practices)
3. Raw material cannot be extracted/produced from lands with high carbon stock (i.e., forested areas, wetlands, peatlands)
4. Compliance with the EU Common Agricultural Policy, relating to biomass production in the EU

On a biannual basis, each Member State must also report on the impacts of biofuels on biodiversity, water resources, water and soil quality, changes in commodity prices (e.g., food), and land use, in addition to GHG emission reduction (Goovaerts et al. 2013). More recently, it was decided that the use of biofuels produced from "food crops" shall be limited to 5%, considering the 10% target previously mentioned (as for 2020). This policy was the result of strong pressure by NGOs to reduce the production of the so-called first-generation biofuels (1G).

The RFS-2 is the second version of a regulation that defines the volume of different biofuels that must be blended with conventional fuels between 2006 and 2022: it is valid throughout the USA, except in California. Biofuels are classified as conventional, bio-based diesel, advanced, and cellulosic, and only one fourth of non-conventional biofuels can be made from food crops. On the other hand, the LCFS is the regulatory standard in the case of California, which aims to reduce GHG emissions from the transport sector in California by at least 10% in 2020 (Goovaerts et al. 2013).

In the context of RFS-2, the minimum GHG thresholds that must be reached for each of the four categories of biofuels are 20% in case of conventional (e.g., ethanol from corn produced in USA), 50% in case of bio-based diesel and advanced biofuels (e.g., ethanol from sugarcane produced in Brazil), and 60% in case of cellulosic biofuels. GHG emissions are evaluated on a life cycle basis, and the emissions resulting from land use (both direct emissions and emissions from indirect impacts) are taken into account. RFS-2 also set restrictions on the type of land used for feedstock production to protect areas with high carbon stock and high biodiversity value (Goovaerts et al. 2013).

Neither RFS-2 nor LCFS are sustainability standards that rely on voluntary certification schemes.

8.3.2 Technical Standards

The Global Bioenergy Partnership (GBEP) was created in 2006, following the commitments taken by the G8 (the group of the largest economies in the world) to support bioenergy, in particular in developing countries. Currently, 23 countries and 14 international organizations (including many United Nations Programmes, the World Bank, and the International Energy Agency) are members of GBEP, and 28 other countries and 12 organizations act as observers (GBEP 2017). To facilitate the sustainable development of bioenergy, GBEP developed a set of voluntary indicators (24, in total, that is, 8 related to each of the three pillars: environment, social, economic) aiming to support and to monitor the results of bioenergy policies and programs (GBEP 2011). The set of GBEP indicators corresponds to a voluntary technical standard, aiming at guiding different stakeholders (e.g., policy makers, regulators, and investors).

ISO 13065, the Sustainability Criteria for Bioenergy, was approved in 2015 as an international standard and aims to facilitate the sustainable production, use, and trade of bioenergy. The standard can facilitate business-to-business communications, the comparison of sustainability information, and can also be a reference for how to provide information regarding sustainability (ISO 2015). In total, the standard has 13 principles (7 environmental, 4 social, 2 economic), 17 criteria, and 62 indicators. ISO 13065 is also a voluntary technical standard that can be used for guidance.

8.3.3 Conformity Assessment Procedures

Examples are the schemes used for verifying compliance of sustainability criteria; in practice, these schemes are used for certifying conformity. In the case of bioenergy, almost all existing certification schemes have been created—or adopted—in response to EU-RED. Next is presented brief information about three certification schemes used for bioenergy sustainability.

ISCC (International Sustainability & Carbon Certification) is able to certify the production of food, feed, bio-based products, and energy (including bioenergy); the standard ISCC EU is fully compatible with EU-RED. According to the company (ISCC 2017), in June 2017, 2750 certificates had been issued in accordance with EU-RED. Worldwide, the number of ethanol plants holding valid certificates at that moment was 365, plus 244 biodiesel plants with ISCC certificates, and 18 HVO (hydrogenated vegetable oil) plants.

RSB, the Roundtable on Sustainable Biomaterials, was created with a focus on bioenergy, but later the organization extended its scope to biomaterials. It is considered the most stringent certification scheme among those that can certify bioenergy and, so far, has issued a relatively small number of certificates. In June 2017, according to RSB (2017), the total number of certificates issued was 32, with 16 being biofuels producers (among which 3 are able to produce bio-jet fuels).

Bonsucro is a certification scheme originally created to verify the sustainability of sugar produced from sugarcane, and it was later adapted to cover sustainability of ethanol. According to the company (Bonsucro 2017), more than 480 companies held a certificate by June 2017, covering 25% of the sugarcane cropped area worldwide, and nearly 3 million tonnes of sugar produced; at that moment the number of certified mills was 63 (39 producing, or able to produce ethanol; 47 in Brazil).

8.4 Requirements of Sustainability Initiatives

This section presents the sustainability aspects covered by one technical regulation (EU-RED) and two technical standards (GBEP and ISO 13065) that are concerned with bioenergy production. The understanding is that they covered, somehow, the general concerns regarding bioenergy production on a large scale. Table 8.1 summarizes environmental aspects covered by the three sustainability initiatives analyzed; in the case of EU-RED and ISO 13065, all aspects should be accomplished by the economic operator (e.g., a biofuel producer), whereas in the case of GBEP the aspects/indicators are guidance for assessing the impacts of bioenergy production at a regional or national level.

For legal reasons, social aspects are not explicitly considered by EU-RED, despite their importance. In a sustainability initiative, social principles and criteria cannot go beyond what is established by the United Nations Declaration on Human Rights and, second, by the ILO Conventions (International Labor Organization).

Table 8.1 Environmental aspects addressed by three selected sustainability initiatives

Aspect	EU-RED	GBEP	ISO 13065
GHG emissions (to be evaluated on life cycle basis)	Set thresholds of required avoided emissions	Assessment of GHG emissions from bioenergy production and use	GHG emissions should be reduced; no threshold is defined
Soil	The Common Agricultural Policy shall be observed in case of production in EU	Soil quality, in terms of soil organic carbon, to be preserved	Soil quality and productivity shall be preserved
Water	The Common Agricultural Policy shall be observed in case of feedstock production in EU	Assessment of impacts on water resources, considering water use and efficiency, and on water quality	Water resources shall be preserved (water availability and water quality shall be observed)
Air	The Common Agricultural Policy shall be observed in feedstock production in EU	Assessment of non-GHG emissions along the whole supply chain; to be compared with other energy sources	Air emissions shall be controlled to maintain air quality
Biodiversity	Define that biomass production cannot occur in areas of high biodiversity value	Address: (a) conversion of high biodiversity value areas for feedstock production; (b) introduction of invasive species; (c) use of recognized conservation methods	Actions to identify potential impacts on biodiversity; Actions for protecting biodiversity; biomass removal from areas designated as biodiversity-protected areas to be reported
Land use and land use change	Define that biomass production cannot occur in areas with high carbon stock; in biannual basis each Member State should report land use changes from bioenergy	Aspects mentioned: areas used for feedstock production; bioenergy production without pressure on agricultural land; land use changes caused by bioenergy feedstock production	Aspect not addressed
Harvest level of wood resources	Aspect not addressed	Annual harvest (volume and as a percentage of net growth or sustained yield), plus the amount used for bioenergy	Aspect addressed in one of the biodiversity criteria
Energy efficiency	Aspect not addressed	Aspect not addressed	Energy consumption to be monitored
Wastes	Aspect not addressed	Aspect not addressed	Waste management is required

Table 8.2 Social aspects addressed by GBEP and ISO 13065

Aspect	GBEP	ISO 13065
Human rights	Aspect not addressed	Respect human rights
Labor rights	One indicator related to unpaid time spent by women and children and another indicator related to occupational injury, illness, and fatalities in the production of bioenergy	Respect labor rights (i.e., avoiding forced and child labor, allowing collective bargaining and assessing working conditions)
Jobs creation	Assessment of net job creation as a result of bioenergy production, plus an indicator about job quality	Aspect not addressed
Changes in income	Contributions of bioenergy production to wages and income	Aspect not addressed
Land use rights	Land title and procedures for change of land title shall be observed	Consent of local people for feedstock production
Water use rights	Aspect not addressed	Identification of potential impacts on water availability and on water quality; consent of people affected
Food price and food supply	Changes in prices (including price volatility), production, imports, and exports should be observed	The criteria related to land use rights and water use rights have specificities for food-insecure regions
Access to modern energy services	Impacts of bioenergy to be assessed	Aspect not addressed
Impacts of phasing out indoor smoke	Impacts of bioenergy to be assessed	Aspect not addressed

Even in this case, aspects addressed by ILO Conventions can only be required when a country has ratified them (Correia 2011).

Table 8.2 summarizes the social aspects covered by GBEP and ISO 13065. Social aspects do not perfectly match when these two technical standards are compared, as they are applicable to different contexts (regional or national, in the case of GBEP, and at the operator level in the case of ISO 13065).

EU-RED does not mention economic sustainability aspects, and ISO 13065 has only two economic criteria that address the economic and financial feasibility of bioenergy production and trade, besides financial risk management. On the other hand, resulting from the motivation of assessing impacts at a regional or even national level, GBEP has an extensive list of economic indicators, although some of the indicators are not economic, sensu stricto:

- Productivity of feedstocks and production costs
- Net energy balance, expressed by the energy ratio and compared to other energy sources
- Gross value added per unit of bioenergy
- Change in the consumption of fossil fuels and traditional use of biomass

- Impacts on training and on re-qualification of workforce
- Change in diversity of total primary energy supply from bioenergy
- Infrastructure and logistics for distribution of bioenergy
- Capacity and flexibility of use of bioenergy

8.5 Sustainability Aspects Addressed by Certification Schemes

8.5.1 Aspects Addressed by Certification Schemes

This section presents the sustainability aspects handled by the three sustainability schemes considered in this chapter. Table 8.3 presents a list of the aspects addressed and a synthesis of principles (ISCC 2016; RSB 2016; Bonsucro 2016).

8.5.2 Environmental Aspects

8.5.2.1 GHG Emissions

The implementation of efficient bioenergy has been considered essential to reduce and stabilize GHG emission levels in the coming decades (Souza et al. 2015). Bioenergy is different from other renewable energy technologies in that it is a part of the terrestrial carbon cycle. The CO_2 emitted by bioenergy use was earlier sequestered from the atmosphere and will be sequestered again if the bioenergy system is managed sustainably, although emissions and sequestration are not necessarily in temporal balance with each other (e.g., long rotation periods of forest stands) (Edenhofer et al. 2012), while other GHG emissions in the supply chain must be considered as well.

Studies of environmental effects, including those focused on energy balances and GHG emissions, usually employ methodologies in line with the principles, framework, requirements, and guidelines of ISO 14040:2006 and 14044:2006 standards for Life Cycle Assessment (LCA). During past years, new standards and methods have also emerged, particularly focused on GHG emissions and removals (also referred to as carbon footprint), such as the GHG Protocol Product Standard, PAS 2050:2011, ISO/TS 14067:2013, but they are all based on the same principles as ISO 14040.

Most bioenergy LCAs are designated as attributional to the defined boundaries, as they describe the impacts of an average unit of a bioproduct. Consequential LCAs, on the other hand, analyze bioenergy systems beyond these boundaries, in the context of the economic interactions, chains of cause and effect in bioenergy production and use, and effects of policies or other initiatives that increase bioenergy production and use. In summary, consequential LCAs investigate the responses

Table 8.3 Aspects addressed by three sustainability initiatives and their principles

Aspects	ISCC	RSB	Bonsucro
GHG emissions	GHG emissions, on a life cycle basis, should be reported; default values can be presented	Biofuels shall significantly reduce life cycle GHG emissions as compared to fossil fuels	Monitoring GHG emissions
No-go areas: high biodiversity value and high carbon stock	No production on land with high biodiversity value or high carbon stock	Topic addressed in the "conservation" principle, constraining feedstock production in such areas	Percentage of land with high biodiversity value, high carbon stock, or peatlands planted to sugarcane
Soil, water, and air	Environmentally responsible production to protect soil, water, and air	In three different principles: (1) to reverse soil degradation and/or maintain soil health; (2) to maintain or enhance the quality and quantity of surface and groundwater resources; (3) to minimize air pollution	Continuously improve key areas of the business (the principle also address practices to protect soil, water, and air)
Working conditions	Safe working conditions	Compliance with national laws and international conventions	Safe and healthy working conditions
Human and labor rights	Compliance with human, labor, and land rights	No violation of human rights or labor rights	Respect human rights and labor standards
Continuous improvement	Good management practices and continuous improvement	Sustainable operations are planned, implemented, and continuously improved	Continuously improve key areas of the business
Biodiversity	Requirements mentioned in the principle "Environmentally responsible production …"	To avoid negative impacts on biodiversity, ecosystems, and conservation values	Actively manage biodiversity and ecosystem services
Production efficiency	Not specifically addressed	Maximization of production efficiency and of social and environmental performance	Manage input, production and processing efficiencies to enhance sustainability (also address GHG emissions)
Rural and social development	Addressed in a broader way	Contribution to the social and economic development of local, rural, and indigenous people	Not specifically addressed

(continued)

Table 8.3 (continued)

Aspects	ISCC	RSB	Bonsucro
Food security	Biomass production shall not replace stable crops or impair the local food security	To ensure food supply and improve food security	Not specifically addressed
Respect of land rights	Land is being used legitimately and traditional land rights have been secured	Operations shall respect land rights and land use rights	To demonstrate clear title to land and water in accordance with national practice and law
Legality	Compliance with laws and international treaties	Operations follow all applicable laws and regulations	Obey the law

to bioenergy expansion and have hence become a common approach in the regulatory context of biofuels (ARB 2017; EPA 2010).

The different focuses of attributional and consequential approaches are reflected in several methodological choices in LCA, which may in turn lead to widely different results. For bio-based supply chains, one important element addressed in consequential LCAs is indirect land use change, which has been the most uncertain and contentious issue in the evaluation of GHG effects of biofuels (see Sect. 8.6).

Apart from land use change effects, bioenergy LCA results vary considerably among the different biofuel types and regions in response to the different agricultural practices, technology performances, as well as according to the study methods. The ranges of GHG emissions for bioenergy systems and their fossil alternatives are illustrated in Fig. 8.2 for different uses (transport, power, heat).

The life cycle performance of biofuels is usually dominated by feedstock cultivation, with significant contributions from diesel for farming and fertilizers (and associated N_2O emissions). For corn ethanol, however, relevant contributions come from the fuel production stage, which are strongly influenced by the process fuel choice (usually fossil, opposed to bagasse used in sugarcane ethanol plants). In the case of advanced biofuels, processes for feedstock pretreatment and conversion also lead to significant differences in GHG emissions. Cellulosic ethanol plants can use residues such as lignin to generate steam and electricity with CHP and export excess electricity to the grid to help improve plant economics and reduce the GHG footprint (Macedo et al. 2015).

It is worth noting that, for any form of bioenergy (current or advanced, solid or liquid), the life cycle performance increases as the efficiency of feedstock production and conversion improves, and as the penetration of renewables increases in the overall energy matrix. Given the relevance of the fossil energy used in the ancillary processes involved in the production cycle (fertilizer manufacture, feedstock cultivation, and logistics), as the energy supply system moves toward low-carbon sources, the emissions associated with these ancillary processes are reduced, thereby lowering the life cycle GHG emissions of bioenergy (Macedo et al. 2015).

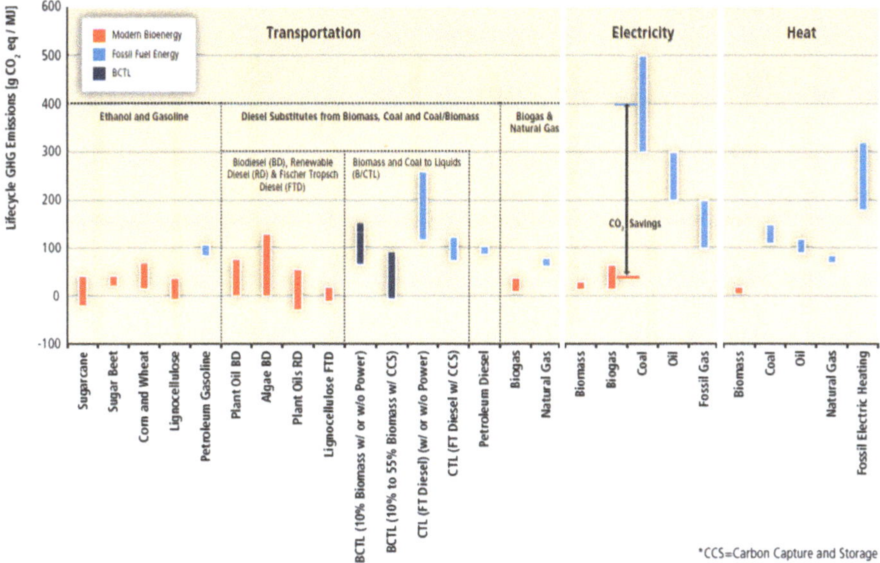

Fig. 8.2 Ranges of GHG emissions per unit energy output (MJ) from major modern bioenergy chains (excluding land use-related effects) compared to conventional and selected advanced fossil fuel energy systems (Edenhofer et al. 2012)

8.5.2.2 Water

Global water withdrawal has sharply increased during past decades, largely dominated by agriculture (FAO 2017a), which brings concerns about the additional pressure on freshwater resources that might be caused by the increase in demand for food in combination with the expansion of the bioeconomy. Water scarcity is seen by many as the world's single biggest water problem, although it is not equally distributed around the globe. The distribution of global freshwater (or river) runoff among the continents is highly uneven and corresponds poorly to the distribution of world population. As a consequence, in some countries the increase in evapotranspiration appropriation for human uses could lead to further exacerbation of an already stressed water resources situation, although there are also countries where such impacts are less likely to occur (Berndes 2002; Hernandes et al. 2014).

In general, the water footprint of fossil energy carriers and derived fuels is much lower than for biomass and biofuels, mostly because plants consume water to grow (Gerbens-Leenes et al. 2009). But the water footprints of biofuels reported in the literature vary by orders of magnitude (Fig. 8.3), as the reporting methods were not standardized. Some reports include rainwater inputs, theoretical transpiration losses from plant growth, and in some cases theoretical use of irrigation water (Berndes et al. 2015). There are also efforts to quantify the potential impacts from depriving human users and ecosystems of water resources, as well as specific potential impacts

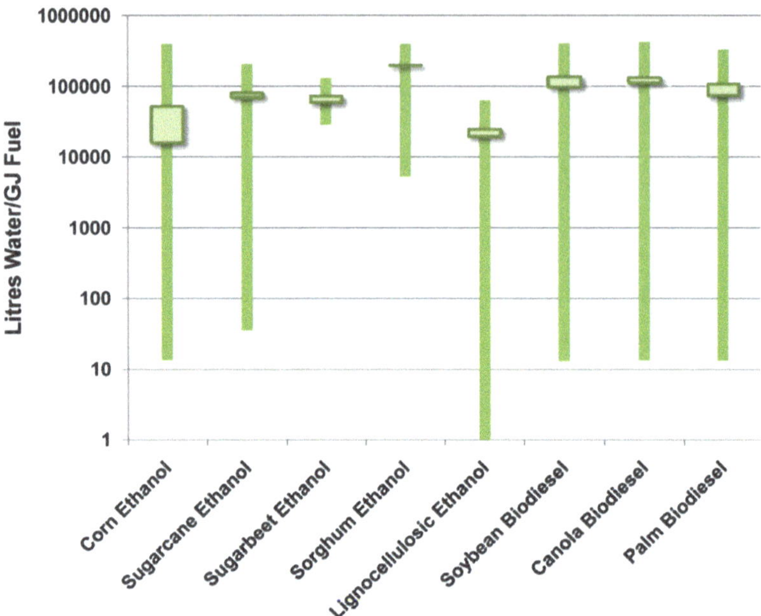

Fig. 8.3 Water intensity indicators of biofuels (Berndes et al. 2015). *Solid bars* indicate the range of reported values; *boxes* represent the differences in median and mean values. The range for lignocellulosic ethanol includes thermochemical and biological conversion pathways

from the emitted contaminants affecting water, through different environmental impact pathways and indicators (Boulay et al. 2011; Pfister et al. 2009, 2011).

The cultivation of conventional annual crops for bioenergy affects water resources in the same way as when such crops are cultivated for food and feed, and effects occur in both positive and negative ways. Feedstock cultivation can lead to leaching and emission of nutrients that increase the eutrophication of aquatic ecosystems, and pesticide emissions that reach water bodies may also negatively impact aquatic life (Diaz-Chavez et al. 2011; Edenhofer et al. 2012). However, semi-perennial crops and perennial grasses, or even trees grown in rotation, tend to have lower water quality impacts than conventional crops because of their extensive root systems, long-term soil cover and protection, and reduced need for tillage and weed suppression (Berndes et al. 2015). Additionally, perennial grasses may increase seasonal evapotranspiration (ET) compared to maize because these grasses have access to deeper soil moisture, which could lead to higher humidity, lower surface temperature, greater precipitation, and more cloud cover with lower solar radiation. In turn, soil moisture would increase, affecting soil metabolism and finally carbon sequestration. The same trend was observed for sugarcane in parts of the Central Brazil savannas and in Southeast Brazil (Berndes et al. 2015).

Consumptive water use during the production of cultivated feedstocks dominates the water demand of bio-based supply chains, whereas feedstock processing for fuels and electricity requires much less water, even though water must be extracted from lakes, rivers, and other water bodies. Water demand for bioenergy and bioproducts can be reduced substantially through process changes and recycling. For instance, water withdrawal in sugarcane mills in Southeast Brazil has been reduced from 15–20 m^3 t^{-1} cane to about 1.85 m^3 t^{-1} cane in past decades through water recycling and other measures to improve water use efficiency. Further improvements of wastewater treatment systems allowing increased reuse of water move the sector toward the water withdrawal goal of 1.0 m^3 t^{-1} cane (Neto et al. 2009; Yeh et al. 2011).

8.5.2.3 Soil

With the possibility of intense agricultural practices, a main concern from the sustainability aspect is maintaining soil quality and its productivity. The most important drawbacks of feedstock production, frequently quoted in the literature, are soil erosion, loss of soil carbon (with impacts on its physical, chemical, and biological characteristics), the risk of borne diseases (that may reduce the frequency of cropping), and eutrophication (Vries et al. 2010).

Essentially, ISO 13065 and the three certification schemes analyzed in this chapter mention the same actions aiming at keeping soil quality and productivity, such as the use of best practices for avoiding erosion, the implementation of practices to protect soil structure (e.g., preventing compaction), the adoption of practices to at least maintain soil nutrient balance (reducing nitrate pollution), to recover agricultural residues as much as possible, to control soil pH value, and to prevent salinization.

8.5.2.4 Air

The main concerns regarding the impacts on air quality from non-GHG emissions can be separated into four groups: (1) first, related to open burning in agricultural activities; (2) second, also in agricultural sites, related to the application of agrotoxics; (3) in regard with the transport of feedstocks and inputs, with the emissions of pollutants and air deterioration from dust; and (4) finally, related to emissions from industrial activities.

Particularly in Brazil, because for decades sugarcane was burned before manual harvesting, many researchers have focused on assessing the health impacts from this activity (Arbex et al. 2007; Cançado et al. 2006). This problem has been minimized with the phasing out of sugarcane in recent years.

The potential impacts of feedstock and biofuels production on air quality are addressed in the sustainability initiatives. In the certification schemes, for instance, RSB has a specific principle on air quality, whereas in the other schemes mentioned in this chapter (ISCC and Bonsucro) the issue is addressed by a more general principle. In practice, a possible approach is the development of a management plan in which actions to allow compliance with air quality standards are listed.

8.5.2.5 Biodiversity

Sustainable agriculture requires the preservation of natural resources (TEEB 2015). Biodiversity and the services derived from ecosystems are essential for the secure production and supply of food, fibers, biofuels, and freshwater, and also for maintaining the resilience of production systems (CBD 2014). However, bioenergy production can be a significant driver of ecosystem degradation, biodiversity loss, and social and cultural externalities (TEEB 2015), mostly allied with the conversion of natural habitats to agricultural land. On the other hand, benefits can be derived when bioenergy policies promote planning and more sustainable land management systems (Kline et al. 2015). In this sense, the environmental advantages of bioenergy depend largely on whether the production contributes to improved management of previously cleared and disturbed lands or, on the contrary, causes additional clearing and degradation on lands that have high conservation value for biodiversity and carbon storage (Dale et al. 2015).

By far, the major concern regarding large-scale production of bioenergy is the conversion of natural habitats to agricultural land, and this explains the EU-RED requirement that feedstock production cannot occur in areas of high biodiversity value, such primary forests, in protection areas, and in highly biodiverse grasslands, excepting when good management practices require the extraction of biomass. Also, because EU-RED imposes this condition, the three certification schemes analyzed in this chapter specifically mention that sustainable feedstock production cannot occur in such areas. In ISO 13065, a criterion states that the economic operator shall indicate how biomass is removed from areas designated as protected areas from a biodiversity aspect.

Other concerns are related to the introduction of alien species, mainly potentially invasive species, with the possible fragmentation of habitats caused by land use change, and with changes in habitat as the consequence of alterations in hydrology, biomass removal, biomass burning, etc. Aligned with these concerns, the certification schemes explicitly mention the maintenance or restoration of ecological corridors, the conservation (or restoration) of natural vegetation around springs and watercourses, the adoption of buffer zones to protect protected areas, restrictions on burning, and actions to minimize the use of fertilizers and agro-chemicals. It has been mentioned that the cultivation of invasive species and of genetically modified species should be avoided or, at least, a management plan shall be prepared to avoid negative impacts. In this sense, some certification schemes describe environmental management plans (EMP) to document actions taken for reducing impacts on biodiversity and ecosystem services.

The standard ISO 13065 has a criterion that demands the identification of potential impacts on threatened ecosystems and/or habitats, as well as on vulnerable and threatened species.

In conclusion, biofuel certification schemes address the main biodiversity concerns about agricultural activities in two different ways. First, based on the precautionary principle, biomass production is restricted in risky areas. Second, the schemes demand specific actions for minimizing relevant impacts (Kline et al. 2015).

8.5.2.6 Wastes

Concerns regarding wastes are related to the substances of which the economic operator is required to dispose, involving such actions as storing, reusing, recycling, and disposal. The aim is to reduce the potential impacts on human health and on the environment (soil, water, air, biodiversity).

The certification schemes address this topic in more or less detail. In regard to agricultural waste, the aim is reducing waste generation or fostering reuse. In a general sense, the principal requirement is the use of best practices of waste management and waste disposal.

8.5.3 Social Aspects

8.5.3.1 Human Rights, Labor Rights, and Labor Conditions

The most basic legal document on human rights is the UN Declaration of Human Rights and, for this reason, both ISO 13065 and the three certification schemes analyzed in this chapter have a principle, or at least one criterion, in which the Declaration is addressed: any violation of human rights is unacceptable.

The concerns regarding labor rights and working conditions involve issues such compulsory labor, child labor, working journey, collective bargaining rights, and safety and health aspects. The three schemes mentioned in this text address these aspects in one or more criteria, in some cases with focus only on compliance of applicable legislation, in some cases requiring more pro-active actions (see Table 8.3).

On the subject of labor conditions, examples of the concerns frequently presented in the literature are related to sugarcane cropping, in Brazil, and this can be explained by three main reasons: first, until recently, the sugarcane sector employed a large working force on manual harvesting; second, because of the heterogeneity of this large sector, it was relatively easy to find examples of bad practices; and third, because of the old governance practices that have been gradually eliminated. Examples of papers criticizing working conditions in sugarcane cropping are the publications by Martinelli and Filoso (2008) and Azadi et al. (2012).

It is worth mentioning that more recently, with mechanization of sugarcane harvesting, the number of employees working under tough conditions has decreased, although other problems have been reported (e.g., unemployment, and extensive working periods for those who operate the machines).

8.5.3.2 Land Use and Water Use Rights

In some cases, land use rights are not supported by law or by titles of ownership, but rather are customary. This difference is an important aspect, and its importance has increased with what has been called land grabbing. Land grabbing is an expression used to describe the large growth of transnational land transactions that has been occurring in recent years (Borras and Franco 2010). The practice is more common where land tenure is more malleable and is exacerbated by both the food price crisis (2007–2008) and the broader financial crisis. It has been said that biofuels production has boosted land grabbing, mainly in Asia and Africa (Borras et al. 2011; Scheidel and Sorman 2012).

Indeed, Ravagnani et al. (2016) studied data from the Land Matrix (2015), for a period of 10 years, and concluded that land acquisitions for biofuels production occurred, with a significant amount of contracts and land extensions (estimated at 0.7% of total agricultural area in Africa, i.e., 2.3 million hectares, and 0.32 million hectares in Asia). However, land acquisitions for biofuels production have recently diminished.

The sustainability initiatives are concerned with land rights demanding that local people, and the potentially affected community, would be informed about intended land use, and also requiring proof of consent; obviously these actions have not eliminated the risks of grabbing. ISO 13065 has an indicator that demands respect of traditional land use rights and the necessity of evidence of previous consultation and permission for using the land. In the case of water use rights, the same technical standard also mentions the necessity for evidence showing that the local population has been consulted about the enterprise.

Equally, the three certification schemes analyzed in this text address land tenure in greater or less detail. Basically, the recommended or required actions encompass (a) the identification of potentially affected local communities, and of their traditional land uses, (b) the verification of existence of statutory or customary land use rights, and (c) the negotiation of agreements for use of the land, a process that shall be free and result in formal documents, with fair compensation for the local people.

8.5.3.3 Food Security

The debate that is known as food versus fuel gained momentum with the food supply crisis in 2007–2008, as previously mentioned, when biofuels production was pointed out as the main reason for rise in food prices. Although afterward it was demonstrated that the impact from biofuels production was limited (Hochman et al. 2010; Zilberman et al. 2012), the issue is still very contentious and, as a consequence, there is aversion to biofuels production using food feedstocks and arable land. One of the consequences was the constraint imposed by EU-RED to the so-called first-generation biofuels, which are produced from edible feedstocks.

One possible negative impact is the direct competition between food and biofuels production, but another potential negative impact is the pressure on food supply resulting from the limited availability of arable land. Both aspects have led to the support for biofuels that are produced from nonedible feedstocks and that do not demand high-quality land (e.g., second- and third-generation biofuels, 2G and 3G), such as from waste, cellulosic materials, or algae. So far, however, throughout the world there is not a single biofuels plant that commercially produces 2G and 3G. On the other hand, a possible positive impact of biofuels production would be higher incomes in rural areas, minimizing the food supply problem.

In this sense, sustainability initiatives such as GBEP and EU-RED require monitoring of the impact of biofuels production on food supply, including impacts on prices, production, and consumption. This approach is in line with the assessment of macroeconomic impacts. On their side, the certification schemes demand an evaluation of impacts at the level the activity takes place, mainly in poor regions, or in regions that are recognized as food-insecure regions, as stated in RSB ("Operations ensure the human right to adequate food and improve food security in food insecure regions") and ISCC ("Biomass production shall not replace stable crops or impair the local food security"). However, the Bonsucro scheme does not address food supply in any way.

The great importance assigned to potential impacts on food supply predicts that this issue will also be the focus of debates on future large-scale production of biomaterials.

8.5.3.4 Rural Development

The potential social and economic development of the regions where the production of feedstocks and biofuels take place (and, in the future, the production of biomaterials) is one of the main reasons for fostering bioproduction. In particular, the aim is enhancing the development of rural areas, by the creation of jobs, enhancement of the working force, larger income, improved infrastructure, etc. Some social and one of the economic (training and requalification of the workers) indicators proposed by GBEP induces the continuous monitoring of parameters that can attest the effectiveness, from a local development perspective, of policies aiming at fostering biofuels production.

The papers by Machado et al. (2015, 2016) are examples of the effort aimed at evaluating the average local and regional socioeconomic impacts of large-scale ethanol production from sugarcane in Brazil. The results show that, on average, the impacts were positive so long as municipalities of the same size (on a population basis), where sugarcane is or is not relevant, are compared. The conclusion is based on indicators such as income, wealth distribution, social health, the Human Development Index, and infrastructure. The same author is completing a research project assessing the social and economic potential impacts of future production of biochemicals in Brazil (Machado 2017); the assessment is based on the use of a general equilibrium model. Also regarding socioeconomic impacts of the sugarcane sector in Brazil, a number of papers were published in recent years (Moraes et al. 2015, 2016; Gilio and Moraes 2016) assessing regional impacts.

In the general sense, it is a difficult task to assess the impacts of a single economic activity at the local level, first because a longer period is required, and second because other economic activities have influence in the same region. As the regional economy becomes more diversified, the more difficult is this assessment. Thus, the authors of this chapter believe the aim of assessing the socioeconomic impacts of a single enterprise is senseless, as is the attempt of some certification schemes. However, one certification scheme analyzed along this text (RSB) has a principle dedicated to this issue, and another scheme (ISCC) has a criterion that addresses the issue.

8.5.4 Economic Aspects

From a microeconomic perspective, the aspects to be considered regarding biofuels/biomaterials production are limited to ensure that the enterprise is economically and financially viable and that the economic operator must act to minimize financial risks.

However, the most important economic issues are related to the impacts on regional and on national economy, and these are determined by the results for jobs, farm income (probably raised, diversified), workers' wages, etc., as well as by the indirect impacts caused by reducing imports and dependence, developing technology, supporting local industry, improving capacity building, and so on. These impacts are rarely related to a single enterprise and cannot be assessed directly. In this sense, economic models such as the input/output matrix and general equilibrium models (computable general equilibrium, CGE) have been used for this purpose (Channing et al. 2008; Moschini et al. 2012; Herreras-Martínez et al. 2013).

The resulting impacts are site specific, and depend on the local reality, on the way supply chains are developed, on the investments for improving local infrastructure, and on whether basic rights have been assured and how conflicts have been managed. Thus, results from previous experiences, whether positive or negative, cannot be generalized. In the general sense, the macroeconomic impacts are closely related to rural/regional development, as discussed previously.

8.6 Land Use Change

Land use activities for human use, whether through conversion of natural landscapes or changing management practices, have transformed a large proportion of the planet's land surface (Foley et al. 2005). Today, approximately 38% of the world's 13 billion ha of land is dedicated to pasture and crops; forests retain about 30% of the total area (FAO 2017b). Looking ahead, based on population and dietary trends, the FAO projects a net increase of about 70 Mha in land requirements to grow food crops by 2050, resulting from an expansion of 130 Mha in developing countries combined with a loss of 60 Mha to infrastructure, recreation, etc. in developed countries (Alexandratos and Bruinsma 2012). For the case of modern bioenergy, estimations of gross land demand vary between 50 and 200 Mha by 2050, which could deliver between 44 and 135 EJ year^{-1} (Woods et al. 2015). Those values are well below the land availability for bioenergy, once most estimations from global studies exceed 500 Mha for 2050, after allowing for food production, protected areas, urban expansion, and increased biodiversity protection.

One major opportunity to compensate for growth in biomass resource use is to intensify the use of low productivity pastureland and make use of (part of) the available area of pasture, which is estimated to be around 950 Mha, for multipurpose agriculture (Woods et al. 2015). However, land potentially available for bioenergy also includes degraded areas, as well as those areas suitable for rainfed agriculture that are not expected to be needed for other purposes, in addition to those not suitable for rainfed agriculture but potentially suitable for energy crops.

Even though land is not a constraint for bioenergy at a global level (Souza et al. 2015), land use change (LUC) is expected to play a central role in the sustainability of bioeconomy, as a common denominator among food, energy, and environment. Given the continuous changes in human society, LUC can be seen as an inevitable consequence, with likely socioeconomic and environmental implications. However, the effects do not have to be negative.

Although unintended consequences to food security may surge at specific regional contexts when bioenergy expansion is not well implemented, bioenergy, when well planned, can actually enhance access to food through productivity gains enabled by technology deployment, or by promotion of higher household income, or improved infrastructure. Modern bioenergy can also allow better food quality/stability if directly used for drying, cooking, and water purification (Nogueira et al. 2015). With respect to the environment, it is widely recognized that converting land planted with row crops to perennial grasses can be accompanied by significant benefits in terms of biodiversity, habitat, soil carbon, and fertility, whereas worst-case scenarios such as clearing rainforests or draining peatland to make land available for bioenergy must be avoided (Karp et al. 2015).

When addressing the impacts from LUC, the distinction between direct and indirect LUC is important. Direct LUC (dLUC) involves "changes in land use on the site used for bioenergy feedstock production, such as the change from food or fiber production (including changes in crop rotation patterns, conversion of pasture land, and changes in forest management) or the conversion of natural ecosystems." Indirect LUC (iLUC) refers to "the changes in land use that take place elsewhere as a consequence of the bioenergy project" (for example, the conversion of natural ecosystems to agricultural land caused by the reestablishment of displaced food producers) (Berndes et al. 2011).

Land use changes can influence global warming through CO_2 emissions or sequestration from carbon stock changes in biomass, dead organic matter, and soil organic matter.[1] Indicators such as "carbon debt" (Fargione et al. 2008) and carbon

[1] LUC may also influence the extent to which the land surface reflects incoming sunlight (i.e., the albedo), and thereby global warming. As pointed out by Berndes et al. (2011), in regions with seasonal snow cover or a seasonal dry period (e.g., savannahs), reduction in albedo caused by the introduction of perennial green vegetative cover can counteract the climate change mitigation benefit of bioenergy. Conversely, albedo increases associated with the conversion of forests to energy crops (e.g., annual crops and grasses) may counter the global warming effect of CO_2 emissions from deforestation.

payback time (Gibbs et al. 2008) have been proposed to reflect the amount of CO_2 released from the direct land conversion to bioenergy production, respectively, and how many years are required to repay those debts by avoiding fossil fuel emissions with biofuels. As shown in Fig. 8.4, all biofuel options have significant payback times when dense forests or peatlands are converted into bioenergy plantations. However, payback times are relatively low for the most productive systems when grasslands and woody savannas are used (Edenhofer et al. 2012), and practically zero when degraded land or cropland is used (not considering the iLUC that can arise). Actually, studies have found that growing perennial grasses in lieu of row crops increases soil carbon stocks at a rate of 1 Mg C ha^{-1} $year^{-1}$ or more (Woods et al. 2015), which is similar to the outcome of replacing annual crops with sugarcane in Brazil (Mello et al. 2014).

Even though there are large uncertainties about the overall carbon stock changes, the dLUC effects can be measured and observed with time. iLUC implications, conversely, result from projections using economics models, which are only able to capture both effects (dLUC and iLUC) together.[2] When it is not empirically observable, iLUC estimations depend on assumptions of cause–effect relationships that will attribute responsibility of land conversion to individual agricultural land use (Macedo et al. 2015). Quantifying the indirect effects of bioenergy policies is therefore a complex exercise that requires a combination of energy, agro-economic, global land use, and emission modeling approaches.

As illustrated in Fig. 8.5, the more recent studies of iLUC report a lower effect than did earlier studies. Estimates for new land brought into cultivation by the expansion of corn ethanol have been reduced by an order of magnitude and by threefold for sugarcane ethanol. Recent modeling exercises therefore suggest that the land use sectors are able to accommodate a significant part of the projected bioenergy expansion without claiming new land (Macedo et al. 2015).

Furthermore, bioenergy does not always result in LUC. Lignocellulosic feedstocks can be produced in combination with food and fiber, avoiding land use displacement and improving the productive use of land (Edenhofer et al. 2012). Bioenergy can also make use of by-products from the agricultural and forest industries, or even organic municipal wastes, according to principle 7 of green chemistry.

8.7 Final Remarks

Bioenergy has been promoted in different countries with the aim of reducing dependence on nonrenewable resources, supporting local economies, and offering a better quality of life in rural areas. However, to effectively contribute to sustainable

[2] As there is no information about how the feedstock from a newly cultivated land parcel will be used, the concept of dLUC is not applicable within the models. For this reason, some authors refer to the land use changes modeled in economics models as induced land use change.

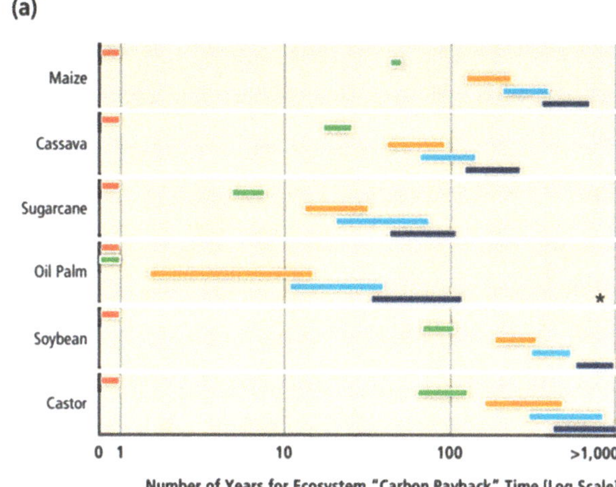

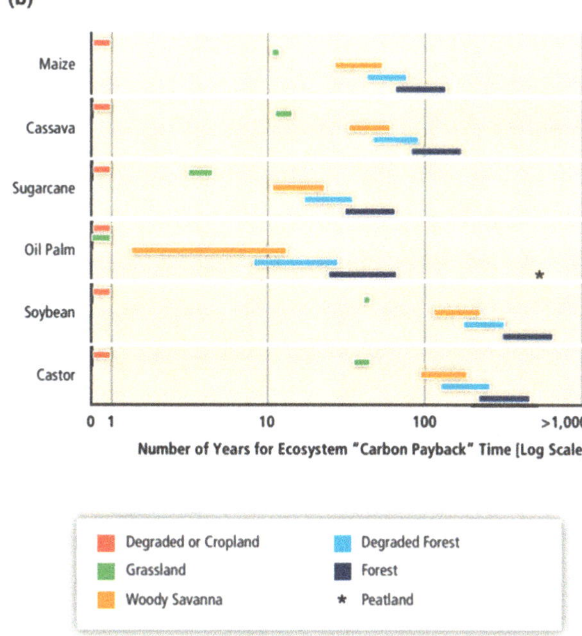

Fig. 8.4 Ecosystem carbon payback time for potential biofuel crop expansion pathways across the tropics, as based on crop yields circa 2000 (**a**), and if all crops achieved the top 10% global yields through gradual or abrupt improvements in agricultural management or technology (**b**). (Adapted from Edenhofer et al. 2012)

development, bioenergy deployment needs to be well planned and carefully implemented to avoid environmental and social risks.

As a result of interest group advocacy, a diversity of sustainability initiatives has emerged in recent years in the bioenergy context, which may soon be extended to renewable chemicals and biomaterials to accommodate principle 7 of green chemistry (see Chap. 1). Even though these initiatives address the main sustainability

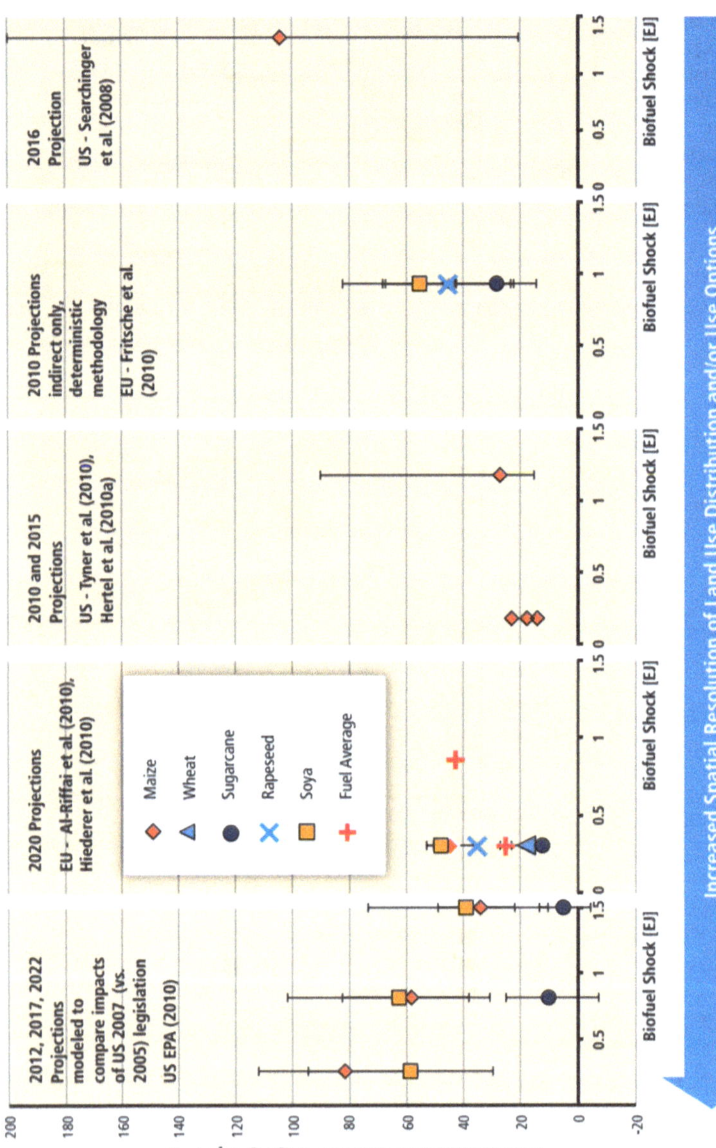

Fig. 8.5 Select model-based estimates of LUC emissions (30-year accounting framework) for major biofuel crops given a certain level of demand (Edenhofer et al. 2012)

concerns, compliance with certification schemes does not necessarily translate into sustainable production. Risks of green-washing exist, and further investigation is needed to gauge the implications on trade, new producers, and ultimately the effective promotion of sustainable development.

Nevertheless, there is evidence supporting that, if implemented correctly, the bio-based economy can indeed improve energy security, generate jobs, spur economic activity (especially in rural areas), enhance food supply, decrease air and water pollution, protect biodiversity, and counteract climate change. To that end, however, research and development, good governance (helped by appropriate certification schemes), and innovative business models are essential to address knowledge gaps and foster innovation across the value chain. As argued by Souza et al. (2015), with these measures, a sustainable future would be more easily achieved with the contributions from biomass than without that input.

References

Alexandratos N, Bruinsma J (2012) World agriculture towards 2030/2050: the 2012 revision (No. 12-03). ESA Working Paper, FAO Agricultural Development Economics Division, Rome

ARB (2017) Low carbon fuel standard [WWW Document]. https://www.arb.ca.gov/fuels/lcfs/lcfs.htm. Accessed 20 May 2017

Arbex M, Martins L, Oliveira R, Pereira L, Arbex F, Cançado J, Saldiva P, Braga A (2007) Air pollution from biomass burning and asthma hospital admissions in a sugar cane plantation area in Brazil. J Epidemiol Community Health 61:395–400

Azadia H, de Jong S, Derudder B, De Maeyer P, Witlox F (2012) Bitter sweet: how sustainable is bio-ethanol production in Brazil? Renew Sust Energ Rev 16:3599–3603

Berndes G (2002) Bioenergy and water—the implications of large-scale bioenergy production for water use and supply. Glob Environ Change 12:253–271. https://doi.org/10.1016/S0959-3780(02)00040-7

Berndes G, Bird N, Cowie A (2011) Bioenergy, land use change and climate change mitigation: background technical report (No. ExCo:2011:04). Background Technical Report. IEA Bioenergy

Berndes G, Youngs H, Ballester MVR, Cantarella H, Cowie AL, Jewitt G, Martinelli LA, Neary D (2015) Soils and water In: Bioenergy and sustainability: bridging the gaps. Scientific Committee on Problems of the Environment (SCOPE), Paris Cedex, pp 618–659

Bonsucro (2016) Bonsucro production standard – version 4.2. Available at https://www.bonsucro.com/

Bonsucro (2017) Information available at https://www.bonsucro.com/. Accessed 12 June 2017

Borras SM Jr, Franco J (2010) Towards a broader view of the politics of global land grab: rethinking land issues, reframing resistance. ICAS Working paper series 001

Borras SM Jr, Fig D, Suárez SM (2011) The politics of agrofuels and mega-land and water deals: insights from the ProCana case, Mozambique. Rev Afr Polit Econ 38(128):215–234

Boulay A-M, Bouchard C, Bulle C, Deschênes L, Margni M (2011) Categorizing water for LCA inventory. Int J Life Cycle Assess 16:639–651. https://doi.org/10.1007/s11367-011-0300-z

Cançado J, Saldiva P, Pereira L, Lara L, Artaxo P, Martinelli L, Arbex M, Zanobetti A, Braga A (2006) The impact of sugar cane-burning emissions on the respiratory system of children and the elderly. Environ Health Perspect 114(5):725–729

CBD (Convention on Biological Diversity) (2014) How sectors can contribute to sustainable use and conservation of biodiversity. CDB Technical Series 79. PBL Netherlands Environmental Assessment Agency, The Hague

Channing Arndt C, Benfica R, Tarp F, Thurlow J, Uaiene R (2008) Biofuels, poverty, and growth: a computable general equilibrium analysis of Mozambique. IFPRI Discussion paper 00803

Correia BB (2011) Sustainability requirements for biofuels and the rules of the international law. Masters Dissertation, University of Campinas, Unicamp, Campinas

Crutzen PJ, Mosier AR, Smith KA, Winiwarter W (2007) N_2O release from agro-biofuel production negates global warming reduction by replacing fossil fuels. Atmos Chem Phys Discuss 7:11191–11205

Dale VH, Parish ES, Kline KL (2015) Risks to global biodiversity from fossil-fuel production exceed those from biofuel production. Biofuels Bioprod Biorefin 9(2):177–189

Delucchi MA (2010) Impacts of biofuels on climate change, water use, and land use. Ann NY Acad Sci 1195(1):28–45

Diaz-Chavez R, Berndes G, Neary D, Elia Neto A, Fall M (2011) Water quality assessment of bioenergy production. Biofuels Bioprod Biorefin 5:445–463. https://doi.org/10.1002/bbb.319

Dominguez-Faus R, Powers S, Burken J, Alvarez P (2009) The water footprint of biofuels: a drink or drive issue? Environ Sci Technol 43(9):3005–3010

Edenhofer O, Pichs Madruga R, Sokona Y, United Nations Environment Programme, World Meteorological Organization, Intergovernmental Panel on Climate Change, Potsdam-Institut für Klimafolgenforschung (eds) (2012) Renewable energy sources and climate change mitigation: special report of the intergovernmental panel on climate change. Cambridge University Press, New York

Endres JM, Diaz-Chaves R, Kaffka SR, Pelkmans L, Seabra JEA, Walter A (2015) Sustainability certification. In: Bioenergy & sustainability: bridging the gaps. Scientific Committee on Problems of the Environment (SCOPE), Paris Cedex, pp 660–681

EPA (2010) Renewable fuel standard program (RFS2). Regulatory impact analysis (No. EPA-420-R-10-006). Assessment and Standards Division, Office of Transportation and Air Quality, U.S. Environmental Protection Agency, Washington, DC

Eriksen SHE, Watson HK (2009) The dynamic context of southern African savannas: investigating emerging threats and opportunities to sustainability. Environ Sci Pol 12(1):5–22

European Commission (2017) Energy – voluntary schemes. Information available at http://ec.europa.eu/energy/node/74. Accessed 26 June 2017

FAO (2017a) AQUASTAT [WWW Document]. Food Agric Organ. U. N. URL http://www.fao.org/faostat/en/#home. Accessed 17 May 17

FAO (2017b) FAOSTAT [WWW Document]. Food Agric Organ. U. N. URL http://www.fao.org/faostat/en/#home. Accessed 30 May 17

Fargione J, Hill J, Tilman D, Polasky S, Hawthorne P (2008) Land clearing and the biofuel carbon debt. Science 319:1235–1238. https://doi.org/10.1126/science.1152747

Foley JA, DeFries R, Asner GP, Barford C, Bonan G, Carpenter SR, Chapin FS, Coe MT, Daily GC, Gibbs HK, Helkowski JH, Holloway T, Howard EA, Kucharik CJ, Monfreda C, Patz JA, Prentice C, Ramankutty N, Snyder PK (2005) Global consequences of land use. Science 309:570–574. https://doi.org/10.1126/science.1111772

GBEP (Global Bioenergy Partnership) (2011) The global bioenergy partnership sustainability indicators for bioenergy. FAO, Rome

GBEP (Global Bioenergy Partnership) (2017) Information available at http://www.globalbioenergy.org/. Accessed June 2017

Gerbens-Leenes PW, Hoekstra AY, van der Meer T (2009) The water footprint of energy from biomass: a quantitative assessment and consequences of an increasing share of bio-energy in energy supply. Ecol Econ 68:1052–1060. https://doi.org/10.1016/j.ecolecon.2008.07.013

Gibbs HK, Johnston M, Foley JA, Holloway T, Monfreda C, Ramankutty N, Zaks D (2008) Carbon payback times for crop-based biofuel expansion in the tropics: the effects of changing yield and technology. Environ Res Lett 3:034001. https://doi.org/10.1088/1748-9326/3/3/034001

Gilio L, Moraes M (2016) Sugarcane industry's socioeconomic impact in São Paulo, Brazil: a spatial dynamic panel approach. Energ Econ 58:27–37
Goovaerts L, Pelkmans L, Goh CS, Junginger M, Joudrey J, Chum H, Smith CT, Stupak I, Cowie A, Dahlman L, Englund O, Goss A (2013) Monitoring sustainability certification of bioenergy. Report of task 1: examining sustainability certification of bioenergy. IEA Bioenergy, IEA, Paris
Hennenberg K, Dragisic C, Haye S, Hewson J, Semroc B, Savy C, Wiegmann K, Fehrenbach H, Fritsche U (2010) The power of bioenergy-related standards to protect biodiversity. Conserv Biol 24(2):412–423
Hernandes TAD, Bufon VB, Seabra JEA (2014) Water footprint of biofuels in Brazil: assessing regional differences. Biofuels Bioprod Biorefin 8:241–252. https://doi.org/10.1002/bbb.1454
Herreras-Martínez SH, van Eijck J, Cunha MP, Guilhoto JMM, Walter A, Faaij A (2013) Analysis of socio-economic impacts of sustainable sugarcane–ethanol production by means of inter-regional input–output analysis: demonstrated for Northeast Brazil. Renew Sust Energ Rev 28:290–316
Hochman G, Zilberman D, Rajagopal D (2010) Are biofuels the culprit? OPEC, food, and fuel. Am Econ Rev 100(2):183–187
ISCC (2016) Sustainability requirements – vesioSSS n 3.0. Available at www.iscc-system.org/
ISCC (International Sustainability & Carbon Certification) (2017) Information available at www.iscc-system.org/. Accessed 12 June 2017
ISO (2015) ISO 13065: sustainability criteria for bioenergy. ISO, Geneva
Karp A, Artaxo Netto PE, Berndes G, Cantarella H, El-Lakany H, Estrada TEMD, Faaij A, Fincher GB, Huntley BJ, Ravindranath NH, Van Sluys M-A, Verdade LM, Youngs H (2015) Environmental and climate security. In: Bioenergy and sustainability: bridging the gaps. Scientific Committee on Problems of the Environment (SCOPE), Paris Cedex, pp 138–183
Kline K, Martinelli F, Mayer AL, Medeiros R, Oliveira COF, Sparovek G, Walter A, Venier LA (2015) Bioenergy and biodiversity: key lessons from the Pan-American region. Environ Manag 56(6):1377–1396
Koh LP, Ghazoul J (2008) Biofuels, biodiversity, and people: understanding the conflicts and finding opportunities. Biol Conserv 141(10):2450–2460
Koh LP, Wilcove DS (2007) Cashing in palm oil for conservation. Nature (Lond) 448(7157):993
Kullander S (2010) Food security: crops for people not for cars. Ambio 39(3):249–256
Land Matrix (2015) Land Matrix is an online public database on land deals. Available at: http://www.landmatrix.org/en/
de Macedo IC, Nassar AM, Cowie AL, Seabra JEA, Marelli L, Otto M, Wang MQ, Tyner WE (2015) Greenhouse gas emissions from bioenergy. In: Bioenergy and sustainability: bridging the gaps. Scientific Committee on Problems of the Environment (SCOPE), Paris Cedex, pp 582–617
Machado PG (2017) Sustainable development potentials and pathways for biobased economy options: an integrated approach on land use, energy system, economy and greenhouse gases emissions. PhD Thesis, University of Campinas, Unicamp, Campinas
Machado PG, Picoli MCA, Torres LJ, Oliveira JG, Walter A (2015) The use of socioeconomic indicators to assess the impacts of sugarcane production in Brazil. Renew Sust Energ Rev 52:1519–1526
Machado PG, Walter A, Picoli MC, João CG (2016) Potential impacts on local quality of life due to sugarcane expansion: a case study based on panel data analysis. Environ Dev Sustain 18:1–24
Martinelli L, Filoso S (2008) Expansion of sugarcane ethanol production in Brazil: environmental and social challenges. Ecol Appl 18(4):885–898
Mello FFC, Cerri CEP, Davies CA, Holbrook NM, Paustian K, Maia SMF, Galdos MV, Bernoux M, Cerri CC (2014) Payback time for soil carbon and sugar-cane ethanol. Nat Clim Chang 4:605–609. https://doi.org/10.1038/nclimate2239
Moraes MAFD, Oliveira FCR, Diaz-Chavez RA (2015) Socio-economic impacts of Brazilian sugarcane industry. Environ Dev 16:31–43

Moraes MAFD, Bacchi MRP, Caldarelli CE (2016) Accelerated growth of the sugarcane, sugar, and ethanol sectors in Brazil (2000–2008): effects on municipal gross domestic product per capita in the south-central region. Biomass Bioenergy 8(91):116–125
Moschini GC, Cui J, Lapan H (2012) Economics of biofuels: an overview of policies, impacts and prospects. Bio-based Appl Econ 1(3):269–296
Nantha H, Tisdell C (2009) The orangutan–oil palm conflict: economic constraints and opportunities for conservation. Biodivers Conserv 18:487–502
Neto AE, Shintaku A, Pio AAB, Conde AJ, Donzelli JL (2009) Manual de conservação e reuso de água na agroindústria sucroenergética. ANA, Brasília
Nogueira LAH, Leal MRLV, Fernandes E, Chum HL, Diaz-Chaves R, Endres JM, Mahakhant A, Otto M, Seebaluck V, van der Wielen L (2015) Sustainable development and innovation. In: Bioenergy and sustainability: bridging the gaps. Scientific Committee on Problems of the Environment (SCOPE), Paris Cedex, pp 184–217
Pfister S, Koehler A, Hellweg S (2009) Assessing the environmental impacts of freshwater consumption in LCA. Environ Sci Technol 43:4098–4104. https://doi.org/10.1021/es802423e
Pfister S, Bayer P, Koehler A, Hellweg S (2011) Environmental impacts of water use in global crop production: hotspots and trade-offs with land use. Environ Sci Technol 45:5761–5768. https://doi.org/10.1021/es1041755
Phalan B (2009) The social and environmental impacts of biofuels in Asia: an overview. Appl Energ 86:21–29
Pilgrim S, Harvey M (2010) Battles over biofuels in Europe: NGOs and the politics of markets. Sociol Res Online 15(3):1–16
Pimentel D, Marklein A, Toth MA, Karpoff MN, Paul GS, McCormack R, Kyriazis J, Krueger T (2009) Biofuel impacts on world food supply: use of fossil fuel, land and water resources. Hum Ecol 37:1–12
Ravagnani RM, Walter A, Matuguma, CA (2016) Background review on social impacts of land grabbing associated to biofuels production. 5th IAEE Asian Conference, Perth, 24–17 February 2016
REN21 (2016) Renewables 2016 global status report. REN21 Secretariat, Paris
Richardson B (2012) From a fossil-fuel to a biobased economy: the politics of industrial biotechnology. Environ Plan C 30(2):35
Robertson B, Pinstrup-Andersen P (2010) Global land acquisition: neo-colonialism or development opportunity? Food Security 2(3):271–283
RSB (2016) RSB principle and criteria. Available at http://rsb.org/
RSB – Roundtable on Sustainable Biomaterials (2017) Information available at http://rsb.org/. Accessed 12 June 2017
Scheidel A, Sorman AH (2012) Energy transitions and the global land rush: ultimate drivers and persistent consequences. Global Environ Change 22:588–595
Searchinger T, Heimlich R, Houghton RA, Dong F, Elobeid A, Fabiosa J, Tokgoz S, Hayes D, T-H Y (2008) Use of U.S. croplands for biofuels increases greenhouse gases through emissions from land-use change. Science 319(5867):1238–1240
Semino S, Paul H, Tomei J, Joensen L, Monti M, Jelsoe E (2009) Soybean biomass produced in Argentina: myths and realities. IOP Conference Series. Earth Environ Scie 8(1)
Souza GM, Victoria RL, Joly CA, Verdade LM (2015) Bioenergy and sustainability: bridging the gaps. Scientific Committee on Problems of the Environment (SCOPE), Paris Cedex
TEEB – The Economics of Ecosystems and Biodiversity (2015) TEEB for Agriculture & Food: an interim report. United Nations Environment Programme, Geneva
Venter O, Meijaard E, Wilson K (2008) Strategies and alliances needed to protect forest from palm-oil industry. Nature (Lond) 451(7174):16
de Vries SC, van de Ven GWJ, van Ittersum MK, Giller KE (2010) Resource use efficiency and environmental performance of nine major biofuel crops, processed by first-generation conversion techniques. Biomass Bioenergy 34:588–601

Woods J, Lynd LR, Laser M, Batistella M, Victoria D de C, Kline K, Faaij A (2015). Land and bioenergy. In: Bioenergy and sustainability: bridging the gaps. Scientific Committee on Problems of the Environment (SCOPE), Paris Cedex, pp 258–301

Yeh S, Berndes G, Mishra GS, Wani SP, Elia Neto A, Suh S, Karlberg L, Heinke J, Garg KK (2011) Evaluation of water use for bioenergy at different scales. Biofuels Bioprod Biorefin 5:361–374. https://doi.org/10.1002/bbb.308

Zilberman D, Hochman G, Rajagopal D, Sexton S, Timilsina G (2012) The impact of biofuels on commodity food prices: assessment of findings. Am J Agric Econ 95(2):275–281

Erratum to: Lignocellulosic Biomass for Energy, Biofuels, Biomaterials, and Chemicals

Abla Alzagameem, Basma El Khaldi-Hansen, Birgit Kamm, and Margit Schulze

Erratum to
Chapter 5 in: S. Vaz Jr. (ed.), *Biomass and Green Chemistry*, https://doi.org/10.1007/978-3-319-66736-2_5

The original version of this chapter was published without printing the 2nd affiliation for one of the chapter authors – Abla Alzagameem. The correct affiliation for Abla Alzagameem should be:

1 Faculty of Environment and Natural Sciences, Brandenburg University of Technology BTU Cottbus-Senftenberg, Cottbus, Germany

2 Department of Natural Sciences, Bonn-Rhein-Sieg University of Applied Sciences, Rheinbach, Germany

The original version of this chapter was inadvertently published without printing Acknowledgements. The acknowledgement section is added in the current version of this chapter as below:

Acknowledgements Financial support (scholarship) was given to Abla Alzagameem by the Avempace-II Erasmus-Mundus Programme and the Graduate Institute of the Bonn-Rhein-Sieg University of Applied Sciences.

The updated online version of the chapter can be found at https://doi.org/10.1007/978-3-319-66736-2_5

S. Vaz Jr (ed.), *Biomass and Green Chemistry*,
https://doi.org/10.1007/978-3-319-66736-2_9

A Glossary of Green Terms

Sílvio Vaz Jr., Birgit Kamm, Claudio Mota, Marcelo Domine, German Aroca, Mônica Caramez Triches Damaso, Paulo Marcos Donate, and Roger Sheldon

Abstract To achieve an economy based on bioproducts instead of petrochemicals and one that is more sustainable, it is relevant to share scientific knowledge promoting the development and application of renewable chemistry. We can observe renewable molecules obtained from renewable and nonfood sources of biomass for applications such as polymers, solvents, biofuels, bioenergy, and pharmaceuticals. These molecules can be obtained by means of innovative processes that take into account reduction in the generation of residues, energy economy, development and

S. Vaz Jr. (✉) • M.C.T. Damaso
Brazilian Agricultural Research Corporation, National Research Center for Agroenergy (Embrapa Agroenergy), Embrapa Agroenergia, Parque Estação Biológica, Brasília, DF, Brazil
e-mail: silvio.vaz@embrapa.br

B. Kamm
Brandenburg University of Technology BTU Cottbus-Senftenberg, Cottbus, Germany

Kompetenzzentrum Holz GmbH, Linz, Austria

C. Mota
Federal University of Rio de Janeiro, Rio de Janeiro, Brazil

M. Domine
Instituto de Tecnología Química, Universitat Politècnica de València/Consejo Superior de Investigaciones Científica (UPV-CSIC), Valencia, Spain

G. Aroca
Pontificia Universidad Católica de Valparaiso, Valparaiso, Chile

P.M. Donate
University of São Paulo, São Paulo, Brazil

R. Sheldon
Department of Biotechnology, Delft University of Technology, Delft, The Netherlands

Molecular Science Institute, School of Chemistry, University of the Witwatersrand, Johannesburg, South Africa

S. Vaz Jr. (ed.), *Biomass and Green Chemistry*,
https://doi.org/10.1007/978-3-319-66736-2

application of new catalysts and biocatalysts, and improved microorganisms and plants. Thus, a glossary that compiles the terms and significances of this new chemical area has, as its main objective, dissemination of the concept of renewable chemistry.

In this glossary were considered contributions from the several areas closely related to the renewable chemistry concept and its application: analytical chemistry, biochemistry, environmental chemistry, organic chemistry, inorganic chemistry, and chemical engineering.

Keywords Bio-based economy; Biomass; Green chemistry; Renewable resources

Abbreviations

1G	First generation (for biofuels)
2G	Second generation (for biofuels)
3G	Third generation (for biofuels)
AFEX	Ammonia fiber explosion
APR	Aqueous phase reforming
ATP	Adenosine triphosphate
B100	Biofuel mixture with 100% v/v biodiesel
Bio-ETBE	Ethyl *tert*-butyl ether or 2-ethoxy-2-methyl-propane, $C_6H_{14}O$
Bio-MTBE	Methyl *tert*-butyl ether or 2-methoxy-2-methyl-propane, $C_5H_{12}O$
C5	Pentoses from carbohydrate polymers
C6	Hexoses from carbohydrate polymers
CDH	Cellobiose dehydrogenase
CHP	Combined heat and power
FAME	Fatty acid methyl ester
FSCW	Food supply chain waste
GBR	Green biorefinery
HTS	High-throughput screening
LCA	Life cycle analysis
LC-MS	Liquid chromatography–mass spectrometry
LPMO	Lytic polysaccharide monooxygenase
MSW	Municipal solid waste
NMR	Nuclear magnetic resonance
PHA	Polyhydroxyalkanoate
PHB	Polyhydroxybutyrate
PMO	Polysaccharide monooxygenase

Acid Pretreatment

Process catalyzed by an acid (Brønsted acid or Lewis acid) to promote the separation of lignocellulosic fractions, such as hemicellulose (*see Hemicellulose*). Sulfuric, hydrochloric, phosphoric, and nitric acids are commonly used.

Advanced Fuels
Liquid fuel derived from sustainable or renewable sources of organic matter (*see Organic matter*) that do not typically compete with food production, such as agricultural wastes (*see Waste*), forest and wood residues, certain oilseeds, and algae; these produce low CO_2 emissions from their use.

Aerobic Digestion
Digestion process occurring by means of the activity of bacteria and fungi that requires molecular oxygen. This process has a high energy yield (*see Yield in biotechnology*) obtained when molecular oxygen reacts with organic matter (*see Organic matter*), releasing heat (e.g., composting). It can be applied to hazardous wastes (*see Waste*) such as chemical process wastes and landfill leachates, with microorganisms (*see Microorganisms*) degrading the hazardous molecules.

Agricultural Biomass
All biomass (*see Biomass*) generated from crop production in the field, be it soil or an aquatic environment.

Agricultural Residues
Residues from crop production, such as bagasse (*see Bagasse*), straw, or liquid effluents.

Alternative Fuels
Alternative fuels for transportation applications include methanol, denatured ethanol, and other alcohols; fuel mixtures containing 85% or more by volume of methanol, denatured ethanol, and other alcohols with gasoline or other fuels; natural gas; liquefied petroleum (*see Petroleum*) gas (propane); hydrogen; coal-derived liquid fuels; fuels (other than alcohol) derived from biological materials [biofuels (*see Biofuels*) such as soy diesel fuel]; and electricity [including electricity from solar energy (*see Solar energy*)].

Alternative Solvents
Solvents that have low toxicity, are easy to recycle, are inert, and do not contaminate the product. They are also called green solvents (*see Green solvent*). Biodegradability, biocompatibility, and sustainability (*see Sustainability*) are desirable qualities for an alternative solvent.

Reference: (Paiva et al. 2014).

Ammonia Fiber Explosion (AFEX)
Ammonia-based process for pretreatment of lignocellulosic biomass; AFEX improves the susceptibility of lignocellulosic biomass to enzymatic attack.

Amylases
Starch-degrading enzymes (*see Enzymes*) of hexose monomers such as glucose.

Anaerobic Digestion
Series of biological processes [hydrolysis (*see Hydrolysis*), acidogenesis, acetogenesis, and methanogenesis] in which microorganisms (*see Microorganisms*) convert organic material, in the absence of oxygen, to a mixture of methane (CH_4) and

carbon dioxide (CO_2) and other gaseous compounds in small amounts, called biogas (*see Biogas*), for bioelectricity generation (*see Bioelectricity*). It produces also a liquid fraction (rich in inorganic species in aqueous medium, which can be precipitated) and a solid fraction that when dewatered can be used as a fertilizer (*see Fertilizers*), depending on its composition.

Aqueous Effluent
Discharge of liquid waste (*see Waste*) from a factory, refinery, or biorefinery (*see Biorefinery*) containing a large proportion of water and other compounds, which commonly must be treated before its disposal.

Aqueous Phase Reforming (APR)
Catalytic process to obtain H_2, CO, CO_2, and other volatile compounds by means of reforming (*see Reforming*) in an aqueous medium, normally using supported metal heterogeneous catalysts (*see Heterogeneous catalyst*). When the process is carried out by feeding reactants at high temperature (>350–400 °C) in the gas-phase system, it is called steam reforming.

Basic Pretreatment
Alkaline agents such as sodium, potassium, calcium, and ammonium hydroxides are added in the pretreatment of lignocellulosic biomass (see *Lignocellulosic biomass*) to release the cellulosic fraction (see *Cellulosic fraction*). This pretreatment promotes delignification and removal of hemicellulose (*see Hemicellulose*) from lignocellulosic biomass, allowing, in a later step, the enzymatic hydrolysis (*see Hydrolysis*) to release sugars. Beyond that, this has the advantage of causing less sugar degradation than acid treatment.

Bagasse
Fibrous fraction of the sugarcane after juice extraction constituting a lignocellulosic biomass (*see Lignocellulosic biomass*). It can be used to generate bioelectricity (*see Bioelectricity*), second-generation ethanol, chemicals, and materials.

Bio-Based Economy
Model of economy based on the use of renewable sources instead of nonrenewable sources such as petroleum (*see Petroleum*). In a bio-based economy, renewable resources (*see Renewable resources*) are used for industrial purposes and energy production, without compromising food and feed provision.

Reference: (Hennig et al. 2015)

Bio-Based Polymers
Polymers (*see Polymers*) obtained from renewable sources such as sugars, starch, or oleaginous or lignocellulosic biomass. Examples are the 'green polyethylene' from sugarcane ethanol, and polyhydroxyalkanoates (PHAs) and polyhydroxybutyrate (PHB) produced by a variety of microorganisms (*see Microorganisms*) as a reserve of carbon and energy. Cellulose is the major natural and bio-based polymer constituent of the vegetal cell wall.

Bio-Based Chemicals
Chemicals obtained from renewable sources such as sugars, starch, or oleaginous or lignocellulosic biomass by means of chemocatalytic, biochemical, or thermochemical processes; examples include furfural, succinic acid, and levulinic acid.

Biocatalysis
Use of enzymes (*see Enzymes*), isolated or present in whole cells, to promote an increase in the rate of chemical and biochemical reactions to produce, for instance, biofuels (*see Biofuels*) and chemicals.

Biocatalyst
Enzyme or a group of enzymes (*see Enzymes*) consisting of, or derived from, an organism or cell culture (in cell-free or whole-cell forms) that catalyzes metabolic reactions in living organisms or substrate conversions in one or various biochemical reactions.

Reference: (PAC 1992a)

Biochar
Product obtained from fast pyrolysis of biomass, when the biomass is heated quickly (<2 s) to 700 °C in the absence of oxygen, generating a solid with good fuel properties.

Biochemical Conversion Process
Processes based on enzymatic or fermentation technologies to convert a substrate into a product, for instance, biofuels (*see Biofuels*) or chemicals from sugars, or oleaginous, starch, or lignocellulosic feedstock.

Biodegradable Polymer
Polymer susceptible to being degraded by biological activity, with the degradation accompanied by a decrease of its molar mass.

Notes:

1. See also Note 2 to polymer degradation.
2. In the case of a polymer, its biodegradation (*see Biodegradation*) proceeds not only by the catalytic activity of enzymes (*see Enzymes*), but also by a wide variety of biological activities.

Reference: (PAC 2004)

Biodegradation
Breakdown of a substance catalyzed by enzymes (*see Enzymes*) in vitro or in vivo: this may be characterized for purposes of hazard assessment as follows.

1. Primary. Alteration of the chemical structure of a substance, resulting in loss of a specific property of that substance.
2. Environmentally acceptable. Biodegradation to such an extent as to remove undesirable properties of the compound; this often corresponds to primary biodegradation but it depends on the circumstances under which the products are discharged into the environment.

3. Ultimate. Complete breakdown of a compound to either fully oxidized or reduced to simple molecules (such as carbon dioxide/methane, nitrate/ammonium, and water. It should be noted that the products of biodegradation can be more harmful than the substance that was degraded.

   Reference: (PAC 1992b)

Biodiesel
Biofuel derived from oleaginous biomass (see *Oleaginous biomass*) for using in compression ignition engines (diesel-type engines). It can be usually obtained by various processes, such as transesterification, cracking (*see Cracking*), hydrotreatment, and the Fischer–Tropsch process (see *Fischer–Tropsch process*): transesterification is the most common process. Technically, biodiesel is defined as a fuel composed of mono-alkyl esters of long-chain fatty acids derived from vegetable oils or animal fats, designated B100.
Reference: (American Society for Test and Methods 2015)

Bioeconomy
Economics model based on renewable sources (*see Renewable resources*), such as biomass, instead of an economics model based on nonrenewable resources (*see Nonrenewable resources*), such as petroleum (*see Petroleum*) and coal, according to sustainable metrics (*see Sustainable metrics*). Products of a bioeconomy system are biofuels (*see Biofuels*), chemicals, bioenergy (*see Bioenergy*), and biomaterials (*see Biomaterials*).

Bioelectricity
Production of electricity from biomass, for example, sugarcane bagasse (*see Bagasse*) as a feedstock, biomass derivatives by means of the co-generation process.

Bioenergy
Energy generated from biomass for industrial or domestic uses, commonly as electricity (for power) and heat.

Bioethanol
Ethanol (C_2H_6O) produced from sucrose or starch (first generation) or lignocellulosic biomass (second generation) by fermentation, mainly by the yeast (*see Yeast*) *Saccharomyces cerevisiae*, to be used as biofuel or renewable feedstock (*see Renewable feedstock*) for the chemical industry.

Biofuels
Fuels obtained from different biomass sources (sucrose, starch, oleaginous, or cellulosic) by a chemical, biochemical, or thermochemical process to substitute green fossil fuels for petroleum (*see Petroleum*). Examples of biofuels: bioethanol (*see Bioethanol*), biodiesel (*see Biodiesel*), biogas (*see Biogas*), biomethanol, biodimethylether, bio-ETBE (ethyl *tert*-butyl ether, $C_6H_{14}O$), bio-MTBE (methyl *tert*-butyl ether, $C_5H_{12}O$), biokerosene (*see Biokerosene*), or biohydrogen (H_2) (*see Biohydrogen*), and pure vegetable oil.

Biogas

Mixture of gaseous compounds produced by anaerobic digestion (*see Anaerobic digestion*) of organic materials that can contain CH_4 (ϕ = 48–75%), CO_2 (ϕ = 25–47%), N_2 (ϕ = 0–20%), O_2 (ϕ = 0–5%), H_2S (ϕ = 100–2000 × 10^{-6}), and mercaptans (0–100 × 10^{-6}). Biogas can be used for electricity and heat generation, and therefore by industrial or rural properties, bringing economic and environmental gains.

Note: ϕ = volume fraction.

Biohydrogen

Hydrogen produced biologically from organic materials, most commonly by algae, Bacteria, or Archaea, and hydrogen produced chemically from organic materials or biomass in general.

Biokerosene

Biofuel that meets standards for being used in the aviation sector as an alternative fuel or in mixtures with kerosene from petroleum (*see Petroleum*). It can be produced from fatty acid methyl ester (*see Fatty acid methyl ester*), when it comes from an oleaginous feedstock (e.g., soybean), or by hydrocarbons obtained by the Fischer–Tropsch process of syngas (*see Syngas*) from biomass (*see Biomass*) gasification, as well as from the fermentation of sugars using recombinant microorganisms (*see Microorganisms*) (e.g., farnesene).

Biomass

Material produced by the growth of microorganisms, plants, or animals.

Reference: (PAC 1992b)

In plants, we can consider four types: lignocellulosic (*see Lignocellulosic biomass*), oleaginous (*see Oleaginous biomass*), starch (*see Starch biomass*), and saccharide (*see Saccharide biomass*).

Biomass Waste

Organic non-fossil material of biological origin that is a by-product or a discarded product. Biomass waste (*see Waste*) includes municipal solid waste (*see Municipal solid waste*) from biogenic sources, landfill gas, sludge waste, agricultural crop by-products, straw, and other biomass (*see Biomass*) solids, liquids, and gases, but excludes wood and wood-derived fuels, biofuels feedstock, biodiesel (*see Biodiesel*), and fuel ethanol.

Note:

Biomass waste data also include energy crops grown specifically for energy production, which would not usually constitute waste.

Biomass Chemistry

Study of the chemical composition and properties of biomass (*see Biomass*) and its conversion processes.

Biomaterials

Any material originated from biological and renewable resources (*see Renewable resources*) or applied to biological purposes, such as polymers (*see Polymers*), composites (*see Composites*), and metallic alloys.

Bio-Oil

Oil or liquid bio-crude commonly obtained from the pyrolysis of biomass (*see Biomass*), as, for example, fast pyrolysis of lignocellulosic biomass (*see Lignocellulosic biomass*) by a rapid heating (<2 s) to about 500 °C in the absence of oxygen. The major compounds identified in its composition are carboxylic acids, aldehydes, ketones, furfurals, sugar-like material, and lignin-derived compounds.

Bioproducts

Products obtained from biomass (*see Biomass*) by means of conversion processes (*see Conversion process*), such as biofuels (*see Biofuels*), chemicals, and biomaterials (*see Biomaterials*).

Biorefinery

Industrial plant dedicated wholly to the use of biomass (*see Biomass*). As a petroleum (*see Petroleum*) refinery, it would be able to produce biofuels (*see Biofuels*), bioenergy (*see Bioenergy*), chemicals, and biomaterials (*see Biomaterials*). A biorefinery can be based on a number of processing platforms using mechanical, thermal, chemical, and biochemical processes.

Biotechnology

Integration of natural sciences and engineering sciences to achieve the application of organisms, cells, parts thereof, and molecular analogues for products and services.

Reference: (PAC 1992b)

Black Liquor

By-product (*see By-product*) of the paper production process; alkaline spent liquor that can be used as a source of energy or solubilized lignin (*see Lignins*). Alkaline spent liquor is removed from the digesters in the process of chemically pulping wood. After evaporation, the residual "black" liquor is burned as a fuel in a recovery furnace that permits the recovery of certain basic chemicals.

Building-Block Molecules

Molecules from renewable sources, for example, phenolic monomers, fatty acids, and carbohydrates, which act as raw materials (*see Raw material*) for other products with high value such as polymers (*see Polymers*); several synthetic molecules, among others, for a large number of applications.

Carbohydrate Chemistry

Chemical aspects related to production, analysis, and conversion of C5 (pentoses) and C6 (hexoses) carbohydrates.

Carbon Capture and Storage
Family of methods and technologies to capture CO_2 emitted by combustion processes (*see Combustion process*), for example, burning of fuels to generate energy, burning of coal by the cement and steel industries, or gasification processes (*see Gasification process*). An example is the CO_2 captured and pressurized to about 100 bar or more to be transported and stored in a site by means of injection into geological features, to prevent its emission into atmosphere for thousands of years.

Carbon Fiber
Fibers (filaments, tow, yarns, rovings) (*see Fibers*) consisting of w at least 92% carbon, usually in the non-graphitic state.

Note 1:

Carbon fibers are fabricated by pyrolysis of organic precursor fibers or by growth from gaseous hydrocarbons. Use of the term graphite fibers instead of carbon fibers, as is often observed in the literature, is incorrect and should be avoided. The term graphite fibers is justified only if three-dimensional crystalline order is confirmed, such as by X-ray diffraction measurements.

Reference: (PAC 1995a)

Note 2:

w = mass fraction.

Carbon Fixation
Carbon dioxide is converted into sugar that is found only in plants by means of photosynthesis (*see Photosynthesis process*).

Carbon Footprint
Metric to measure the global warming (*see Global warming*) potential of a process or a product, mainly by means of the consideration of the consumption of fossil fuels during the production and the CO_2 produced or released (input and output) by the product or process. For renewable chemicals (*see Renewable chemicals*), their carbon footprint can be calculated by means of life cycle assessment tools (see *Life cycle assessment*) and compared against petrochemicals to prove their advantage or disadvantage.

Carbon Cycle
All carbon sinks and exchanges of carbon from one sink to another by various chemical, physical, geological, and biological processes.

Carbon Dioxide Equivalents
Amount of carbon dioxide by weight emitted into the atmosphere that would produce the same estimated radiative forcing as a given weight of another relatively active gas. Carbon dioxide equivalents are computed by multiplying the weight of the gas being measured (for example, methane) by its estimated global warming (*see Global warming*) potential (which is 21 for methane). "Carbon equivalent units" are defined as carbon dioxide equivalents multiplied by the carbon content of carbon dioxide (e.g., 12/44).

Carbon Stocks

Quantity of carbon stored in biological and physical systems including trees, products of harvested trees, agricultural crops, plants, wood and paper products, and other terrestrial biosphere sinks, soils, the oceans, and sedimentary and geological sinks.

Carbonization

Process by which solid residues with increasing content of the element carbon are formed from organic material, usually by pyrolysis in an inert atmosphere.

Note:

As with all pyrolytic reactions, carbonization is a complex process in which many reactions take place concurrently such as dehydrogenation, condensation, hydrogen transfer, and isomerization. It differs from coalification in that its reaction rate is faster by many orders of magnitude. The final pyrolysis temperature applied controls the degree of carbonization and the residual content of foreign elements, for example, if the carbon content of the residue exceeds w of 90%, whereas at more than w of 99%, carbon is found.

Reference: (PAC 1995b)

Note:

w = mass fraction.

Catalysis

Application of a catalyst (*see Catalyst*) in a conversion process (*see Conversion process*).

Catalyst

Substance that increases the rate of a reaction without modifying the overall standard Gibbs energy change in the reaction; the process is called catalysis (*see Catalysis*).

Reference: (PAC 1996a)

Alternatively, we can consider organic and inorganic molecules as chemical catalysts, in homogeneous or heterogeneous phase, according to the substrate, and biological catalysts as isolated enzymes (*see Enzymes*) or whole cells, in solution (homogeneous) or immobilized (heterogeneous).

Cellulase

Enzyme or a group of enzymes (*see Enzymes*), mainly hydrolytic, used to catalyze the depolymerization (*see Depolymerization*) of cellulose to obtain glucose. The cellulase system in both bacteria and fungi constitutes four different classes of enzymes: (1) *endo*-1,4-β-glucanases; (2) *exo*-1,4-β-D-glucanases, including both 1,4-β-D-glucan cellobiohydrolases and 1,4-β-D-glucan glucohydrolases; (3) 1,4-β-D-glucosidases, also referred to as cellobiases; and (4) lytic polysaccharide monooxygenases (LPMOs), which are oxidoreductases.

Cellulose

The major constituent of plant cell walls and the most abundant organic compound in nature with glucose as the monomer, linked by 1,4-β-D bonds. Generally,

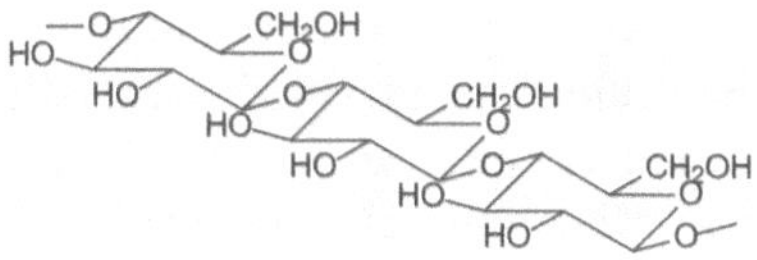

Fig. 1 Cellulose structure

cellulose occurs in two forms: amorphous and crystalline. It is the framework substance in wood, accounting for *w* of 40–50% of the wood in the form of cellulose microfibrils. Cellulose is a source of sugar for second-generation ethanol and chemicals after a hydrolysis (*see Hydrolysis*) step (Fig. 1).

Note:
w = mass fraction

Cellulosic Fraction
Fraction obtained after the lignocellulosic biomass (*see Lignocellulosic biomass*) pretreatment by means of the use of chemicals in aqueous medium, commonly alkali or acid, separating it from hemicellulose (*see Hemicellulose*) and lignin fractions (*see Lignin fraction*) for subsequent conversion to glucose.

Chemical Analysis
Application of chemical reactions to obtain quantitative or qualitative data about chemical composition.

Chemical Conversion Process
Transformation process (organic or inorganic) that occurs with or without the use of catalysts (*see Catalyst*), used to obtain biofuel, bioenergy (*see Bioenergy*), chemicals, or materials from a certain feedstock.

Chemocatalytic Platform
Set of technologies to promote the transformation of biomass (*see Biomass*) into a large number of molecules with a high value using catalysts (homogeneous or heterogeneous) (*see Catalyst*).

Cogeneration (or Combined Heat and Power, CHP)
Use of lignocellulosic biomass (see *Lignocellulosic biomass*) [e.g., sugarcane bagasse (*see Bagasse*)] as fuel in boilers for heating water and generating steam, which will turn a turbine that produces electricity, or bioelectricity (see *Bioelectricity*), for power.

Combustion Process
Rapid exothermic process in the presence of oxygen that liberates substantial energy as heat together with CO_2 and water. This process can be used to generate energy from biomass (*see Biomass*), as the co-generation process applied to the sugarcane bagasse (*see Bagasse*) to produce bioelectricity (*see Bioelectricity*).

Conversion Process
Certain process used to transform a biomass (*see Biomass*) into product(s), such as a biofuel, chemical, or material. It can be subdivided into noncatalytic processes, chemocatalytic processes [heterogeneous and homogeneous catalysis (*see*

Heterogeneous catalysis and *Homogeneous catalysis*)], biochemical processes [fermentation and enzymatic catalysis or biocatalysis *(see Biocatalysis)*], and thermochemical processes (combustion, gasification, pyrolysis). Each process involves its own technological approach.

Composites
Systems formed by different materials in the same state; there are interactions between the materials near the surface. Wood is a composite.

Compositional Analysis
Chemical analysis (*see Chemical analysis*) or chemical analyses set used to determine the chemical composition of a certain sample, as biomass (*see Biomass*), based on analytical techniques such as gravimetry, thermogravimetry, and elemental analysis. It normally involves fractionation of the sample and it is fundamental to the establishment of quality criteria for feedstock and products.

Co-products
Secondary products from a biomass (*see Biomass*) conversion process (*see Conversion process*) that could be used to add value in an economic chain. An example is glycerine obtained from biodiesel (*see Biodiesel*) production.

Co-products Valorization
Exploitation of possible by-products (*see By-product*) to generate fuels, energy, or value-added chemicals (*see Value-added chemicals*) to improve the process of economic viability.

Cracking (or Catalytic Cracking)
Process commonly used in refinery or biorefinery (*see Biorefinery*) for breaking down the larger, heavier, and more complex hydrocarbon molecules into simpler and lighter molecules with the help of a catalyst (*see Catalyst*).

Dehydration
Removal of water from feedstock or products. It is very useful for biomass (*see Biomass*) and its products because of their high oxygen content.

Deoxygenation
Removal of one or more oxygen atom from a molecule or group of molecules. It is very useful to obtain aromatic and alkyl compounds from biomass (*see Biomass*) components.

Depolymerization
Process of converting a polymer into a monomer or a mixture of monomers.

Note:

Unzipping is depolymerization occurring by a sequence of reactions, progressing along a macromolecule (*see Macromolecule*) and yielding products, usually monomer molecules at each reaction step, from which macromolecules similar to the original can be regenerated.

Reference: (PAC 1996b)

Detoxification
Process, or processes, of chemical modification that make a toxic molecule less toxic.

Reference: (PAC 1993a)

Downstream Processing
Refers to the recovery and purification of products, including the recycling or valorization of usable components and proper treatment or disposal of waste (*see Waste*). It is an essential step to maintain the economic viability of industries based on biomass (*see Biomass*) and renewable feedstock (*see Renewable feedstock*).

Drop-in
Addition of a renewable chemical (reactant) to make the final product 'greener.' Furthermore, a drop-in solution to a process is one where the process operation/configuration does not have to be altered. For instance, in the case of drop-in biofuels (*see Biofuels*) there is no modification of the engine or final device used.

Eco-Friendly Process
Process with a minimal negative impact on the environment.

Ecology
Interdisciplinary field that studies the interactions among organisms and their environment. Ecology also studies the environmental impact (*see Environmental impact*) from the use of natural resources.

Economic Impact
Impact of a technology, a process, or a product on the economy of a society. It is part of the sustainability (*see Sustainability*) measurement. It can be positive or negative.

E-factor
Measurement of sustainability (*see Sustainability*) for conversion processes (*see Conversion process*) based on the ratio of the amount of waste (*see Waste*) generated divided by the amount of desired product produced. The E-factor is lower for sustainable processes and the ideal E-factor is zero. It can be calculated by

$$E\text{-}factor = m_{(\text{waste})} / m_{(\text{product})}$$

Reference: (Sheldon 2007)

Energy
In mechanics, the sum of potential energy and kinetic energy. In thermodynamics, the internal energy or thermodynamic energy increase, ΔU, is the sum of heat and work brought to the system. Only changes in energy are measurable. For photons:

$$E = h\nu$$

where h is the Planck constant and ν the frequency of radiation. In relativistic physics:

$$E = mc^2$$

which is the speed of light and the mass.

Reference: (Green Book)

Environmental Chemistry

Branch of the chemical sciences dedicated to studying the relationship between chemistry and the environment, involving analyses, remediation processes, and the understanding of the cycle of elements.

Environmental Impact

Impact of a technology, a process, or a product on the environment. It is part of the sustainability (*see Sustainability*) measurement and can be positive or negative.

Enzymes

Macromolecules (*see Macromolecule*), mostly of protein nature, that function as (bio)catalysts (*see Biocatalyst* and *see Catalyst*) by increasing reaction rates. In general, an enzyme catalyzes only one reaction type (reaction specificity) and operates on only one type of substrate (substrate specificity). Substrate molecules are attacked at the same site (regiospecificity) and only one or preferentially one of the enantiomers of chiral substrates or of racemic mixtures is attacked (stereospecificity).

Reference: (PAC 1992c)

Fatty Acid Methyl Esters (FAME)

Large-chain aliphatic carboxylic acids (fatty acids) that constitute the commonly named biodiesel (*see Biodiesel*) obtained, together with glycerine, by means of the transesterification of triglycerides with methanol.

Fast Pyrolysis Process

Rapid thermal decomposition (400–600 °C) of organic compounds [e.g., lignocellulosic biomass (see *Lignocellulosic biomass*)] in the absence of oxygen to produce liquids (bio-oil) (*see Bio-oil*), gases, and biochar (*see Biochar*).

Fertilizers

Mineralized fractions rich in N, O, C, and S, applied to improve soil functionality for agriculture. It is a by-product (*see By-product*), for example, of a sugarcane biorefinery.

Fibers

Polymers (*see Polymers*) with high molecular symmetry and strong cohesive energies between chains that usually result from the presence of polar groups. Fibers can be obtained from natural or synthetic sources with a length-to-diameter ratio of at least 100.

Fine Chemicals

Large family of chemicals with a high added value: pharmaceuticals, intermediates, nutraceuticals, flavors and fragrances, agrochemicals, food and feed additives, and catalysts (*see Catalyst*). These chemicals can increase the economic potential of the vegetable biomass (*see Biomass*).

First-Generation Biofuel
Biofuel (*see Biofuel*) obtained from the direct conversion—chemical or biochemical—or direct preparation of biomass (*see Biomass*) or biomass component without a previous chemical step [e.g., bioethanol (*see Bioethanol*) from sugarcane and biodiesel (*see Biodiesel*) from soybean].

Fischer–Tropsch Process
Conversion technology, which produces liquid products such as fuels, light olefins, and phenols, by means of an organic synthesis catalyzed by metallic catalysts (*see Catalyst*), for example, Co and Ru species, using syngas (*see Syngas*) from gasified biomass (*see Biomass*) as feedstock.

Food Supply Chain Waste (FSCW)
Waste (*see Waste*) generated from food production (animal or vegetable source) that can be used, for example, as a feedstock for chemicals and fuels, among others.

Gasification Process
High-temperature (750–850 °C) conversion of solid carbonaceous fuels into flammable gas mixtures, sometimes known as synthesis gas or syngas (see *Syngas*), consisting of a mixture of CO and H_2 (major components), CO_2, methane (CH_4), nitrogen (N_2), and smaller quantities of higher hydrocarbons.
Reference: (Reed 1981)

Global Warming
Effect from a compound's ability to absorb infrared radiation in the atmosphere. Methane (CH_4) and carbon dioxide (CO_2) are the major contributors.

Green Analytical Chemistry
Application of the green chemistry (*see Green chemistry*) principles to analytical chemistry. Generally, it produces (i) reduction of the amount of solvents required in sample pretreatment; (ii) reduction in the amount and the toxicity of solvents and reagents employed in the measurement step, especially by automation and miniaturization; and (iii) development of alternative direct analytical methodologies not requiring solvents or reagents.
Reference: (Armenta et al. 2008)

Green Biomass
Biomass (*see Biomass*) produced by the growth of plants.

Green Biorefinery (GBR)
Biorefinery (*see Biorefinery*) based on the use of renewable biomass (*see Renewable biomass*) as feedstock (plants).

Green Chemistry
Design of chemical products and chemical processes in a way that minimizes the generation of waste (*see Waste*) and avoids the use of toxic and hazardous materials.

Green chemistry is based on 12 principles that go beyond concerns over hazards from chemical toxicity and include energy (*see Energy*) conservation, waste reduction, and life cycle considerations such as the use of more sustainable or renewable feedstocks (*see Renewable feedstock*) and designing for the end of life or the final disposition of the product.

References: (American Chemical Society 2016; Anastas and Warner 1998)

Green Engineering

Green engineering is the design, commercialization, and use of processes and products in a way that minimizes pollution, promotes sustainability (*see Sustainability*), and protects human health without sacrificing economic viability and efficiency.

Reference: (United States Environmental Protection Agency 2016)

Green Metrics

Mathematical parameters, such as atom economy and the E-factor (*see E-factor*), for any chemical transformation, applied to determine the 'greenness' of the experiment, process, or products in a rigorous quantitative way according to green chemistry (*see Green chemistry*) principles.

Greenhouse Gases

Gases that produce global warming (see *Global warming*) from their capacity to absorb infrared radiation, highlighting methane (CH_4) and carbon dioxide (CO_2) as the main species. Their content can be calculated in *kilogram carbon dioxide equivalents per kilogram for each of the gaseous species*.

Green Polymer

Polymer obtained from a renewable source (e.g., sugars) instead of a petrochemical source, such as the green polyethylene produced from ethanol from sugarcane.

Green Solvent

Solvent that promotes the substitution of solvent(s) derived from petroleum (*see Petroleum*) with others obtained from renewable resources (*see Renewable resources*), and the substitution of hazardous solvents with those that show better environmental, health, and safety properties.

Reference: (Espino et al. 2016)

Hemicellulases

Group of hydrolytic enzymes (*see Enzymes*) used to catalyze the depolymerization (*see Depolymerization*) of the hemicellulose (*see Hemicellulose*) structure to obtain sugars, mainly xylose. The complexity of hemicellulose requires using a complex mixture of enzymes for complete hydrolysis (*see Hydrolysis*) of this heteropolymer, such as *endo*-1,4-β-xylanases, β-1,4 xylosidase, α-L-arabinofuranosidase, α-glucoronosidase, and acetylxylane esterase.

Hemicellulose

Natural polymer present in the plant cellular wall. Their oligomers are pentoses–hexoses type, mainly xylose–glucose, linked by means of a 1,4-β-D bond. Xylans are the most abundant kind of hemicellulose (Fig. 2).

Fig. 2 Hemicellulose structure

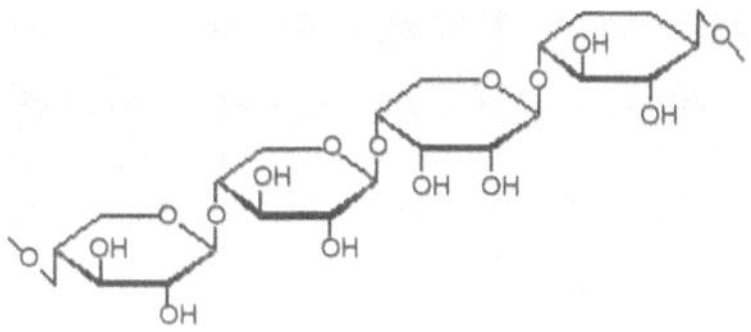

Hemicellulosic Fraction
Fraction obtained after the lignocellulosic biomass (*see Lignocellulosic biomass*) pretreatment by means of the use of chemicals in aqueous medium, commonly acid or alkaline, separating it from cellulose (*see Cellulose*) and lignin fractions (*see Lignin fraction*).

Heterogeneous Catalysis
Catalysis (*see Catalysis*) in which the reaction occurs at or near an interface between phases, and the catalyst (*see Catalyst*) is commonly used in the solid phase.
Reference: (PAC 1996a)

Heterogeneous Catalyst
Catalyst (*see Catalyst*) applied for a heterogeneously catalyzed reaction. Generally, it is in solid state (e.g., zeolites).

High-Throughput Screening (HTS)
Study of the performance of a large number of catalysts (*see Catalyst*) [inorganic, organometallic, or enzyme (*see Enzymes*)] using reduced quantities of reactants, catalysts (*see Catalyst*), and products (milli- or micro-quantities) in parallel mode and under the same experimental conditions. It can be applied also for synthetic chemistry or biology assays.

Holocellulose
Total carbohydrate fraction in the plant, considering α-cellulose plus the hemicellulose (*see Hemicellulose*).

Homogeneous Catalysis
Catalysis (*see Catalysis*) in which only one phase is involved and the catalyst (*see Catalyst*) is miscible in the reaction medium.
Reference: (PAC 1996a)

Homogeneous Catalyst
Catalyst (*see Catalyst*) applied for a homogeneously catalyzed reaction. Generally, it is in liquid state (e.g., mineral acids).

Hydrocracking Process (or Catalytic Hydrocracking)
Process effected under hydrogen pressure and on a catalyst (*see Catalyst*) containing an ingredient with a hydrogenating function.
Reference: (PAC 1976)

Hydrogen Energy
Energy (*see Energy*) generated from the cracking (*see Cracking*) of the H_2 molecule by means of an electrochemical process. Hydrogen-containing compounds such as fossil fuels, biomass (*see Biomass*), or water can be a source of hydrogen.

Hydrogenation
Addition of hydrogen atom(s) in an unsaturated molecule by the use of H_2 (reactant) and a metallic catalyst (*see Catalyst*), such as Pt, Pd, or Ni. As an example, it is applied to produce certain saturated molecules from oleaginous oil.

Hydrogenolysis (or Hydrocracking)
Type of pyrolysis with the addition of hydrogen, which assists in the cleavage of bonds. This process creates more monomeric liquid molecules from biomass (*see Biomass*) (e.g., phenols) because of the lower reaction temperature.

Hydrolysis
Reaction with water involving the rupture of one or more bonds in the reacting solute.

Reference: (PAC 1994)

Cleavage of chemical bonds by the addition of water.

Klason Lignin
Lignin (*see Lignins*) isolated from extracted wood material after treating with cold concentrated sulfuric acid. It is very useful for analytical purposes.

Kraft Process
Alkaline degradation of lignin (*see Lignins*) using, mainly, sodium hydroxide (about 0.8 mol l^{-1}) plus sodium sulfide (about 0.2 mol) at elevated temperatures (160–180 °C) for 1–2 h. It is the most widely used process in the production of chemical pulps for papermaking and cellulose (*see Cellulose*) applications.

Intermediate (of Industrial Synthesis)
Reactant added during an intermediary step of the synthesis to obtain the final product, commonly used in the pharmaceutical industry. An example of intermediate use is to solve the problem of regioselectivity in organic synthesis.

Ionic Liquids
Alternative solvents (*see Alternative solvents*) with unique properties such as low vapor pressure, good thermal stability, and high ion conductivity; these are salts that are liquid or close to liquid at room temperature. Quaternary onium salts are the main ionic liquids (e.g., 1-butyl-3-methylimidazolium hexafluorophosphate).

Reference: (Vafaeezadeh and Alinezhad 2016)

Life Cycle Analysis (LCA)
Conceptual analysis that comprises the consideration of all stages along the life cycle of a chemical [e.g., raw material (*see Raw material*) extraction, preproduction, production, use, recycling, and disposal] as well as the consideration of environmental impacts (*see Environmental impact*) caused by the production of other consumables (such as solvents and additives), by-products (*see By-product*), and

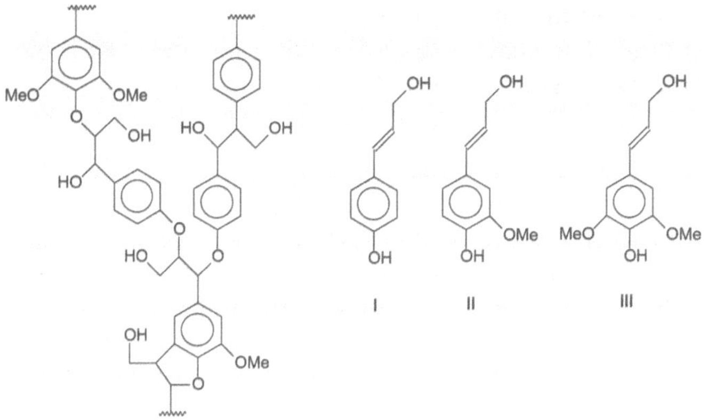

Fig. 3 Lignin structure (*left*) and its precursors (*right*): (I) *p*-coumaryl alcohol, (II) coniferyl alcohol, (III) sinapyl alcohol

auxiliaries (facilities that have to be provided to produce the green chemical). A life cycle conceptual approach also addresses and compares different environmental impacts such as global warming (*see Global warming*), ozone depletion, acidification, and ecotoxicity.

Lignans
Plant products of low molecular weight formed primarily from oxidative coupling of two *p*-propylphenol moieties at their β-carbon atoms; products with units coupled in other ways are neolignans.
Reference: (PAC 1995c)

Lignins
Macromolecular constituents of plants related to lignans (*see Lignans*), composed of phenolic propylbenzene skeletal units, linked at various sites and apparently randomly (Fig. 3).
Reference: (PAC 1995d)

Lignin Fraction
Fraction obtained after the lignocellulosic biomass (*see Lignocellulosic biomass*) pretreatment by use of chemicals in aqueous medium, commonly alkali, separating it from cellulose (*see Cellulose*) and hemicellulose (*see Hemicellulose*) to later phenolic monomers release.

Lignocellulosic Biomass
Biomass (*see Biomass*) from plants formed by a three-dimensional polymeric composite as structural material with cellulose (*see Cellulose*), hemicellulose (*see Hemicellulose*), and lignin (*see Lignins*) as the major constituents. It is the most abundant biomass (*see Biomass*) on Earth. The approximate w is cellulose (*see Cellulose*), 35–50%; hemicellulose, 30–45%; lignin, 15–25%. This content depends on type of plant, soil composition, water availability, and climate conditions.
Note:
w = mass content.

Lignocellulosic Feedstock Biorefinery
Biorefinery (*see Biorefinery*) based on the lignocellulosic biomass (*see Lignocellulosic biomass*) as raw material (*see Raw material*).

Macroalgae
Marine algae: a source of functional ingredients in the food industry for use as emulsifying agents and to enhance viscosity and gelation. Furthermore, macroalgae can produce molecules (e.g., lipids and proteins) for pharmaceuticals, cosmetics, and biofuels (*see Biofuels*), and biomass (*see Biomass*) for bioenergy (*see Bioenergy*). The chemical composition of macroalgae species is significantly different from that of terrestrial plants: they possess lower content of carbon, hydrogen, and oxygen and higher content of nitrogen and sulfur than that of land-based, lignocellulosic biomass (*see Lignocellulosic biomass*).

Reference: (Ghadiryanfar et al. 2016)

Macromolecule
Molecule of high relative molecular mass, the structure of which essentially comprises the multiple repetition of units derived, from molecules of low relative molecular mass.

Notes:

1. In many cases, especially for synthetic polymers (*see Polymers*), a molecule can be regarded as having a high relative molecular mass if the addition or removal of one or a few of the units has a negligible effect on the molecular properties. This statement fails in the case of certain macromolecules for which the properties may be critically dependent on fine details of the molecular structure.
2. If a part or the whole of the molecule has a high relative molecular mass and essentially comprises the multiple repetitions of units derived, actually or conceptually, from molecules of low relative molecular mass, it may be described as either macromolecular or polymeric, or by polymer used adjectivally.

Reference: (PAC 1996c)

Metabolomics
Study of metabolics from biochemical routes by means of analytical technologies, as LC-MS and NMR. It is very useful to understand the formation of products by bioprocesses (e.g., fermentation) and the improvement of microorganisms (*see Microorganisms*).

Metabolic Process
Process that occurs by means of a biochemical route in an organism or microorganism and generates metabolites (*see Metabolite*) as products and by-products (*see By-product*).

Metabolic Route
Synthetic route (*see Synthetic route*) that occurs in a living organism (e.g., plants) by means of a metabolic process (*see Metabolic process*).

Metabolism

Entire physical and chemical processes involved in the maintenance and reproduction of life in which nutrients are broken down to generate energy (*see Energy*) and to give simpler molecules (catabolism), which by themselves may be used to form more complex molecules (anabolism).

Reference: (PAC 1992d)

Metabolite

Any intermediate or product resulting from metabolism (*see Metabolism*).

Reference: (PAC 1992d)

Microalgae

Microscopic algae, particularly unicellular algae such as *Chlamydomonas* and *Chlorella*, and diatoms, which can produce metabolites (*see Metabolite*), oil, surfactants, and nutraceuticals, etc.

Microfiltration

Pressure-driven membrane-based separation process in which particles and dissolved macromolecules (*see Macromolecule*) larger than 0.1 μm are separated.

Reference: (PAC 1996d)

Microfluid

Any fluid in which the local motion of contained particles affects the behavior of the fluid as a whole. A monodispersive emulsion is an example.

Microorganisms

Microscopic organisms within the categories of algae, Archaea, Bacteria, Fungi (including lichens), Protozoa, viruses, and subviral agents.

Microreactor

Device to produce chemical transformations in micro-scale. It is very useful for the development of processes in the laboratory and to reduce waste (*see Waste*) generation.

Microwave Reaction (or Assisted-Microwave Reaction)

Reaction conducted by microwave instead of common temperature application. Microwave hydrothermal/solvothermal synthesis provides higher reaction rates at low reaction temperatures in a very short reaction time, along with good yield.

Reference: (Riaz et al. 2016)

Municipal Solid Waste (MSW)

Waste (*see Waste*) rich in organic matter (*see Organic matter*) produced from tree residues and food residues, among others, collected systematically in urban areas and disposed in landfills. MSW are the largest single source of lignocellulosic materials available for utilization in modern society.

Nanomaterials
Materials from synthetic or natural sources obtained in the nano-scale of size. Examples are solids containing metal nano-particles or based on nano-crystals, porous materials with channels and pore sizes at the nano-scale.

Nanoscience
Study of the substances and their properties in the nano-scale size (10^{-9} m).

Nanotechnology
Application of nanoscience (*see Nanoscience*) for real situations.

Natural Fibers
Fibers (*see Fibers*), essentially cellulosic in nature, and chemically described as poly-(1,4-β-D-anhydroglucopyranose), with about 3000 units of repetition. Natural fibers may be derived from plant sources such as cotton and flax.

Natural Polymers
Polymers (*see Polymers*) present in nature, such as polysaccharides (*see Polysaccharides*): cellulose (*see Cellulose*), hemicellulose (*see Hemicellulose*), and starch, and protein-based polymers (e.g., polyamides and polynucleotides).

Natural Products
Chemical compounds or substances produced by a living organism, which are found in nature; for example, products obtained from plants.

Nonfood Crops
Crops for usages other than food in a biorefinery (*see Biorefinery*) [e.g., crops for natural fibers (*see Natural fibers*)].

Nonedible Crops
Crops and plants, in general, inappropriate to be used as human food or to be integrated in the food chain or industry.

Nonhydrolytic Proteins for Biomass Deconstruction
Expansins, polysaccharide monooxygenase (PMO), and cellobiose dehydrogenase (CDH) are nonhydrolytic proteins known as biomass (*see Biomass*) saccharification stimulators. These proteins can improve the degradation of cellulose (*see Cellulose*) in combination with cellulases (*see Cellulase*) by the induction of the cellulose (*see Cellulose*) relaxation, reducing the crystallinity and stimulating the enzyme action.

Nonrenewable Resources
Industrial resources that are not renewable, such as oil and ores, antagonistic to renewable resources (*see Renewable resources*). Here, the idea of a cyclic dynamic from feedstock to residues for formation and consumption cannot be applied.

Oleaginous Biomass
Biomass (*see Biomass*) from plants that produce oil (fatty acids and esters) inside their seeds or fruits, such as soya, cotton, and palm. The approximate *w* value of some fatty acids in oleaginous plants is palmitic, 10–45%; stearic, 5%; oleic,

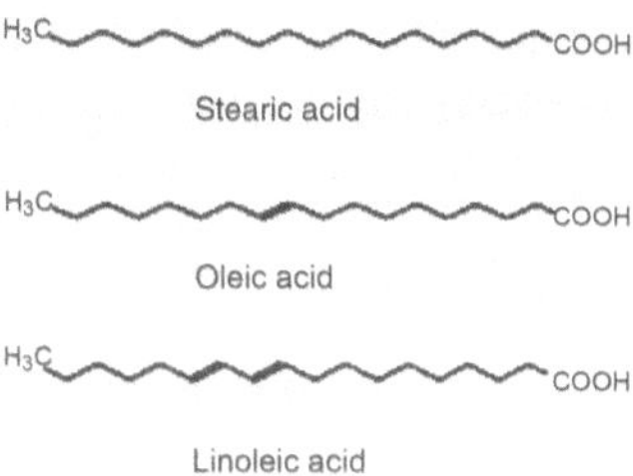

Fig. 4 Structures of some fatty acids from oleaginous plants

20–40%; and linoleic, 10%. This content depends on the type of plant, soil composition, water availability, and climate conditions (Fig. 4).

Note:

w = mass fraction.

Organic Matter
Material generated from the decomposition of biomass (*see Biomass*) (plant and animal) by means of chemical or biological processes. It is rich in carbon, hydrogen, and oxygen, as the humic substances in the soil.

Organic Solar Cell
Solar energy (*see Solar energy*) conversion device with low cost, flexibility, lightweight, and easier fabricability than inorganic cells. An example is a bilayer device using copper phthalocyanine as an electron donor and perylene tetracarboxylic derivatives as an electron acceptor.

Reference: (Sathiyan et al. 2016)

Organosolv Lignin
Lignin fraction (*see Lignin fraction*) obtained by treating lignocellulosic biomass (*see Lignocellulosic biomass*) with the Organosolv process (*see Organosolv process*).

Organosolv Process
Delignification process using an organic solvent (frequently ethanol) in the presence or absence of acid, at 180–200 °C, generating a liquor rich in lignin (*see Lignins*). It is very useful for analytical purposes because the lignin structure suffers less interference.

Oxygenated Compounds Recovery
Attainment of potentially valuable substances from unusable sources containing chemicals with high levels of oxygen, as waste (*see Waste*) biomass (*see Biomass*).

Pellets
Material with reduced dimensions obtained from the densification of biomass that can be treated by thermal process to improve its energetic content.

Petroleum
Mixture of relatively nonreactive hydrocarbons with variable amounts of nonhydrocarbons with a large variety in molecular weight (aliphatic and aromatic

compounds). It is formed by a reductive process of degradation of biomass (*see Biomass*), chemical or biological, during millions of years. It exists in the liquid phase and is called crude oil (*see Crude oil*).

Photochemical Reaction
Generally used to describe a chemical reaction caused by absorption of ultraviolet, visible, or infrared radiation. There are many ground-state reactions, which have photochemical counterparts. Among these are photochemical nitrogen extrusions, photo-cycloadditions, photo-decarbonylations, photo-decarboxylations, photo-enolizations, photo-Fries rearrangement, photo-isomerizations, photo-oxidations, photo-rearrangements, photo-reductions, and photo-substitutions.

Note:

Photochemical paths offer the advantage over thermal methods of forming thermodynamically nonfavored products, overcome large activation barriers in a short period of time, and allow reactivity otherwise inaccessible by thermal methods.

Reference: (PAC 2007)

Photosynthesis Process
Metabolic process (*see Metabolic process*) involving plants, microalgae (*see Microalgae*), and some types of bacteria (e.g., *Chromataceae*, *Rhodospirillaceae*, *Chlorobiaceae*) in which light energy (*see Energy*) absorbed by chlorophyll and other photosynthetic pigments results in the reduction of CO_2 followed by the formation of organic compounds. In plants, the overall process involves the conversion of CO_2 and H_2O to carbohydrates (and other plant material) and the release of O_2.

Reference: (PAC 1992e)

Photovoltaic Cell
Solid-state device, usually a semiconductor such as silicon, which absorbs photons with energies higher than or equal to the bandgap energy (*see Energy*) and simultaneously produces electric power.

Reference: (PAC 1996e)

Platform Chemicals
Molecule(s) used as precursors to obtain a large amount of other molecules for industrial purposes.

Reference: (Gunukulaa et al. 2016)

Polymers
Macromolecular structures (*see Macromolecule*) formed by repeating units (monomers). Cellulose (*see Cellulose*) is the main natural polymer from biomass (*see Biomass*), where the monomeric unities are glucose molecules.

Polysaccharides
Compounds consisting of a large number of monosaccharides linked glycosidically. This term is commonly used only for those containing more than ten monosaccharide residues. Also called glycans.

Reference: (PAC 1995e)

Pretreatment Process
A crucial step used to deconstruct the lignocellulosic biomass (*see Lignocellulosic biomass*) into cellulose (*see Cellulose*), hemicellulose (*see Hemicellulose*), and lignin (*see Lignins*). Steps are physical (e.g., densification), chemical (e.g., acid or alkali), biological [e.g., fungi, bacteria, enzymes (*see Enzymes*)], or a combination of these steps, depending on the feedstock composition and the final products to be obtained.

Pyrolysis Process
Thermal decomposition (above 500 °C) of organic compounds in the absence of oxygen, when fast pyrolysis, with a short residence time, is the most interesting for biomass (*see Biomass*). Products are bio-oil (*see Bio-oil*) and biochar (*see Biochar*).

Raw Material
Feedstock for a transformation or conversion process: it can be renewable [e.g., biomass (*see Biomass*)] or nonrenewable (e.g., oil).

Reforming (or Catalytic Reforming)
Process using controlled heat and pressure together with catalysts (*see Catalyst*) to rearrange certain medium to high molecular weight hydrocarbon molecules into high-quality fuel stocks suitable for blending into finished gasoline or naphtha. In the case of vapor-phase reforming, water is added in the system to transform a wide range of feedstocks [including biomass (*see Biomass*) -derived fractions] into light gases useful for syngas (*see Syngas*) and hydrogen production.

Renewable Biomass
Biomass (*see Biomass*) from a renewable source (*see Renewable source*), as plants, which can be cultivated and processed for various purposes.

Renewable Chemistry
Concept related to the development of products and processes from renewable sources to promote a sustainable chemistry (*see Sustainable chemistry*): involves all areas from chemical sciences: analytical chemistry, biochemistry, organic chemistry, inorganic chemistry, physical chemistry, environmental chemistry, and chemical engineering.

Renewable Chemicals
Chemicals obtained from renewable feedstock (*see Renewable feedstock*), as several biomass (*see Biomass*) types, such as bioethanol (*see Bioethanol*), and organic acids such as succinic and lactic acid. It can be a building block (*see Building-block molecule*), an industrial intermediate (see intermediate for industrial synthesis), or an end product. It is desirable that a renewable chemical can substitute for a petrochemical in an ideal situation.

Renewable Energy Resources
Energy (*see Energy*) resources that are naturally replenishing but flow limited: these are virtually inexhaustible in duration but limited in the amount of energy that is available per unit of time. Renewable energy resources include biomass (*see*

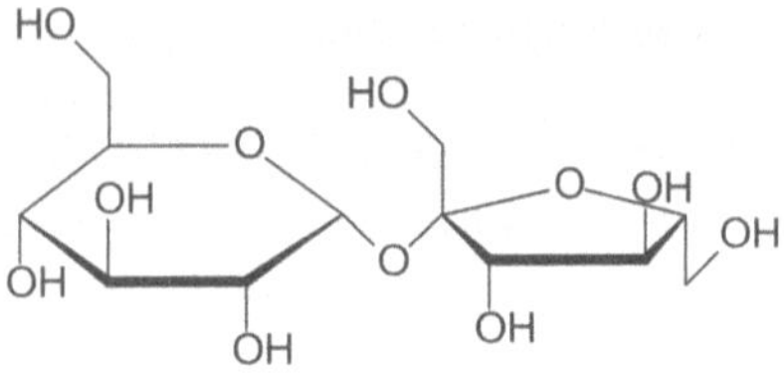

Fig. 5 Structure of sucrose present in sugarcane juice

Biomass), hydro-energy, geothermal, solar, wind, and ocean thermal energy, and energy from wave action and tidal action.

Renewable Feedstock

Feedstock for industrial processes from a renewable resource (*see Renewable resource*) as biomass (*see Biomass*).

Renewable Fuels

Fuels from renewable resources (*see Renewable resources*) such as biomass (*see Biomass*) with a low CO_2 emission.

Renewable Materials

Materials obtained from renewable feedstock (*see Renewable feedstock*), such as lignocellulosic biomass (*see Lignocellulosic biomass*); for example, natural fibers (*see Natural fibers*), resins, composites (*see Composites*).

Renewable Resources

Industrial resources obtained and used by means of a cycle mode. There is a cyclic dynamic from the feedstock to residues for formation and consumption. On the other hand, nonrenewable biomass—oil, coal, and natural gas—take millions of years to produce in geological reservoirs [from renewable biomass (*see Renewable biomass*)], whereas renewable biomass is produced in months or years; that is, in the same time period that it is consumed. Plant biomass is the most representative renewable resource.

Residues

Surplus from a transformation process without economic relevance when compared with the product(s). Sometimes, such surplus has a negative impact on the environment.

Saccharide Biomass

Biomass (*see Biomass*) from plants that produce sucrose (a disaccharide from glucose and fructose), such as sugarcane and sweet sorghum. The D-glucose is linked to the D-fructose moiety by α-β-D-disaccharide bonds. The w of sucrose in sugarcane juice can reach as much as 85%. This content depends on the type of plant, soil composition, water availability, and climate conditions (Fig. 5).

Note:

w = mass fraction.

Saccharification Process
Production of simple sugars (hexoses) by hydrolysis (*see Hydrolysis*) of polysaccharides (*see Polysaccharides*) contained in starch or lignocellulosic biomass (*see Starch biomass*) (*see Lignocellulosic biomass*).

Second-Generation Biofuel
Biofuel (*see Biofuel*) derived from purpose-grown nonfood feedstocks containing cellulose (*see Cellulose*), hemicellulose (*see Hemicellulose*), lignin (*see Lignins*), or pectin as agricultural or forestry wastes (*see Waste*), by means of a previous chemical or biochemical step to release sugars and other monomers to a second chemical or biochemical conversion step (e.g., cellulosic ethanol).

Second-Generation Bioproducts
Bioproducts (*see Bioproducts*) obtained from the deconstruction of lignocellulosic biomass (*see Lignocellulosic biomass*); for example, second-generation ethanol (2G ethanol), furfural.

Slow Pyrolysis Process
Pyrolysis process (*see Pyrolysis process*) with only two products at the output: solid biochar (*see Biochar*) and heat in the form of hot media such as water. Heat generated by direct burning of all gaseous products from pyrolysis runs the process and is also an output, a second final product. Temperature reaches nearly 1000 °C at a residence time up to 20 min.

Reference: (Klinar 2016)

Societal Impact
Generation of wealth, poverty, and employment, etc., from a product or from a process in the society. It is part of the sustainability (*see Sustainability*) analysis and can be positive or negative.

Solar Energy
Energy (*see Energy*) generated from the sun. Plants can convert small quantities, with a low conversion rate (nearly 1%); photovoltaic surfaces can achieve nearly 15%.

Solar Conversion Efficiency
Ratio of the Gibbs energy (*see Energy*) gain per unit time per square meter of surface exposed to the sun and the solar irradiance, E, integrated between $\lambda = 0$ and $\lambda = \infty$.

Reference: (PAC 1996f)

Steam Explosion Pretreatment
Unit operation in which biomass (*see Biomass*) is processed at high-pressure saturated steam (160–290 °C), for a period between 2 min and 10 min, followed by rapid decompression to atmospheric pressure. The process increases the potential of cellulose (*see Cellulose*) hydrolysis (*see Hydrolysis*) mainly because it can remove hemicellulose (*see Hemicellulose*). It is the most commonly used method for the pretreatment of lignocellulosic materials.

Fig. 6 Structure of starch

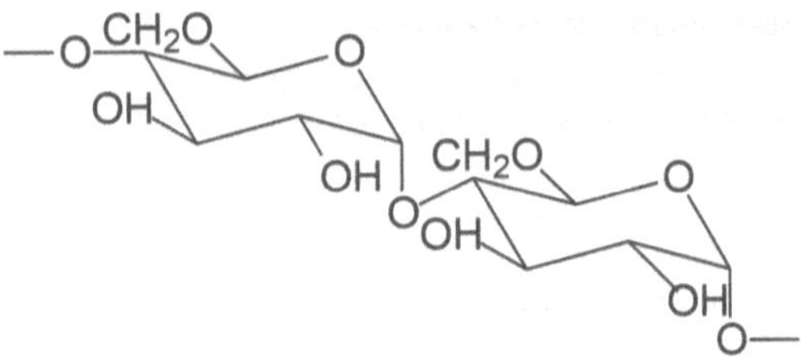

Starch Biomass

Biomass (*see Biomass*) from plants that produce starch (a polymer of glucose), such as corn, sugar beet, and cassava. The glucose units (monomers) are linked by α-1-4-D-disaccharide bonds. The approximate *w* of starch can reach 70–85%. This content depends on the type of plant, soil composition, water availability, and climate conditions (Fig. 6).

Note:

w = mass fraction.

Supercritical Fluid

Defined state of a compound, mixture, or element above its critical pressure (p_c) and critical temperature (T_c).

Reference: (PAC 1993b)

Supercritical Fluid Extraction

Extraction process using supercritical fluid (*see Supercritical fluid*) as the solvent. It can be used as an alternative green extraction process for oleaginous.

Sustainability

Concept and metric set that considers the following components for an analysis of a product or a process, from the feedstock to the end product: the economic impact (*see Economic impact*), the environmental impact (*see Environmental impact*), and the societal impact (*see Societal impact*). Each impact has its own metric for measurement [e.g., life cycle assessment (*see Life cycle assessment*)] to quantify and to qualify (negative or positive) the environmental impact.

Sustainable Chemistry

Chemistry based on sustainability (*see Sustainability*) components: economic impact (*see Economic impact*), environmental impact (*see Environmental impact*), and societal impact (*see Societal impact*). It is desirable to have a high economic impact [e.g., valorization of a biomass (*see Biomass*) chain], a low or nonexistent environmental impact [e.g., process without residues generation and with energy (*see Energy*) economy], and a high societal impact (e.g., distribution of wealth and job creation).

Sustainable Metrics

Set of parameters based on sustainability (*see Sustainability*) criteria (e.g., costs and economic aspects, environmental impact (*see Environmental impact*), and water and land use, among others, taken into account for the evaluation of a chemical process.

Reference: (Sheldon and Sanders 2015)

Sucrochemistry
Production of chemicals from sucrose (a disaccharide from glucose and fructose) by chemical or biochemical processes.

Syngas (or Synthesis Gas)
Gaseous mixture rich in CO and H_2 (the major components) obtained by means of a gasification process (*see Gasification process*). It is an important feedstock for the chemical industry to produce fuels, chemicals, polymers (*see Polymers*), and other materials.

Synthetic Biology
Division of biology that studies the engineering of biological systems using genetic strategies to modify metabolic routes (*see Metabolic routes*) in organisms and plants for specific purposes, such as to produce chemicals.

Synthetic Route
Logical sequence of transformations (organic or inorganic) to obtain a product based on in one or in a set of reactions.

Thermal Cracking
Process in which heat and pressure are used to break down, rearrange, or combine hydrocarbon molecules, for example, those present in lignocellulosic biomass (*see Lignocellulosic biomass*).

Thermochemical Conversion Process
Conversion process (*see Conversion process*) based on the use of thermochemical technologies, such as gasification and fast pyrolysis, to obtain fuels and chemicals.

Third-Generation Biofuel
Biofuel (*see Biofuel*) derived from marine biomass (*see Biomass*) as microalgae (e.g., diesel-type fuel).

Transesterification Process
Conversion process (*see Conversion process*) based on reaction of an ester with an alcohol (e.g., methanol) to form a new ester and a new alcohol.

Two-Platform Biorefinery
Biorefinery (*see Biorefinery*) in which all carbohydrates constituting the vegetable biomass (*see Biomass*) [e.g., cellulose (*see Cellulose*), starch, sucrose] are enzymatically or chemically saccharified and unified to one sugar platform (hexose or pentose). As a sugar, they will continue processing to methane or carbon monoxide and hydrogen consumed [syngas (*see Syngas*) platform]. From this syngas, products such as methanol or higher hydrocarbons could be obtained by means of the Fischer–Tropsch process (*see Fischer–Tropsch process*).

Waste
Residue generated from a transformation process. Generally, it has an environmentally negative impact. It could be used in a biorefinery (*see Biorefinery*) as a feedstock.

Whole-Crop Biorefinery
Biorefinery (*see Biorefinery*) based on, for example, cereals with potential to be used for the production of not only traditional foods but also novel functional foods and nonfood products (e.g., biodegradable plastics, chemicals, fuels).

Yeast
Unicellular eukaryotic microorganism belonging to the kingdom Fungi. *Saccharomyces cerevisiae* is one of the best known yeast genera used in the fermentation process.

Yield (in Chemistry)
Amount of product obtained from a chemical reaction. The absolute yield can be given as the weight in grams or in mol (molar yield). The percentage serves to measure the effectiveness of a synthetic procedure; it is calculated by dividing the amount of the desired product obtained by the theoretical yield (amount predicted by a stoichiometric calculation based on the number of moles of all reactants present).

Yield (in Biotechnology)
Ratio expressing the efficiency of a mass conversion process. The yield coefficient is defined as the amount of cell mass (kg) or product formed (kg, mol) related to the consumed substrate (carbon or nitrogen source or oxygen in kilograms or moles or to the intracellular ATP production (moles).

Reference: (PAC 1992f).

References

American Chemical Society (2016) Green chemistry definition. http://www.acs.org/content/acs/en/greenchemistry/what-is-green-chemistry/definition.html

American Society for Test and Methods (2015) Standard specification for biodiesel fuel blend stock (B100) for middle distillate fuels. ASTM D6751

Anastas PT, Warner JC (1998) Green chemistry: theory and practice. Oxford University Press, New York, p 30

Armenta S, Garrigues S, de la Guardia M (2008) Green analytical chemistry. Trends Anal Chem 27:497–511

Espino M, Fernandez MA, Gomez FJV, Silva MF 2016) Natural designer solvents for greening analytical chemistry. Trends Anal Chem 76:126–136

Ghadiryanfar M, Rosentrater KA, Keyhani A, Mahmoud OM (2016) A review of macroalgae production, with potential applications in biofuels and bioenergy. Renew Sust Energ Rev 54:473–481

Green Book, 2nd edn, p 12

Gunukulaa S, Keelingb PL, Anex R (2016) Risk advantages of platform technologies for biorenewable chemical production. Chem Eng Res Des 107:24–33

Hennig C, Brosowski A, Majer S (2016) Sustainable feedstock potential – a limitation for the bio-based economy? J Cleaner Prod 123:200–202

Klinar D (2016) Universal model of slow pyrolysis technology producing biochar and heat from standard biomass needed for the techno-economic assessment. Bioresour Technol 206:112–120

PAC, 1976, *46*, 71 (Manual of Symbols and Terminology for Physicochemical Quantities and Units – Appendix II. Definitions, Terminology and Symbols in Colloid and Surface Chemistry. Part II: Heterogeneous Catalysis) on page 86
PAC, 1992a, *64*, 143 (Glossary for chemists of terms used in biotechnology (IUPAC Recommendations 1992)) on page 147
PAC, 1992b, *64*, 143 (Glossary for chemists of terms used in biotechnology (IUPAC Recommendations 1992)) on page 148
PAC, 1992c, *64*, 143 (Glossary for chemists of terms used in biotechnology (IUPAC Recommendations 1992)) on page 152
PAC, 1992d, *64*, 143 (Glossary for chemists of terms used in biotechnology (IUPAC Recommendations 1992)) on page 160
PAC, 1992e, *64*, 143 (Glossary for chemists of terms used in biotechnology (IUPAC Recommendations 1992)) on page 162
PAC, 1992f, *64*, 143 (Glossary for chemists of terms used in biotechnology (IUPAC Recommendations 1992)), on page 168
PAC, 1993a, *65*, 2003 (Glossary for chemists of terms used in toxicology (IUPAC Recommendations 1993)) on page 2038
PAC, 1993b, *65*, 2397 (Nomenclature for supercritical fluid chromatography and extraction (IUPAC Recommendations 1993)) on page 2399
PAC, 1994, *66*, 1077 (Glossary of terms used in physical organic chemistry (IUPAC Recommendations 1994)) on page 1123
PAC, 1995a, *67*, 473 (Recommended terminology for the description of carbon as a solid (IUPAC Recommendations 1995)) on page 480
PAC, 1995b, *67*, 473 (Recommended terminology for the description of carbon as a solid (IUPAC Recommendations 1995)) on page 484
PAC, 1995c, *67*, 1307 (Glossary of class names of organic compounds and reactivity intermediates based on structure (IUPAC Recommendations 1995)) on page 1347
PAC, 1995d, *67*, 1307 (Glossary of class names of organic compounds and reactivity intermediates based on structure (IUPAC Recommendations 1995)) on page 1348
PAC, 1995e, *67*, 1307 (Glossary of class names of organic compounds and reactivity intermediates based on structure (IUPAC Recommendations 1995)) on page 1360
PAC, 1996a, *68*, 149 (A glossary of terms used in chemical kinetics, including reaction dynamics (IUPAC Recommendations 1996)) on page 155
PAC, 1996b, *68*, 2287 (Glossary of basic terms in polymer science (IUPAC Recommendations 1996)) on page 2309
PAC, 1996c, *68*, 2287 (Glossary of basic terms in polymer science (IUPAC Recommendations 1996)) on page 2289
PAC, 1996d, *68*, 1479 (Terminology for membranes and membrane processes (IUPAC Recommendations 1996)) on page 1487
PAC, 1996e, *68*, 2223 (Glossary of terms used in photochemistry (IUPAC Recommendations 1996)) on page 2265
PAC, 1996f, *68*, 2223 (Glossary of terms used in photochemistry (IUPAC Recommendations 1996)) on page 2274
PAC, 2004, *76*, 889 (Definitions of terms relating to reactions of polymers and to functional polymeric materials (IUPAC Recommendations 2003)) on page 898
PAC, 2007, *79*, 293 (Glossary of terms used in photochemistry, 3rd edition (IUPAC Recommendations 2006)) on page 386
Paiva A, Craveiro R, Aroso I, Martins M, Reis RL, Duarte ARC (2014) Natural deep eutectic solvents: solvents for the 21st century. ACS Sustain Chem Eng 2:1063–1071
Reed T (ed) (1981) Biomass gasification: principles and technology. In: Energy Technology Review No. 67, Noyes Data Corp., Park Ridge
Riaz U, Ashraf SM, Budhiraja V, Aleem S, Kashyap J (2016) Comparative studies of the photocatalytic and microwave-assisted degradation of alizarin red using ZnO/poly(1-naphthylamine) nanohybrids. J Mol Liq 216:259–267

Sathiyan G, Sivakumar EKT, Ganesamoorthy R, Thangamuthu R, Sakthivel P (2016) Review of carbazole based conjugated molecules for highly efficient organic solar cell application. Tetrahedron Lett 57:243–252

Sheldon RA (2007) The E-factor: fifteen years on. Green Chem 9:1273–1283

Sheldon RA, Sanders JPM (2015) Comparison of the sustainability metrics of the petrochemical and biomass-based routes to methionine. Catal Today 239:44–49

United States Environmental Protection Agency (2016) Green engineering. https://www.epa.gov/green-engineering

Vafaeezadeh M, Alinezhad H (2016) Brønsted acidic ionic liquids: green catalysts for essential organic reactions. J Mol Liq 218:95–105

Zeitfracht Medien GmbH
Ferdinand-Jühlke-Straße 7
99095 Erfurt, Deutschland
produktsicherheit@kolibri360.de